T0185676

Vascular Differentiation and Plant Hormones

Roni Aloni

Vascular Differentiation and Plant Hormones

 Springer

Roni Aloni
School of Plant Sciences and Food Security
Tel Aviv University
Tel Aviv, Israel

ISBN 978-3-030-53204-8   ISBN 978-3-030-53202-4   (eBook)
https://doi.org/10.1007/978-3-030-53202-4

© Springer Nature Switzerland AG 2020
This work is subject to copyright. All rights are reserved by the Publisher, whether the whole or part of the material is concerned, specifically the rights of translation, reprinting, reuse of illustrations, recitation, broadcasting, reproduction on microfilms or in any other physical way, and transmission or information storage and retrieval, electronic adaptation, computer software, or by similar or dissimilar methodology now known or hereafter developed.
The use of general descriptive names, registered names, trademarks, service marks, etc. in this publication does not imply, even in the absence of a specific statement, that such names are exempt from the relevant protective laws and regulations and therefore free for general use.
The publisher, the authors, and the editors are safe to assume that the advice and information in this book are believed to be true and accurate at the date of publication. Neither the publisher nor the authors or the editors give a warranty, expressed or implied, with respect to the material contained herein or for any errors or omissions that may have been made. The publisher remains neutral with regard to jurisdictional claims in published maps and institutional affiliations.

This Springer imprint is published by the registered company Springer Nature Switzerland AG
The registered company address is: Gewerbestrasse 11, 6330 Cham, Switzerland

For my late wife Orna

Acknowledgments

It is my great pleasure to acknowledge and warmly thank the important contributions of my mentors and colleagues with whom I have been fortunate to work and learn from their knowledge, experience, and wisdom. I acknowledge my PhD supervisors, Tsvi Sachs and Avraham Fahn (Hebrew University, Jerusalem), and my Postdoctoral mentor, William P. Jacobs (Princeton University). I am thankful to the following people for fruitful collaboration during sabbaticals in their laboratories: Martin H. Zimmerman (Harvard University), Carol A. Peterson (University of Waterloo, Canada), Melvin T. Tyree (University of Vermont, Burlington), John R. Barnett (University of Reading, England), Lewis J. Feldman (University of California, Berkeley), Cornelia I. Ullrich (Technische Universität Darmstadt, Germany), and Jim Mattsson (Simon Fraser University, Canada).

I am also thankful to my graduate students, who run research in my laboratory for their important research contributions, and the numerous undergraduate and graduate students in my courses for their stimulating questions.

I am extremely grateful to Varda Wexler for her professional help in the preparation of the high-quality illustrations of the book.

I am thankful to the colleagues and friends who very kindly contributed their original micrographs from their research work, Tsvi Sachs, Ryo Funada, Vikram Singh, Nurit Firon, and Marco R. Pace.

It is my pleasure to thank my friends who stimulated me to write the book, Ray F. Evert, Ulrich Lüttge, Jim Mattsson, and Eilon Shani.

I thank the following publishers for their permission to reproduce micrographs: Oxford University Press (for Figs. 3.7 and 5.14) and Cold Spring Harbor Laboratory Press (for Fig. 3.3B).

I warmly thank Springer for publishing my book and the excellent professional team that were very helpful along the process, especially to Christina Eckey, Zuzana Bernhart, and Bibhuti B. Sharma, resulting in the highest quality book.

Finally, I thank my children, Erez Aloni, Miri Aloni Rivlin, and Daphna Rothschild, and my grandchildren, Gur, Romi, Gal, Jonathan, Yuval, Tom Isaac, Maya, and Danielle, for making me happy and inspiring my life.

Tel Aviv University, Israel   Roni Aloni
April 2020

Contents

1	**Introduction**.	1
	References.	5
2	**Structure, Development, and Patterns of Primary, Secondary,**	
	and Regenerative Vascular Tissues	7
	2.1 Primary and Secondary Vascular Meristems	7
	2.2 The Two Conducting Tissues of Vascular Plants	12
	2.2.1 Longitudinal Vascular Bundles of Phloem and Xylem	12
	2.2.2 Phloem Anastomoses.	24
	2.3 Patterns of Primary Vascular Differentiation in Roots	
	and Shoots	27
	2.4 Discontinuous Vascular Patterns During Bundle Maturation	31
	2.5 Parenchyma Cells in the Vascular System	34
	2.6 Usual, Uncommon, and Regenerative Tracheids and Fibers	36
	2.7 Vascular Differentiation in Organ Junctions	40
	2.8 Reaction Wood and Gelatinous Fibers.	40
	2.8.1 Compression Wood	43
	2.8.2 Tension Wood.	43
	2.8.3 Gelatinous Fibers in Climbing Shoots.	45
	2.9 Vascular Differentiation Under Metabolic and Environmental	
	Stresses.	45
	References and Recommended Reading.	48
3	**The Hormonal Signals that Regulate Plant Vascular**	
	Differentiation	55
	3.1 Auxin (IAA) Is the Young Leaf Signal	57
	3.1.1 Free and Conjugated Auxin	60
	3.1.2 Sachs' *Canalization Hypothesis*	62
	3.1.3 Subcellular Mechanisms of Polar Auxin Transport:	
	Cycling and Vesicle Trafficking of PIN1 and AUX1	64
	3.1.4 IAA Transport Pathways	67
	3.1.5 Polar Movement of IAA in a Wave-Like Pattern.	72
	3.1.6 Sensitivity to Auxin	73
	3.1.7 Sugar from Photosynthesis Promotes Auxin Synthesis	73

3.2   Gibberellins (GAs) Are the Mature Leaf Signals   74
3.3   Cytokinins (CKs) Are the Root Cap Signals   76
3.4   Ethylene (C_2H_4) the Gaseous Signal   78
3.5   Abscisic Acid (ABA) Regulates Stress Responses   79
3.6   Jasmonates: Activate Defense Responses   81
3.7   Brassinosteroids (BRs).   82
3.8   Strigolactones (SLs) Are Root Hormones and Attracting
      Signals of Symbiotic Fungi and Root-Parasitic Weeds   83
3.9   Florigen Is the Universal Trigger for Flowering and Associated
      Fiber Differentiation   84
3.10 Hormonal Cross Talk and Interactions.   84
References and Recommended Readings.   85

4   **Importance of NO_3^- and PO_4^- for Development, Differentiation,
    and Competition**   97
    References and Recommended Reading.   98

5   **Phloem and Xylem Differentiation** 101
    5.1   Low- and High-Concentration Polar Auxin Streams Induce
          Phloem and Xylem. 101
    5.2   Hormonal Control of Xylem Cell Differentiation 107
          5.2.1   Tracheid Differentiation. 108
          5.2.2   Vessel Differentiation. 111
          5.2.3   Fiber Differentiation 113
          5.2.4   Insights from Transdifferentiation of Isolated Single
                  Parenchyma Cells into Tracheary Elements in Liquid
                  Cultures 122
    References and Recommended Readings. 124

6   **Apical Dominance and Vascularization** 131
    6.1   Organ Communication. 131
    6.2   Regulation of Shoot Apical Dominance 132
    6.3   Cytokinin-Dependent Root Apical Dominance 134
    References and Recommended Readings. 137

7   **Leaf Development and Vascular Differentiation** 141
    7.1   Auxin-Dependent Leaf Initiation and Phyllotaxis. 141
    7.2   Hormonal Control of Venation Pattern Formation in Leaves. 142
          7.2.1   The *Leaf Venation Hypothesis* and Evidence. 145
          7.2.2   Hydathodes Are Induced by Auxin Maxima that
                  Suppress the Development of Lower Hydathodes. 155
          7.2.3   Tertiary Veins and Freely Ending Veinlets Are Induced
                  by Minor Auxin-Producing Sites 157
    References and Recommended Readings. 160

8   **Flower Biology and Vascular Differentiation** 163
    8.1   The FT Protein Is the Leaf-Inducing Signal for Flowering
          and Fiber Differentiation 163

8.2 The ABC Model of Flower Development 164
8.3 High-IAA-Producing Anthers Synchronize Flower Development . . 165
8.4 Vascular Differentiation in Floral Organs and Fruit Development . . 169
References and Recommended Readings 175

9 Can a Differentiating Vessel Induce Lateral Root Initiation? 177
References and Recommended Readings 182

10 Vascular Regeneration and Grafting 185
10.1 Vascular Regeneration 185
10.2 Grafting 192
References and Recommended Reading. 196

11 Regulation of Cambium Activity 199
11.1 Hormonal Control of Cambial Activity and Secondary
Vascular Differentiation 199
11.2 Cambial Dormancy 207
11.3 Cambial Activity Reflects the Social Status of a Forest Tree 208
11.4 Woodiness 209
References and Recommended Reading. 211

12 Regulation of Juvenile-Adult Transition and Rejuvenations 215
References and Recommended Readings 221

**13 The Control of Tracheid Size, Vessel Widening and Density
Along the Plant Axis** 223
References and Recommended Reading. 233

**14 Circular Vascular Tissues, Vessel Endings and Tracheids
in Organ Junctions** 237
References and Recommended Readings 242

15 Ray Differentiation: The Radial Pathways 245
References and Recommended Readings 249

16 Environmental Adaptation of Vascular Tissues 251
References and Recommended Readings 257

17 Resin Glands and Traumatic Duct Formation in Conifers 259
References and Recommended Readings 262

18 Hormonal Control of Reaction Wood Formation 265
References and Recommended Readings 270

19 Hormonal Control of Wood Evolution 273
19.1 The Development of Perforations 273
19.2 Evolution of Vessels and Fibers from Tracheids 276
19.3 Evolution of Ring-Porous Wood from Diffuse-Porous Xylem
in Limiting Environments 277
References and Recommended Reading. 288

**20 How Vascular Differentiation in Hosts Is Regulated by Parasitic
    Plants and Gall-Inducing Insects**. 293
    20.1 Xylem and Phloem Connections Between Parasitic
          and Host Plants 293
    20.2 Vascular Differentiation Controlled by Gall-Inducing Insects 298
    References and Recommended Readings. 305

21 Cancer and Vascular Differentiation. 309
    21.1 Role of Vascular Differentiation in Plant Tumor
          Development 309
    21.2 Comparison Between Plants and Animal Tumors 321
    21.3 The Plant Hormone Jasmonate, Antibiotics and Bioactive
          Secondary Metabolites Are Promising Candidates for
          Human Cancer Therapy. 323
    References and Recommended Readings. 326

Index. 331

About the Author

Roni Aloni, completed his studies at the Hebrew University of Jerusalem, Israel. He graduated in Biology (1968), obtained MSc in Plant Ecology (1970), and completed his PhD in Plant sciences (1973). Dr. Aloni became Lecturer in Biology at the Technion, Israel Institute of Technology (1973–77). He completed his Postdoctoral research as Visiting Fellow in Biology at Princeton University, USA (1974–76). In 1977, he moved to Botany at Tel Aviv University, Israel, and became Professor in 1992 and Professor emeritus in 2013. He continues teaching and runs active research. Dr. Aloni is author of numerous scientific publications, including Springer's book *Vascular Differentiation and Plant Growth Regulators* by: LW Roberts, PB Gahan, R Aloni (1988). His research goal is to understand the hormonal mechanisms that induce and regulate plant vascular differentiation, adaptation, and evolution in organized tissues, galls, and tumors. Since 2005, he is a Member of Leopoldina, The German National Academy of Science, and since 1991, a Fellow of the International Academy of Wood Science, and on the Academy Board from 2014 to 2020. He served as the President of the Botanical Society of Israel (1996–98). Dr. Aloni coordinated the working parties on "Biological control of wood quality" (5.01.01) and "Xylem physiology" (2.01.10) of the International Union of Forestry Research Organizations (IUFRO) (1990–2012). He was an Adjunct Professor of Biology at the University of Waterloo, Canada (1989–99). Dr. Aloni has spent sabbaticals at Harvard University, USA (1982–83); University of

Waterloo, Canada (1989); University of Vermont, USA (1991); University of Reading, England (1995); University of California, Berkeley, USA (1999–2000); Technische Universität Darmstadt, Germany (2004); and Simon Fraser University, Canada (2009–10). He has been an Editor of *Planta* since 2008, was Editor of *Trees, Structure and Function* (1990–2013), and was an editor and editorial board member of a few international plant science journals.

Abbreviations

ABA	Abscisic acid
ABC	ATP-binding cassette
ACC	1-Aminocyclopropane-1-carboxylic acid
ACS	1-Aminocyclopropane-1-carboxylate synthase
AHK	ARABIDOPSIS HISTIDINE KINASE
APL	ALTERED PHLOEM DEVELOPMENT
ARR	ARABIDOPSIS RESPONSE REGULATOR
ARR5::GUS	Cytokinin-sensitive response element
AtHB8	HOMEOBOX GENE8
AUX1	AUXIN-RESISTANCE1
AUX/LAX	AUXIN-RESISTANCE/LIKE-AUX
AuxRE	Auxin-responsive elements
AVG	Aminoethoxyvinylglycine
BR	Brassinosteroid
CC	Companion cell
CF	Methyl-2-chloro-9-hydroxyfluorene-9-carboxylic acid
C_2H_4	Ethylene
CK	Cytokinin
CKK	CYTOKININ OXIDASES/DEHYDROGENASES
CLSM	Confocal laser-scanning microscope
CLV3	CLAVATA3
CSC	Cancer stem cell
DAG	Days after grafting
DR5::GUS	Synthetic auxin-sensitive response element
FT	FLOWERING LOCUS T
GA	Gibberellin
G-fiber	Gelatinous fiber
GFP	Green fluorescent protein
GH3	Gretchen-*Hagen* 3
G-layer	Gelatinous layer
GN	GNOM
GTR1	Influx carrier of JA-Ile and gibberellin
GUS	β-glucuronidase

HD-ZIP III	CLASS III HOMEODOMAIN-LEUCINE ZIPPER
IAA	Indole-3-acetic acid, auxin
IAAld	Indole-3-acetaldehyde
IAM	Indoleacetamide
IAOx	Indole-3-acetaldoxime
ICA	INCREASED CAMBIAL ACTIVITY
iP	Isopentenyladenine
iPA	Isopentenyladenosine
IPT	Isopentenyltransferase
Irx	Irregular xylem
JA	Jasmonic acid
JA-Ile	Jasmonoyl-l-isoleucine
JAT1	Efflux carrier of jasmonic acid
KAN	KANADI
LeFRK	Fructose-phosphorylating enzyme
LRi	Lateral root initiation
MeJA	Methyl ester of jasmonic acid
MP (ARF5)	MONOPTEROS (AUXIN RESPONSE FACTOR 5)
mRNA	microRNA
NAA	Naphthaleneacetic acid
NPA	1-N-naphthylphthalamic acid
NPF	Nitrate transporter peptide
Nr	Never ripe
OPDA	cis-12-oxo-phytodienoic acid
oxIAA	2-oxindole-3-acetic acid
PCD	Programmed cell death
PDR	PLEIOTROPIC DRUG RESISTANCE
Pe	Pericycle
PEAR	PHLOEM EARLY DOF
PILS	PIN-LIKES
PIN	PIN-FORMED
P-protein	Phloem protein
PX	Protoxylem vessels
QC	Quiescent center
RAM	Root apical meristem
REV	REVOLUTA
SA	Salicylic acid
SAM	Shoot apical meristem
SCN	Stem cell niche
SCWB	Secondary cell wall biogenesis
SE	Sieve element
SFT	SINGLE FLOWER TRUSS
SL	Strigolactone
STS	Silver thiosulphate
TAA	TRYPTOPHAN AMINOTRANSFERASE OF ARABIDOPSIS

TAA/YUC	Auxin biosynthesis pathway in plants
T-DNA	Transferred DNA of the tumor-inducing Ti plasmid
TE	Tracheary element
TIBA	2,3,5-triiodobenzoic acid
Ti-plasmid	Tumor inducing plasmid of *Agrobacterium tumefaciens*
Trp	Tryptophan
VND	VASCULAR-RELATED NAC-DOMAIN
WOX	WUSCHEL-related HOMEOBOX
WT	Wild-type
XP	Xylem parenchyma
YUC	YUCCA
Z	Zeatin
ZR	Zeatin riboside

List of Figures

Fig. 1.1   Micrographs of cleared leaf primordia of DR5::GUS transformed
           Arabidopsis thaliana demonstrating how the bioactive auxin
           hormone, namely, indole-3-acetic acid (IAA), induces vascular
           veins, visualizing the process by the blue staining of *DR5::GUS*
           expression during leaf morphogenesis. (**A**) At the lobe of a young
           leaf primordium, where a hydathode will develop, showing a
           center of strong expression (*arrow*) marking the synthesis site of
           high-auxin concentration, from which the auxin starts to flow
           downward in a diffusible pattern that gradually becomes canalized
           to a narrow stream which induces the vascular vein (*arrowhead*).
           (**B**) In a more developed primordium, *GUS* expression is detected
           at the base of a few auxin-producing trichomes (*arrows*), showing
           how these trichomes induce their water-supporting vascular veins.
           Note that two of the trichomes are associated with freely ending
           veinlets (*arrowheads*) that they induced. Bar = 50 μm (**A**), 250 μm
           (**B**). (From Aloni et al. 2003) 2

Fig. 1.2   Photographs of coast redwoods (*Sequoia sempervirens*) grown in
           north California. (**A**) Showing three giant trees of about 100 m
           height. The two trunks on the right have straight bark patterns,
           while the left tree has a spiral bark pattern, likely induced by
           spiral movement of auxin in the phellogen, which is the cork
           cambium (see Chap. 3, Fig. 3.11A). (**B**) An upward photograph
           made from the inside of a continuously growing tree, showing a
           living redwood tree where its inside dry wood was burnt by a
           strike of lightning, while the outer moisturized wood layers
           remain alive, and the tree is covered by its own many green active
           leaves and growing branches, demonstrating that only the outer
           vascular layers are the living tissues of a tree and the dead core of
           a tree can be burnt and the tree continues to grow. *Arrow* marks
           some annual wood rings that were not burnt; T, top of tree
           damaged by the lightning 4

Fig. 2.1   Longitudinal view of the shoot apical meristem (SAM) of *Coleus blumei*, showing the primary vascular meristem, the procambium (*arrows*), descending from the leaf primordia tips. Primary phloem and later also primary xylem differentiate at the node (*arrowhead*). Scale Bar = 200 μm   8

Fig. 2.2   Longitudinal views of both the SAM and the root apical meristem (RAM) in a 3-day-old seedling of free-cytokinin-responsive ARR5::GUS transformant of *Arabidopsis thaliana*, comparing the plant apices regarding the distribution of the active root hormone cytokinin in both cleared plant's poles (see Aloni et al. 2004, 2005). (**A**) The SAM of a 3-day-old seedling is cytokinin-free. (**B**) The RAM is protected by the root cap that produces high quantities of free cytokinin (detected by blue *ARR5::GUS expression*). The *arrows* mark the quiescent center (QC) at both apices, where the frequency of cell divisions is the lowest. The dotted line marks the border of the vascular cylinder with the cortex and root cap. Bars = 25 μm. (Aloni et al. 2004)   9

Fig. 2.3   Longitudinal tangential view of the cambium cells in *Populus maximowiczii*, showing the short ray initials (r) that build the vascular rays and the long fusiform initials (f). Bar = 50 μm, Courtesy of R. Funada and Y. Murakami   10

Fig. 2.4   Cambial zone pattern in a cross section of *Nicotiana tabacum*, showing similar undifferentiated cambial derivatives on both sides of the cambial initials (c). An early stage of vessel (v) expansion is observed on the xylem side. Bar = 50 μm   11

Fig. 2.5   Schematic diagrams of longitudinal and its cross sections illustrating the developmental patterns of the meristematic (pro)cambium (yellow), the phloem (red), primary xylem (light blue), and secondary xylem (dark blue) in a dicotyledonous plant. Besides the longitudinal drawing, the cross sections 1–5 show vascular development along the stem, while cross sections I–IV illustrate vascular development in the root. Note that the phloem starts to differentiate before the xylem along the stem and root. In the stem, the vascular bundles differentiate at the periphery, and wide pith develops in the center. In the root, the vascular tissues occupy the center and are surrounded by a wide cortex. These patterns are primarily induced by polar auxin streams, originating in leaves, which merge into the root   13

Fig. 2.6   Cross section in a young internode of *Coleus blumei*, showing phloem (*arrows*) and collateral (when vessels develop) vascular bundles at the stem periphery. Vessels (v) differentiate in the more developed bundles, while the narrow bundles are built only of phloem. During internode development, additional vessels will differentiate in the larger bundles. Bar = 200 μm. (From Aloni 1988)   14

Fig. 2.7   Cross section in a young internode of corn (*Zea mays*), showing
large collateral vascular bundles in the center of the stem (lower
side) with phloem (*arrows*), functioning narrow protoxylem
vessels (*arrowheads*), and expanding metaxylem vessels (m). The
bundles at the periphery near the epidermis (e) descended from an
upper internode, where they were located in the center of the stem.
Because they are away from their inducing leaf, they receive
low-concentration polar auxin stimulation; therefore, these bundles
are small, and they do not produce protoxylem vessels; their
low-auxin stimulation is enough to produce one or two slow
differentiating metaxylem vessels. Bar = 200 µm   15

Fig. 2.8   Cross sections in well-developed internodes of *Luffa cylindrical*
stained with lacmoid according to Aloni (1979). (**A**) Showing a
bicollateral vascular bundle, with phloem (*arrows*) on both sides
of the xylem. In the xylem, the protoxylem vessels are marked
with *black arrowhead*, and the first produced secondary vessels are
marked with *small white arrowhead*, while the giant vessels (*large
arrowhead*) are the last ones to differentiate by very low-auxin
stimulations descending from far away leaves, which therefore
induce very slow differentiation for a few weeks. (Auxin application to such an internode prevents the formation of these giant
vessels and induces additional regular size secondary vessels; the
experimental evidence is not presented, R. Aloni unpublished).
(**B**) Magnified view of the phloem, showing sieve elements
(*arrows*), and the middle one with a sieve plate covered with
callose (stained sky blue by lacmoid). The companion cells
(*arrowheads*) have dense cytoplasm, which was stained blue.
Bars = 100 µm   16

Fig. 2.9   Longitudinal views of the phloem (which was removed from the
xylem at the cambial zone and is observed from the cambium side)
in intact stems of *Dahlia pinnata*, stained with aniline blue and
observed under epifluorescence microscope. (**A**) Showing sieve
tubes in a bundle with simple (*arrow*) and compound sieve plates
(*arrowhead*) with callose (stained yellow). Callose is also detected
in the plasmodesmata scattered on the longitudinal walls of the
sieve elements (observed as many small yellow points).
(**B**) Patterns of phloem anastomoses (*arrows*) are evident between
the longitudinal vascular bundles. Bars = 100 µm.......................   17

Fig. 2.10   Longitudinal views of the phloem in the stems of *Dahlia pinnata*,
demonstrating the effect of wounding on callose accumulation in
the same studied location, showing (**A**) the zero-time control
before wounding, (**B**) 8 minutes after injury, and (**C**) 15 min after
injury. (To view the phloem from the cambium side, 50-mm-long
portion of the xylem was gently removed at the cambium, while a
wide strip of continuous phloem was kept intact in 0.01% aniline

blue, in 0.05 M K-Na phosphate buffer, pH 7.2, and observed
under epifluorescence microscope.) (**A**) No callose can be detected
in the intact phloem (during spring, in a 6-week-old stem) before
making the horizontal wound. (**B**) 8 min after wounding, large
quantities of callose accumulation are evident immediately above
the wound (w) plugging the damaged sieve tubes. (**C**) 15 min after
wounding, callose accumulation is observed further away at a
greater distance from the wound. *Arrows* mark callose on remote
sieve plates. Bar = 100 μm... 18

Fig. 2.11   Interxylary phloem strands embedded in the secondary xylem of a
Salvadora persica stem stained with aniline blue and observed
under epifluorescence microscope. (**A**) Showing the anatomy of a
well-developed stem with the external phloem outside the xylem
(*arrowhead*) and interxylary phloem strands (*arrow*) of different
sizes embedded inside the secondary xylem, forming safer
pathways for photosynthate translocation than through the external
phloem. (**B**) in a young internode, narrow interxylary phloem
strands built of sieve tubes (the *arrow* marks a sieve plate with
callose accumulation) surrounded by parenchyma cells (*black
color*). The xylem is mainly composed of fibers (F) and vessels
(V). (**C**) A wide (tangentially flattened) interxylary phloem strand
in an older internode, resulting from the merging of narrow
phloem strands (the *arrow* marks phloem parenchyma inside the
strand). Bars = 50 μm (**A**), 100 μm (**B, C**) 19

Fig. 2.12   Longitudinal three-dimensional view of both the xylem and
phloem in a thick section of cucumber (*Cucumis sativus*) stem
cleared with lactic acid and stained with lacmoid according to
Aloni and Sachs (1973). Showing in the xylem, annular rings
(*arrows*) and spiral secondary wall thickenings (*arrowheads*)
inside vessels. Due to early internode elongation, the annular rings
of the first formed narrow protoxylem vessel (at the left side) were
stretched away from each other (*arrows*). Note that the distances
between the annual rings and late formed spiral secondary wall
thickenings become smaller (from left to right), as the internode
slows its elongation. In the phloem, sieve tubes can easily be
followed due to the blue staining of their P protein and the sky
blue staining of their sieve plates (*arrows*). The first induced sieve
tube has very long protophloem elements (on the right side). Late
formed sieve elements are shorter (close to xylem). A phloem
anastomosis (*arrowhead*) differentiated inside the bundle.
Bar = 100 *μm* 20

Fig. 2.13   A drawing of the spiral secondary wall thickenings in a partially
cut leaf of *Vitis vinifera* published in the pioneering *Anatomy of
Plants* book, written by Nehemiah Grew (1682), indicating that
the spiral secondary wall thickenings (*arrows*) along a vessel
might be continuous along a few vessel elements........................ 21

Fig. 2.14 (**A**) Experimentally induced vessel ending in the upper side of a vessel of *Luffa cylindrica*, characterized by a pitted secondary wall pattern. The *arrowheads* mark pits (From Indig and Aloni 1989). (**B**) Early stages of penetration of parenchyma cells (*arrows*) into an air-filled nonfunctional vessel in hops (*Humulus lupulus*) stem. These tyloses, which have balloon-like appearance, tend to expand and prevent invasion and movements of pathogens along the vessel. Bars = 25 μm.. 22

Fig. 2.15 Longitudinal views of basic pit structures in vascular elements. (**A**) Showing the radial and (**B**) The tangential views of bordered pits in earlywood tracheids of *Cupressus sempervirens*. In (**A**) and (**B**), the *black arrows* mark the minimized pit aperture (opening) in the secondary walls of the tracheids, the *white arrows* mark the torus, and the *arrowheads* mark the margin of the relatively porous region of primary wall (margo) inside the pit cavity through which water is transported between adjacent tracheids (see schematically drawn in Fig. 2.16). (**C**) Showing simple pit between adjacent primary phloem fibers in *Coleus blumei*. Their primary cell wall is marked by an *arrow* and the secondary wall by *arrowhead*. Bars = 5 μm 22

Fig. 2.16 Schematic diagrams showing how bordered pit function. (**A**) Pathway of water (w) flow through the primary wall margo around the torus of intact transporting tracheids. (**B**) Air (a) entrance from an embolized tracheid to the water (w)-transporting tracheid is prevented by the torus that seals the opening (pit aperture) at the functioning tracheid side, due to the negative xylem pressure in the water-transporting tracheids. (**C**) Simple pit enables water movement through the primary cell wall of the pit cavity between two secondary walled fibers 23

Fig. 2.17 Longitudinal views of transporting phloem in the stems of *Dahlia pinnata*, showing the pattern of functioning sieve tubes transporting the fluorescent dye, fluorescein, in living phloem tissue observed under an epifluorescence microscope. (**A**) In an intact stem, the fluorescein (yellow stain) is translocated along a longitudinal sieve tube, but not in anastomoses. (**B**) Wound (w) promoted fluorescein translocation in a phloem anastomosis (*arrow*). (**C**) Magnified view of fluorescein-filled sieve tubes (*arrow*) in phloem anastomoses beside a wound. Note that there is a local increase of fluorescein concentration at the sieve plates. (**D**) Diagram showing the transport of fluorescein in the sieve tubes around the wound (w) in a *Dalia* stem. The fluorescein was applied to the internode below the wounded one shown in the diagram. The diagram demonstrates that wounding activated the phloem anastomoses, which serve as an emergency system providing alternative pathways for assimilates around the stem. Bars = 125 μm (**C**), 250 μm (**A, B, D**). (From Aloni and Peterson 1990) 25

Fig. 2.18   Phloem anastomoses originate from parenchyma cells between
             longitudinal vascular bundles in the stem of *Dahlia pinnata*,
             showing longitudinal views of phloem cleared with lactic acid
             and stained with lacmoid. (**A**) Early stage of anastomosis
             differentiation in a very young internode, showing the pattern of
             parenchyma cell divisions (*black arrows*) from them, the sieve
             elements of the anastomosis develop between longitudinal
             bundles. (**B**) Developed anastomosis built of a few sieve tubes
             with callose (*black arrows*) on their sieve plates in a mature
             internode. *White arrows* mark sieve plates in the longitudinal
             bundles. Bars = 1 mm. (From Aloni and Barnett 1996)   26
Fig. 2.19   Primary vascular differentiation in roots is characterized by a
             radial pattern of alternating strands of xylem and phloem,
             showing cross sections of developmental stages of adventitious
             dicotyledonous roots of sweet potato (*Ipomoea batatas*) grown
             in a wet soil, therefore developing large air spaces of aeren-
             chyma (ae) in the cortex. (**A**) Early stage of a young root with
             four xylem (*arrows*) and phloem (ph) strands and early expan-
             sion of a metaxylem (m) vessel. (**B**) More developed adventi-
             tious root with a wider stele that enables the development of five
             alternating xylem and phloem stands with an already functional,
             wide lignified (stained red) metaxylem vessel in the center. Early
             stage of cambium (C) development appears between the primary
             phloem and primary xylem. Bars = 100 μm. (Courtesy of
             V. Singh and N. Firon)..   28
Fig. 2.20   Cross sections showing radial patterns in monocot adventitious
             roots of corn (*Zea mays*), with many alternating protoxylem (*p*,
             or *yellow lines*) archs and protophloem (*s*, or *broken red lines*)
             strands at the periphery of the vascular cylinder, around wide
             metaxylem (m) vessels and parenchyma pith in the center. (**A**)
             Large and rounded metaxylem vessels differentiated in a well-
             developed root grown under favorite water irrigation and root
             space. (**B**) Deformed metaxylem vessels due to thin secondary
             wall formation in a stressed-developed root grown under limited
             root space and irrigation, demonstrating that environmental stress
             decreases secondary wall deposition, resulting in vessels with thin
             secondary cell walls, which become deformed under pressure of
             surrounding cells. The deformed vessels may substantially reduce
             the efficiency of water transport to leaves. Bars = 100 μm.
             (R. Aloni and J. Mattsson, unpublished)   29
Fig. 2.21   Naturally occurring discontinuous vessel patterns in cleared intact
             wild-type *Arabidopsis* leaf primordia observed under Nomarski
             illumination. Under this illumination, the procambium is observed
             as dark lines, and vessels have a white appearance. (**A**)
             Discontinuous xylem (*arrowheads*) at the leaf tip, showing

basipetal (*arrows*) and acropetal (*small arrows*) vessel differentiation. (**B**) Typical discontinuous xylem initiation (*arrow*) in a lobe,
promoted by the high-auxin concentration of a developing
hydathode (asterisk). (**C**) Typical early stages of lateral vascular
bundle maturation, showing two discontinuous vessels (*each
vessel is marked by the same size arrowheads*) progressing
acropetally (a) and basipetally (b) along the procambium.
Bars = 200 μm. (From Aloni 2004) 32

Fig. 2.22   Naturally occurring discontinuous vessel patterns in cleared
intact wild-type *Arabidopsis* flower primordia photographed
under Nomarski illumination (giving the xylem a white appearance). (**A**) and (**B**), showing early discontinuous xylem patterns
immediately beneath the tips of developing flower organs (at the
sites of high-auxin concentrations). (**A**) In the young stamen
primordium, the vein differentiates at the anther (*arrowhead*) and
progresses basipetally. (**B**) Two sites of xylem initiation at the tip
of a very young gynoecium (*arrowheads*), starting beneath the
stigma and progressing downward. (**C**) In mature gynoecium, the
short veinlets (*arrowheads*) originated from the ovules, but could
not connect to the central bundle of the gynoecium (due to
higher-auxin streams descending from the stigma through the
central gynoecium bundle; for explanation, see Fig. 3.7C).
Bars = 200 μm. (From Aloni 2004) 33

Fig. 2.23   Longitudinal views of cleared juvenile needle-like leaves of the
conifer *Thuja plicata*. (**A**) Showing primary tracheids (*arrowhead*) (originated from procambium) in an early stage of bundle
differentiation in a young leaf primordium (before any transfusion tracheid development). (**B**) and (**C**), showing short tracheids
(*arrows*), which form a vessel-like pattern in well-developed
leaves. Water moves from one transfusion tracheid to the other
which occurs via bordered pits (*small vertical arrows*).
Bars = 100 μm (**A**), 25 μm (**B, C**). (From Aloni et al. 2013) 36

Fig. 2.24   Longitudinal views of cleared juvenile needle-like leaves of
T. plicata, showing short transfusion tracheids (*arrows*) and
long primary tracheids (*arrowheads*) at the tip of the leaf's
vascular bundle. Bars = 100 μm (**A**), 50 μm (**B**). (From Aloni
et al. 2013) 37

Fig. 2.25   Longitudinal views of phloem fibers (F) in tissues cleared and
stained with lacmoid (according to Aloni and Sachs 1973). (**A**)
The central part of a primary phloem fiber and the lower end
(*black arrow*) of a nearby fiber, after substantial intrusive
elongation in an intact internode of *Coleus blumei*, in comparison
to sieve elements (the callose on their sieve plates was stained
sky blue, marked by *white arrows*), and surrounding parenchyma
cells that remained short (upper and lower ends marked by

arrowheads). (**B**) The upper portions of typical tumorous regenerative phloem fibers with several ramifications (*arrow*) grown intrusively in various direction in the lower region of a 6-week-old *Ricinus* crown gall (tumor tissue experimentally induced by *Agrobacterium tumefaciens*, Aloni et al. 1995), due to interruptions to their growth caused by neighboring dividing and expanding parenchyma cells. (**C**) A magnified view a tumorous regenerative phloem fiber in a *Ricinus* crown gall with three ramifications (*arrows*) developed after the upper tip stop growth due to interruption caused by neighboring parenchyma. Bars = 50 μm. (**B** and **C** are from Aloni et al. 1995) 38

Fig. 2.26   Regenerative phloem fibers induced above wounds in young *Coleus blumei* internodes (cleared and stained with lacmoid). (**A**) Showing a few short isolated fibers (*arrows*) originated from parenchyma cells between the cut bundles. (**B**) 60-day-old long regenerative phloem fibers after intrusive growth (*arrowheads*) and shorter ones that were likely developed later during the regeneration process (*arrow*). Bars = 100 μm. (From Aloni 1976) 39

Fig. 2.27   Gibberellin signaling at leaf-stem junctions, demonstrating the structure and function of the leaf-stem junction. The figures show longitudinal images of cleared *Arabidopsis* plants with rosette leaves. Dotted lines mark petiole-stem junction line. (**A**)–(**C**), the leaf-stem junction (where the abscission zone will develop) is characterized by short cells. Short tracheary elements (marked with blue in **B**) differentiated between the long vessels (marked with red in **B**) of the petiole and stem. An identical vessel element is marked by short-thick arrows in (**A**) and (**B**). (**B**) The cell walls of some of the short vessel elements between the long ones are outlined in blue and red, respectively. (**C**) Longitudinal section through the base of the petiole, showing short vessel elements (*arrowhead*) with high *ProGA2ox2:GUS* expression, due to high local gibberellin concentration. (**D**) Accumulation of *ProGA2ox2:GUS* expression is observed just above the petiole-stem junctions. *Arrows* point to site of GUS accumulation. (**E**) Paclobutrazol (a gibberellin inhibitor) treatment abolishes the *GUS* expression. Bars = 55 μm. (From Dayan et al. 2012) 41

Fig. 2.28   A large cross section revealing the anatomy at the base of a large branch after its removal from a conifer tree, showing the asymmetric pattern of massive compression wood produced in the lower side of the branch (below the *arrow*, which marks the original central pith of the cut branch), demonstrating minimum growth of the opposite wood above the *arrow*, and the massive compression wood produced in the lower side, which supports the weight of the heavy branch. Note that the lower side of the branch develops most of the active living compression sapwood, distinguished by its lighter color (while the dead heartwood in the center has a dark brown color). Bar = 50 mm. (R. Aloni, unpublished) 42

Fig. 2.29 Confocal laser scanning microscopic images of cross sections of tracheids in *Abies sachalinensis*, after staining with safranin. (**A**) Normal wood; (**B**) Compression wood. The compression wood differentiated in the lower side of the inclined stem. Bar = 25 μm. (Courtesy of R. Funada, Y. Sano, Y. Yamagishi, and S. Nakaba)... 43

Fig. 2.30 Light micrograph of a cross section from the upper side of inclined stem of *Betula platyphylla*, showing gelatinous fibers, which differentiated after stem inclination. The gelatinous fibers are characterized by the development of the inner gelatinous layer, stain blue (*arrow*) by safranin and Astra blue according to Nugroho et al. (2012). The continuous region that developed the G-fibers is marked by a line on the right side of the micrograph. Bar = 100 μm. (Courtesy of R. Funada) 44

Fig. 2.31 Cross sections in the phloem (**A**) and xylem (**B**) of the climbing stem of hops (*Humulus lupulus*) stain with lacmoid according to Aloni and Sachs (1973), showing gelatinous fibers in both vascular tissues. Their lignified walls stained blue and their inside G-layer remained white. (**A**) In the phloem, wide primary phloem G-fibers with cellulose and almost no lignin are the first to differentiate (*black arrow*) during internode development. They are followed by the formation of many narrow secondary phloem G-fibers (*white arrows*) among the secondary sieve tubes (s). (**B**) In the xylem, the large secondary xyle fibers (*arrowhead*) differentiate near the pith during internode elongation, but during the early stage of internode elongation, they do not produce the G-layer. They are followed by the differentiation of many narrow secondary G-fibers (*white arrow*) among wide vessels (v). During later internode developmental stage, at the time of longitudinal stem shrinkage, which causes the coiling and twining of the climbing stem, only then the wide xylem fibers (*arrowhead*) produce their G-layer and the stem becomes fixed. Note that the rays (R) have a dark appearance due to winter starch accumulation. Starch is also observed in the pith. Bars = 100 μm. (R. Aloni, K. Sims, J. Mattsson, unpublished) 46

Fig. 2.32 Xylem and phloem structure in *LeFRK2*-antisense stems (**A–C**) metabolically stressed due to reduced fructose production, compared to wild-type (**D–F**) tomato stems. (**A**) Cross section from the middle of a stem showing that suppression of *LeFRK2* reduces cambium (C), vessel (V), and sieve element (*black*

arrow) width and slows fiber differentiation, resulting in a few
layers of differentiating fibers (*white f*) with dense cytoplasm.
Note that the secondary vessels near the cambium differentiate
with a circular shape (*white arrows*) and become deformed (V)
away from the cambium due pressure of neighboring cells on
their thin secondary wall. Mature fibers (*black f*) are narrow and
have thin cell walls. (**B**) Xylem from the base of the stem
characterized by deformed vessels and narrow fibers (f) both
with thin cell walls. (**C**) Longitudinal section from the base of
the antisense stem showing short and narrow sieve elements with
almost no callose on the sieve plates (*arrows*) and a companion
cell (CC). (**D**) Cross section from the middle of a wild-type stem
characterized by wide cambium, large vessels and sieve tubes
(*arrow*), and mature fibers (f). (**E**) Widest vessels and fibers,
both with thick secondary walls, from the stem base. (**F**)
longitudinal section from the base of a wild-type stem showing
normally wide sieve elements with callose deposited on their
sieve plates (*arrows*). *Scale bars* = 50 μm. (From Damari-
Weissler et al. 2009) 47

Fig. 3.1	The molecular structure of the four primary hormonal signals
that control vascular differentiation in plants: auxin, gibberellin,
cytokinin, and ethylene.. 56

Fig. 3.2	Schematic diagram illustrating the upward transport of auxin
from young leaves to the shoot apical meristem (SAM). Auxin
moves via the epidermis, and the outermost meristematic cell
layer, creating auxin maxima at the shoot tip, which induce the
initiation of new leaf primordia. 0, shows the site of auxin
maximum inducing a new leaf primordium. 1, shows auxin
transport in the youngest growing leaf primordium, upward in
the differentiating epidermis and then downward in the center of
the leaf, where the signal induces the midvein............................ 59

Fig. 3.3	(**A**) Schematic diagram illustrating the upward transport of auxin
in the epidermis, from the base of the leaf primordium to its tip,
establishing auxin maximum (*asterisk*) at the tip, from which
auxin continues downward and induces the midvein in the center
of the growing leaf. (**B**) Upward PIN1 orientation during early
development of *Arabidopsis thaliana* rosette-leaf primordium,
directing polar auxin transport from leaf-base to tip, showing
tip-directed polarity of PIN1 in the epidermis (*arrows*), detected
by PIN1:GFP under GFP fluorescence and transmitted light
illumination. Scale bar = 10 μm. Micrograph **B** was reproduced
from Scarpella et al. 2006, *Genes & Development* 20:
1015–1027, with permission of Cold Spring Harbor Laboratory
Press and the authors 59

Fig. 3.4   Concentrated conjugated auxin accumulation in the youngest
             rosette-leaf primordium (*arrowhead*) and in its stipules (*arrows*)
             of a DR5::GUS transformed *Arabidopsis*. Showing auxin
             immunolocalization (stained red with mouse monoclonal red
             Alexa Fluor 568 fluorescence and rabbit polyclonal antibodies)
             viewed with a confocal laser-scanning microscope (CLSM). The
             accumulation of conjugated auxin in the youngest leaf primor-
             dium demonstrates that young leaf primordia are sink for auxin
             arriving from adjacent young leaves (see Sect. 7.1). This
             youngest leaf does not show free auxin (with DR5::GUS
             staining) at this early developmental stage. Bar = 20 µm. (From
             Aloni et al. 2003)   60
Fig. 3.5   Total auxin (mainly conjugated auxin) label in young
             Arabidopsis rosette-leaves stained with green Alexa Fluor 488
             fluorescence and rabbit polyclonal antibodies, viewed with
             CLSM. (**A**) Longitudinal section in a very young leaf primor-
             dium showing strong labelling in its compact cytoplasm (*arrows*)
             of densely packed mesophyll cells, but nearly no label in the
             epidermal cells (*arrowhead*). (**B**) Paradermal section in a more
             developed leaf, showing in its lamina strong auxin label in the
             chloroplasts (*white arrows*) and weaker labeling in the thin
             cytoplasmic layer in the elongated cells around the vascular
             bundles, containing vessels (*black arrow*), emphasizing the
             important role of chloroplasts in auxin metabolism. Bar = 20 µm.
             (From Aloni et al. 2003)   61
Fig. 3.6   Schematic diagrams illustrating the process of how increase in
             auxin flow promotes cell polarization and vascular tissue
             differentiation, as proposed by Sachs' *canalization hypothesis*.
             The arrows indicate the direction of auxin flow and their number
             in each cell the quantity of transported auxin, resulting from the
             distribution and quantity of the auxin efflux carrier in the plasma
             membrane. (**A**) Early stage of tissue development during which
             auxin flows by diffusion. (**B**) Auxin flow increases along a file of
             cells that start to transport auxin more efficiently. (**C**) The
             efficient transporting cell file become more polarized, transports
             more auxin and attracts auxin from neighboring cells. The
             continuous polar auxin transport through the canalized cell file
             will terminate in the formation of vascular conduit   63
Fig. 3.7   Longitudinal views of thick sections of pea (*Pisum sativum*)
             cleared in lactic acid and stained with phloroglucinol. The
             vascular tissues in the micrographs have a white appearance and
             the other cleared tissues are darker. The three micrographs show
             the basal shoot internode after the shoot above it was decapi-
             tated, for demonstrating how a lateral source of auxin applied in
             lanolin (its application sites, on the internode's right side, is

marked by the lateral *asterisk*) artificially induces new xylem
strands: without auxin (**A**), by low (**B**), or high (**C**) auxin
application on the decapitation cut (marked by one (**B**) or two
(**C**) *asterisks* on the decapitation cut). The laterally applied auxin
induced some tissue swelling at the application site. (**A**) The
induced strand jointed directly, in the shortest way, to the central
vascular cylinder, which was not supplied with auxin (and
therefore attracted the lateral auxin stream to be efficiently
transported through its polar well-transporting vascular cells).
(**B**) The auxin-induced xylem strand connected the central
vascular cylinder, which was supplied with auxin in lanolin
(upper *asterisk*), but in a much lower point (due to a moderate
rejection by the auxin-supplied central cylinder). (**C**) The
induced lateral strand did not connect to the central vascular
system (because it was well supplied with high auxin concentration, marked by the upper two *asterisks*), and therefore the
artificially induced strand differentiated quite parallel to the
pre-existing vascular system. All photomicrographs are at the
same magnification, Bar = 1 mm. The figure was prepared from
the original micrographs of Tsvi Sachs, which were published
in: Sachs (1968) *Annals of Botany* 32: 781–790, with permission
of Oxford University Press 64

Fig. 3.8   Schematic diagrams illustrating the flexibility of modifying
                auxin pathway flow, due to the auxin actin-dependent PIN1
                vesicle (V) cycling (PIN1 was stained *blue* in a protein
                complex *stained yellow and red*), which modifies PIN1
                location along the plasma membrane. (**A**) PIN1 cycling
                movement along actin microfilaments between the lower
                plasma membrane and an endosomal (E) component in intact
                cells; and (**B**) around a damaged cell (marked by X) showing
                that the PIN1 is relocated to the lateral membrane (on the *right
                side*, in the cell above the damaged one), directing the auxin to
                the adjacent right cell, thus forming a new bypass of polar
                auxin flow around a damaged cell 65

Fig. 3.9   (**A**) Downward PIN1 orientation (*arrows*) located on the lower
                membrane in cells adjacent to the midvein of intact young
                Arabidopsis rosette-leaf, directing polar auxin transport
                downward. (**B**) Lateral PIN1 orientation (*arrows*) observed
                2 days after wounding of *Arabidopsis* rosette-leaf. PIN1 was
                detected as PIN1:GFP under GFP fluorescence, viewed with a
                CLSM. Bar = 10 μm. (J. Dayan and R. Aloni, unpublished) 66

Fig. 3.10   Cross sections in stems of the *GH3* transformant of white
                clover, *Trifolium repens*, cv. Haifa, with the naturally occurring
                749-bp auxin responsive promoter region of the soybean gene
                GH3 (Hagen et al. 1991), contained at least three auxin-

responsive elements (AuxRE) fused to the *GUS* gene, transformed according to Larkin et al. (1996) and grown as
described by Schwalm et al. (2003). The histochemical blue
staining for GUS activity in the sections shows auxin pathways
in vascular bundles (**A**) and in a differentiating vessel (**B**, **C**).
Auxin transports through two vascular bundles (small bundle,
mainly phloem bundle on the left side, and large bundle on the
right side) occurs in the cambium (*white arrowheads*), in the
sieve tubes (*white arrows*), in a differentiating vessel (*small
black arrowhead*), in the bundle sheath (*black arrow*) and in
xylem parenchyma cells (*large black arrowhead*). Blue stained
vesicles (*arrowheads*) are observed along actin filaments in a
differentiating vessel (**B**) which is enlarged in (**C**). v = vessel.
Bar = 10 μm in (**C**), 50 μm in **A**, **B**. (Photographed by R Aloni
and CI Ullrich; published in Aloni 2013b) 68

Fig. 3.11 Patterns of auxin distribution in the vascular and protective
tissues of *Arabidopsis thaliana* viewed and recorded in CLSM
fluorescence intensity profile of Alexa 568-labeled
monoclonal-IAA antibodies along two (**A** and **B**) radial
straight line in cross sections in two hypocotyls, excited with
568-nm light (emission 600–615 nm) using a 50-mW krypton/
argon laser, revealing high total auxin peaks in the phellogen,
phloem sieve tubes (SE), cambium, and xylem parenchyma
(XP) cells. (Measured by M Langhans, CI Ullrich and R Aloni;
A - was published in Aloni 2013b) 69

Fig. 3.12 Schematic diagrams showing the three major IAA transport
pathways in the primary shoot (**i**), the secondary body (**ii**), and
the primary root (**iii**). In the *internal route*, IAA moves polarly
through the vascular meristem (M) (which is the procambium
in the primary body), and continuously through the cambium
(C) in the secondary body, the (X) xylem parenchyma cells, the
(B) bundle sheath (in primary shoot); and (Pe) pericycle (in
primary root). In the *non-polar route*, IAA moves up and down
in the (S) sieve tubes. In the *peripheral route*, most of the IAA
moves polarly downward through the (E) epidermis and the
phellogen (Ph) (From Aloni 2010) 70

Fig. 3.13 Schematic diagram illustrating the source and translocation of
the leaf-derived gibberellin signal. The GA originates in
developing and maturing leaves (longer than 3 cm in tobacco
(*Nicotiana tabacum*), but not leaves that are emerging at the
shoot apex). GA movement is non-polar in the leaf blade and
becomes polar only in the lower midvein toward the stem
(*arrows* mark flow orientation). The unique anatomy at the
base of the petiole (*blue dotted line*) of short cells, potentially
retards the flow, which induces a local maximum (*star*),

thereby acting as the leaf's elongation driving force. The signal
flows in both directions along the stem; its upward movement
from developing leaves reaches the young internodes (*star*) and
induces stem elongation at the shoot apex. Throughout the flow
along the stem, the signal results in bioactive GA signaling that
controls cambial activity and fiber differentiation. (From
Dayan et al. 2012)... 75

Fig. 3.14   Free-cytokinin distribution in a root (**A**), in young flowers
protected from the wind (**B**), or exposed to gentle wind (**C**), as
detected by free bioactive-CK-dependent *GUS* expression in
the CK-responsive *ARR5::GUS* transformant of 5-week-old
Arabidopsis grown under long-day conditions (described in
Aloni et al. 2005). (**A**) Root of wind-protected plant with
strong *ARR5::GUS* expression reflecting high bioactive CK
concentration in the cap (*arrow*), considerable concentration in
the vascular cylinder (*arrowhead*) and lower concentration in
the cortex. (**B**) Wind-protected inflorescence (by a tight
translucent plastic bag, under 95–100% humidity) showing
young apical flower buds (*arrow*) without *GUS* expression. (**C**)
Plant from the same experiment as in (**A**) and (**B**), but exposed
to gentle wind (of 0.2–0.7 m s^{-1} under lower humidity of
60–70%) only for the last 3 h before harvesting for the GUS
assay, resulting in very strong *ARR5::GUS* expression in the
young apical flower buds (*arrow*). Scale bars = 40 μm (**A**), 500
μm (**B**, **C**). (From Aloni et al. 2006a).. 77

Fig. 3.15   Molecular structure of hormonal signals that regulate vascular
differentiation in plants: abscisic acid, jasmonic acid, brassino-
steroid, and strigolactone... 80

Fig. 5.1   Micrograph of a thick tissue of soybean (*Glycine max*) cleared
and stained with lacmoid (according to Aloni and Sachs 1973),
which was grown on a solid agar medium in culture, on 1 mg/l
IAA and 6% sucrose, for 21 days. The figure shows a typical
pattern of vascular differentiation in a nodule. The vessel
elements (v) are in the center of the nodule (lower center)
surrounded by circular sieve tubes. *Arrows* mark groups of
callose-rich sieve plates (of three sieve tubes) stained blue by the
lacmoid. Bar = 100 μm. (From Aloni 1980)................................. 104

Fig. 5.2   Schematic diagrams of longitudinal median sections in tissue
cultures grown on a solid medium, illustrating stages in the
developmental patterns of phloem and xylem in the growing
tissues. (**A**) Callus tissue consisting of homogenous parenchyma cells with no vascular elements grown on low-auxin
medium (0.03 or 0.05 mg/l IAA). (**B**) Nodules of only sieve
elements with no xylem developed either at early stage of
vascular differentiation after transferring homogenous paren-

chyma to a medium containing a low-auxin concentration
(0.1 mg/l IAA), or in callus tissue kept on this medium. (**C**)
Intermediate stage in which tracheary elements start to appear
in some of the phloem nodules located in the center of the
callus. (**D**) Typical pattern of callus tissue grown on high-auxin
concentration (0.5 or 1.0 mg/l IAA). In the periphery of the
callus, there are either new developed nodules of phloem with
no xylem, or nodules of phloem with little xylem. The center
of the callus comprises nodules or short strands of well-developed phloem and xylem. (From Aloni 1980) 105

Fig. 5.3   Effect of IAA concentration on the differentiation of sieve and
tracheary elements in tissue culture of *Glycine max* after
21 days. Sucrose concentration was 3%. Each point represents
an average of three replicates from five callus tissues. (From
Aloni 1980)... 106

Fig. 5.4   Micrographs of cross sections taken from the middle of young
internodes of *Coleus blumei* stems from which the leaves were
excised, showing the pattern of primary phloem fiber differentiation induced by hormonal treatments during 20 days. (**A**) No
fiber differentiation occurred when only 0.1% GA$_3$ (w/w in
lanolin) (with no auxin) was applied. The *arrows* show the
primary phloem fiber initials that remain unchanged during the
experiment, demonstrating the need for an auxin background
in order to induce fiber differentiation. (**B**) Small bundles of
primary phloem fibers were induced by 0.1% IAA (w/w).
These fibers were very short. (**C**) Fibers induced by the
combination of 1.0% IAA + 0.05% GA$_3$ (w/w), characterized
by thick secondary walls. (**D**) Fibers induced by 0.05%
IAA + 1.0% GA$_3$, showing thin-walled fibers in the largest
primary phloem bundles. These fibers were the longest in the
experiment. *Arrows* mark primary phloem fibers in **B**, **C**,
D. Bar = 100 µm. (From Aloni 1979)... 109

Fig. 5.5   Typical polar pattern of primary phloem fiber differentiation
detected 10 days after wounding in the middle of a mature
internode of *Coleus blumei*. For observation, the phloem was
separated from the xylem at the cambium and stained by the
lacmoid clearing technique (Aloni and Sachs 1973). The
photograph shows strands of fibers (*arrow*) that differentiated
above and beside the wound (W) and absence of fibers
immediately below the wound (separating the tissues above
and below with parafilm), demonstrating that the fiber-inducing signals arrived from the leaves above the wound (there is
no fiber differentiation when the leaves above the wound are
excised; see Fig. 5.6E). Bar = 100 µm. (From Aloni 1976) 114

Fig. 5.6   Schematic diagrams illustrating the differentiation of primary
            phloem fibers in internodes 5 and 6 of *Coleus blumei* stems.
            The numbers beside the internodes are averages (from 5 stems
            per treatments, plus and minus SE) of the primary phloem
            fibers in the adjacent half cross section, whether taken from the
            middle of the internode, or 3 mm above and below the wound
            made in the middle of the internode. The wound (−w) is shown
            in Fig. 5.5. (**A**) was analyzed at 0 time from untreated stems.
            All the other treatments (**B–F**) were harvested after 14 days;
            their youngest internodes, lateral buds and the absent leaves
            illustrated in the experimental design were excised at 0 time.
            (From Aloni 1976SS)   114
Fig. 5.7   Schematic diagrams illustrating the role of auxin (IAA) and
            gibberellic acid (GA$_3$) in differentiation of primary phloem
            fibers around a wound (−W) in the mature internode 5
            (counted downward from the shoot apical bud) of *Coleus
            blumei*, after 10 days. The values beside the internodes are
            the mean ± SE of the number of primary phloem fibers in the
            adjacent half cross section taken 3 mm above and 3 mm
            below the wound in the middle of the internode. Sample size
            was five in all cases. All the plant leaves were removed,
            except for pair 5 in treatment **F**. The lanolin pastes were
            applied to node 5 (the node above internode 5) in the various
            hormonal concentrations (w/w) shown at the top of each
            diagram (in treatment **A**, the Lan. is plain lanolin; and in **F**,
            the leaves were the source of signals). Treatment **D** shows
            that the low hormonal combination was more effective that
            auxin (**B**), or gibberellin (**C**) alone. The high combination (**E**)
            did not have an additional promoting effect. Gibberellin by
            itself was not effective (**C**). (From Aloni 1979)   115
Fig. 5.8   Micrographs of cross-sections taken from the middle of mature
            internodes of *Nicotiana tabacum* stems from which all the
            leaves were excised; the location of the cross-sections are
            marked by blue bars, in the adjacent schematic diagrams
            illustrating the experimental design, showing the long-distance
            effect of hormonal application to a young internode on
            vascular differentiation in a mature internode. (**A**) Showing the
            effect of low auxin concentration (0.04% NAA in lanolin w/w)
            resulting in slow differentiation rate that enable longer time of
            vessel (V) and xylem fiber (F) expansion, resulting in wide
            xylem elements. (**B**) Mixture of gibberellic acid (0.8% GA$_3$ in
            lanolin w/w) with the auxin (0.04% NAA in lanolin w/w)
            promoted cell divisions in the cambium, faster cell differentia-
            tion resulting in narrow vascular elements, and increased
            lignification of fibers and vessels. The *arrows* mark primary

phloem fibers. Cambium (C), fibers (F), vessels (V).
Bars = 50 μm. (J Dayan and R Aloni, unpublished) 116

Fig. 5.9   Micrographs of cross sections taken from the middle of young
internodes of *Nicotiana tabacum* stems from which the leaves
were excised; the location of the cross sections is marked by
blue bars, in the adjacent schematic diagrams illustrating the
experimental design. (**A**) Showing the pattern of vessel
differentiation induced in the young internode by low auxin
concentration (0.04% NAA in lanolin) applied on the upper
end; the results demonstrate that auxin alone induces only
vessels, and the late formed ones are narrow vessels (v), with
no any fiber differentiation. (**B**) The addition of gibberellin
(0.8% GA₃ in lanolin) to a mature internode demonstrates the
nonpolar effect of GA on a young internode, namely, that GA
can move upward and induce primary fibers in both the inside
and outside phloem (*arrows*) as well as secondary xylem fibers
(F) among the vessels (v). Bars = 50 μm. (J Dayan and R
Aloni, unpublished) 117

Fig. 5.10   Photographs of cross sections taken from the middle of mature
internodes of *Helianthus annuus* stems at the same distance
from roots. The intact control is shown on the left side and the
GA-treated stem on the right. Showing the effect of gibberellic
acid (100 ppm GA₃) applied for 6 weeks to the root system by
irrigation. The GA treatment substantially increased stem size
and secondary xylem fiber formation. The different color of
the xylem fibers (*arrows*) stained with lacmoid indicates
modification in lignin composition. Bar = 10 mm 118

Fig. 5.11   Effect of zeatin and kinetin on the differentiation of secondary
xylem fibers in cultured hypocotyl segments of *Helianthus
annuus* after 30 days in the presence of 0.5 μg/ml IAA and
0.1 μg/ml GA₃. (**A**) The maximal promoting effect of both
cytokinins was at 0.2 μg/ml, each point represents an average
of five replicates from 10 explants. (**B**) Photograph of the
zeatin-induced secondary xylem fibers in *H. annuus* cultured
segment, stained with lacmoid. The fibers (F) are characterized
by small simple pits compare with the large perforation
(*arrow*) and reticulated secondary wall thickenings of the
vessel elements (V). Bar = 25 μm. (From Aloni 1982) 119

Fig. 5.12   Photographs of cross sections taken from the middle of the
hypocotyl of sunflower (*Helianthus annuus*), showing the
pattern of secondary xylem fibers induced by kinetin applied to
the roots of intact plants during 21 days. The two photographs
were taken from the same experiment and under the same
magnification. (**A**) The intact control shows the typical bundles
with interfascicular cambium (*arrows*) between them at the

end of the 21-day period of growth in Hoagland medium
(Hoagland and Arnon 1950) without kinetin. (**B**) Intact plants
treated with 0.25 µg/ml kinetin applied in the medium to their
roots show secondary xylem fibers induced between the
bundles (*arrows*). Note that there are also more fibers within
the bundles. Bars = 100 µm. (From Saks et al. 1984)   120

Fig. 5.13 Stem anatomy of wild-type control and tobacco (*N. tabacum*)
plants over-expressing GA 20-oxidase, silencing GA 2-oxi-
dase, and their crosses. Three columns separate between
different regions of the dissected stems (corresponding to
young (High), middle (Mid), and basal (Low) internodes). It is
evident that the silenced plants have higher cambial activity
with thicker xylem zone. In the young cross sections, a delay
in xylem differentiation is depicted in the transgenic lines,
prolonging the developmental stage, and allowing cell elongation. High xylem fiber production is observed in the basal
internodes of silenced lines and their cross-fertilization with
GA 20-oxidase over-expressing plants. The middle internodes
do not exhibit any significant difference in cambial activity and
xylem development. X xylem; P phloem; VC vascular cambium; V vessels; PV primary vessels; F fibers. Bars = 80 µm.
(From Dayan et al. 2010)...   121

Fig. 5.14 Effect of the auxin transport inhibitor 1-N-naphthylphthalamic
acid (NPA) on tracheary element (TE) differentiation in *Zinnia
elegans* liquid culture. (**A**) Showing isolated tracheary elements (*arrows*) of *Zinnia* cells that were cultured on a xylogenic inducing medium for 96 h without NPA. (**B**) TE
differentiation was inhibited by 20 µM NPA. (**C**) The NPA-
induced suppression of TE formation was overcome by
high-auxin concentration (10.7 µM NAA). Scale
bars = 100 µm. (Reproduced from Yoshida et al. (2005) *Plant
and Cell Physiology* 46: 2019–2028, with permission of
Oxford University Press)   123

Fig. 6.1 *GUS* expression patterns visualizing free cytokinin (CK) in the
root tip of CK-responsive *ARR5::GUS* transformant of
Arabidopsis thaliana plants grown protected from wind,
demonstrating the high-CK concentrations in the main root
compared with much lower CK production in the lateral roots.
(**A**) Main root of soil-grown wind-protected plant at 35 days
after germination, with strong *ARR5::GUS* expression reflecting massive CK production and accumulation due to wind
protection in the entire elongating zone (*arrow*), in the vascular
cylinder, cortex, and epidermis. The concentration of *GUS*
expression in the cortex decreases gradually. (**B** and **C**) Lateral
roots and main root grown on MS basal medium in closed
boxes. (**B**) Due to typically low CK production in a lateral root,

ARR5::GUS expression was only observed in the root cap, with CK export restricted to the base of the central vascular tissue (*arrowhead*). The periphery of the root cap was almost free of *GUS* expression (*arrow*). (**C**) CK production in the tip of a lateral root (*short arrow*), and almost absent from its vascular cylinder, cortex and epidermis (*arrowhead*), while strong CK accumulation is observed along the entire axis of the main root (*large arrow*). Bars = 25 μm (**B**), 50 μm (**A, C**). (From Aloni et al. 2005) 136

Fig. 7.1   Typical downward procambium development in wild-type *Arabidopsis thaliana* rosette leaf primordia, photographed under Nomarski illumination. (**A**) Very young primordium with two procambial loops, loop 2 (marked with *arrow*) developed after loop 1. (**B**) More developed primordium with four procambial loops numbered according to their down-ward (basipetal) developmental pattern. The *arrowhead* marks the analogous site in both photographs emphasizing the basipetal developmental pattern of the procambium induced by auxin. The sites of the inducing auxin maxima are marked by asterisks at the tip and lobe. The low-auxin concentration flows that induce the procambial loops tend to merge into the already formed procambium of the midvein, which is more polarized and transport efficiently the signal. Bar = 200 μm. (From Aloni 2004) 143

Fig. 7.2   The directions of polar auxin flow along the midvein and the secondary veins are detected by *DR5::GUS* expression beside a horizontal (**A**) and a longitudinal (**B**) cuts made in growing leaf primordia of the DR5::GUS-transformed *Arabidopsis thaliana*, revealing that the bioactive auxin flows from auxin maxima (detected by blue GUS staining) at the leaf's periph-ery, toward the midvein (**B**) and along the midvein (**A**) downward to the stem. The cuts stopped the auxin flow, which therefore accumulated at the cut surface. The *arrows* show the direction of the IAA flow. Auxin maxima are marked with asterisks. Bars = 1 mm (**A**), 0.65 mm (**B**) 144

Fig. 7.3   Early morning guttation from hydathodes developed at the leaf margins of strawberries (*Fragaria ananassa*) leaves, showing exudation of drops of xylem sap (*arrows*) from young leaves and sepals... 147

Fig. 7.4   *DR5::GUS* gene expression in DR5::GUS-transformed *Arabidopsis thaliana* cleared in lactic acid after GUS staining, showing histochemical localization of GUS activity (blue GUS staining that marks free auxin) during early leaf primordium development. Incipient (low *GUS* expression) auxin activity is observed (where auxin maximum develops) in the tip (*arrow-*

head) at the stage of acropetal (upward) xylem differentiation of the midvein (*arrow*) in a primordium of a rosette leaf. Bar = 250 μm. (From Aloni et al. 2003) 148

Fig. 7.5   Time course of a downward "wave" of auxin maxima develop-
            mental pattern along the periphery of young leaf primordia
            indicating "leaf apical dominance" detected by *DR5::GUS*
            gene expression in DR5::GUS-transformed *Arabidopsis*,
            depicting histochemical localization of GUS activity in a
            young primordium (**A**) and in an older primordium (**B**). (**A**)
            Early stage in basipetal progression of high-IAA production,
            showing strong *GUS* expression in the upper developing
            hydathodes (*large arrowheads*) and low activity in a lower
            hydathode (*small arrowhead*). Note the wide-band GUS
            activity (*arrow*) accumulated in the midvein. (**B**) Shows a later
            stage in the downward progression of free-auxin production,
            with low *GUS* expression (*small arrowheads*) in the upper
            hydathodes and strong GUS activity (*large arrowhead*) in the
            lower hydathode. Bar = 1 mm. (From Aloni et al. 2003)   149

Fig. 7.6   Close up views of auxin maxima and their development into a
            hydathode in DR5::GUS-transformed *Arabidopsis thaliana*.
            (**A**) Demonstrating how, at the lobe, a center of strong
            DR5::GUS expression (*black arrowhead*) reflecting the high
            free-auxin concentration in a developing hydathode. From the
            auxin maximum, the IAA starts to move by diffusion, gradu-
            ally becomes canalized to a narrow continuous stream (*white
            arrowhead*), flowing into a differentiating secondary vein,
            which is induced by this polar auxin movement. In the vein,
            the IAA flows more rapidly through the polarized and more
            efficient auxin-transporting cells, and therefore cannot be
            detected. (**B**) *GUS* expression at the lobe (*black arrowhead*) in
            a more developed hydathode with four freely ending vessels
            (*large arrow*) differentiating toward the margin. Note some
            weaker blue staining near the margin (*small arrowhead*) and
            low *GUS* expression within the veins (*small arrows*).
            Bar = 150 μm (**A**), 250 μm (**B**). (From Aloni et al. 2003)   150

Fig. 7.7   Low-auxin production sites in the central region of the leaf's
            lamina, which induce the tertiary veins and the freely ending
            veinlets. These minor IAA production sites start to show
            DR5::GUS activity at the late stage of vascular differentiation
            in a leaf primordium, after the cessation of the auxin maxima
            along the primordium periphery. (**A**) *GUS* expression in the
            spongy parenchyma cells (*arrow*). (**B**) Reporter-gene activity
            (*arrow*) detecting IAA in a differentiating freely ending
            veinlet. Bar = 150 μm. (From Aloni et al. 2003)   151

Fig. 7.8   Schematic diagrams showing the gradual changes in sites
             (*black spots locations*) and concentrations (*black spots size*) of
             high- versus low-IAA production during leaf primordium
             morphogenesis in *Arabidopsis*. The rising *green lines* illustrate
             the upward polar auxin flow through the differentiating
             epidermis during the early stage of primordium development
             (**A–C**), originating from nearby auxin-producing leaves. The
             red arrows show the experimentally confirmed directions of
             the downward (basipetal) vein-inducing polar IAA movement,
             descending from the differentiating hydathodes in the growing
             tip and lobes (**B–D**). Incisions (Fig. 7.2A) in leaf primordia
             show that although the midvein (*broken line,* in **B**) matures
             acropetally (Fig. 7.4), it is induced (by auxin accumulation
             above the short cells of the future abscission layer) by the
             basipetal (downward) polar IAA flow (*red arrow*) descending
             from the primordium tip (**B**). *Short red arrows* in the lamina
             originate from *small black spots* indicate random possible
             auxin flow directions from minor auxin production sites (**D,
             E**), which induce the tertiary veins and freely ending veinlets.
             The ontogeny of the midvein and secondary veins is illustrated
             by broken lines (marginal and minor veins are not shown). (**A**)
             Early high-auxin production occurs only in the stipules (s) of a
             very young leaf primordium, before free auxin is detectable in
             the tip. (**B**) Auxin maximum development in the tip of a fast
             elongating primordium induces acropetal midvein differentia-
             tion, illustrating "leaf apical dominance." (**C**) Auxin maxima
             development in the fast-expanding upper lobes induces the
             upper secondary veins and matures into hydathodes. (**D**) Auxin
             maxima development in the lower lobes inducing the lower
             secondary veins and mature into hydathodes; randomly
             distribution of minor auxin production sites start first in the
             upper lamina (**D**), and later also in the lower lamina (**E**),
             induce the tertiary veins and freely-ending veinlets, during
             later phase of primordium development 152
Fig. 7.9   Vein pattern formation and xylem regeneration around a
             wound (*w*) induced in *Arabidopsis thaliana* by injury and
             auxin application (1% IAA in lanolin) to leaf primordia shorter
             than 1 mm. The direction of auxin application site is marked
             by an *asterisk*. No xylem regeneration could be induced either
             without the application of exogenous auxin or in longer (older)
             primordia. After one regeneration week, the treated leaves were
             cleared in lactic acid and studied without staining. (**A**) There
             was typically no vein regeneration (*arrows*) around a wound,
             showing discontinuous patterns of veins, in about half of the
             wounded leaves. (**B**) Continuous pattern of xylem regeneration

(*arrow*) occurred around a wound. (**C**) Two delicate loops (*arrowheads*) differentiated near the wound, attached to thicker regenerative vein (*arrow*) which became a major auxin transporting pathway, likely following procambial configuration. (**D**) Relatively rare typical of xylem regeneration in a leaf, showing continuous pattern of xylem regeneration around a wound, with many regenerative vessel elements that differentiated from parenchyma cells (*arrows*). Bars = 500 μm 153

Fig. 7.10   Regenerative tracheary elements induced from parenchymatic cells around a wound (*w* – marks the wound direction) in leaves of *Arabidopsis thaliana*, by wounding and auxin application (1% IAA in lanolin) to leaf primordia shorter than 1 mm. The direction of the applied auxin site is marked by an *asterisk*. (**A**) Magnified view of a regenerative vein built of regenerated xylem elements characterized by helical wall thickenings (*arrow*). (**B**) Portion of a regenerative vein built of xylem elements with helical wall thickenings (*arrow*) and some annular wall thickenings (*arrowhead*) which allow considerable tissue elongation in a rapidly expending leaf primordium. Bars = 25 μm 154

Fig. 7.11   Hydathode differentiation and lowest hydathode inhibition in leaves of *Arabidopsis thaliana*. Micrographs of leaves, cleared in lactic acid, showing the largest hydathode of the tip spread on a large region (*marked by a dotted ellipse* and *arrow*) that contains a few vessels arranged in a wide pattern (**A**), while the smallest hydathode of the lowest lobe contains only one vessel (**B, D**). Excision of the upper parts of the primordium, leaving only the lowest lobe (**C, E**), promoted the differentiation of four vessels in the lowest hydathode (**E**), demonstrating that when the growing tip and upper lobes were removed experimentally from the primordium, their auxin-maxima-inhibition effect was eliminated, allowing more vessel differentiation in the lowest hydathode. **D** and **E** are close-up views of B and C, respectively, in a different angle. Bars = 20 μm (**D, E**), 100 μm (**A, B, C**)............... 156

Fig. 7.12   Differentiation of tertiary veins and freely ending veinlet is inhibited by auxin maxima produced along the leaf primordium periphery. The DR5::GUS-transformed *Arabidopsis thaliana* was used in these experiments, showing intact control leaves (**A, B**); leaves of plants grown for 2 weeks on the auxin efflux inhibitor 1-*N*-naphthylphthalamic acid (20 μM NPA) (**C, D**); and leaves that their upper portion was excised (**E, F**), and high-auxin concentration was applied (**F**). The *arrows* mark freely ending veinlets. (**A**) Close-up view of freely-ending veinlets normally induced during the late phase of leaf mor-

phogenesis by low-auxin producing trichomes in an intact
leaf (see evidence in Chap. 1, Fig. 1.1B). (**B**) Low magnification shows the regular asymmetric pattern of freely ending
veinlets in the lamina, with different vein patterns in the right
versus the left sides of the midvein (m). (**C** and **D**) Show
leaves that the endogenous auxin produced in their periphery
remained continuously high (detected by green-blue
DR5::GUS staining) along the margin, which prevented
tertiary vein formation in the lamina. Note that the high-
auxin produced in the primordium periphery induced more
vessels in the midvein (m), compared with the thin midvein
in the intact control (**B**). (**E**) Normal pattern of tertiary veins
and freely-ending veinlets differentiated after the upper part
of the leaf primordium was excised (**E** is the control of **F**).
(**F**) Tertiary veins and freely ending veinlet differentiation
was inhibited by the exogenous high-auxin (1% IAA in
lanolin) applied on the cut surface (the auxin application site
is marked with *asterisk*). Bars = 100 μm (**A**), 200 μm (**F**),
250 μm (**B, E**), 1 mm (**C, D**) 158

Fig. 8.1   Schematic representation of the ABC model proposed by
Bowman et al. (1991) in a wild-type *Arabidopsis thaliana*
flower, depicting how three classes of floral homeotic genes
could specify the identity of each of the four whorls of floral
organs. The function of *A* alone specifies sepal (se) identity in
the first whorl (1); the combination of *A* + *B* function specifies
petal (pe) identity in the second whorl (2); the combination of
B + *C* function specifies stamen (st) identity in the third whorl
(3); and *C* function alone specifies carpels (ca) in the fourth
whorl (4) 165
Fig. 8.2   Distribution of conjugated auxin detected by immunolocaliza-
tion with rabbit polyclonal antibodies (green Alexa Fluor 488
fluorescence) in DR5::GUS-transformed *Arabidopsis*, viewed
with confocal laser scanning microscope (CLSM), showing
early development of the stamens and gynoecium (**A–C**), and
late development of the petals (**D**). (**A**) Auxin distribution in
three developmental stages of flower primordia. The youngest
flower bud (*arrow*) before any visible organ initiation is
already loaded with conjugated auxin, while the most developed flower (before petal growth) shows differential auxin
patterns with highest auxin concentrations mainly in the
anthers and gynoecium. (**B**) Very young stamen primordia
(*arrow*) with the highest auxin concentrations, higher than in
the promeristem (*arrowhead*) and sepals which already protect
the stamens. (**C**) Elevated auxin in the stigma, developing
ovules in the gynoecium (g) and the differentiating pollen sacs,

before petal initiation. (**D**) Early stage of petal growth
characterized by a short phase of high auxin concentration
(*arrowhead*). The anthers (*arrow*) and gynoecium continue
to maintain high-auxin concentrations during later developmental stages. Bars = 40 μm (**B**), 80 μm (**A, C, D**). (From
Aloni et al. 2006) 166

Fig. 8.3   Auxin-dependent *DR5::GUS* gene expression in the
DR5::GUS transformed *Arabidopsis thaliana* (cleared in lactic
acid) during early stages of inflorescence and flower morphogenesis, detecting free-auxin by the GUS staining. (**A**)
Inflorescence with very weak (*arrow*) or without any GUS
activity in the youngest flower buds (*arrow*), [which are loaded
with conjugated auxin, detected in Fig. 8.2A]; the larger flower
buds show strong *GUS* expression (dark blue-black staining) in
the stamens. (**B**) Close-up view of a very young anther shows
high GUS activity in the tapetum cells (*arrow*) that produce
IAA and supply the pollen grains with auxin and nutrients.
Bars = 25 μm (**B**), 250 μm (**A**). (From Aloni et al. 2006) 168

Fig. 8.4   Free-auxin production detected by *DR5::GUS* expression in
transformed *Arabidopsis* (cleared with lactic acid), showing
localization of GUS activity during flower morphogenesis,
demonstrating how the bioactive auxin induces vascular
differentiation in the gynoecium. (**A**) Young flower bud, prior
to detectable IAA production in the stigma and therefore with
no vascular development (*arrow*) in the gynoecium. At this
developmental stage the stamens produce the highest free-
auxin concentrations (*stained dark blue*), which likely delay
free-auxin production in the stigma. (**B**) In a young primordium, free-auxin production (*detected by blue GUS expression*)
in the stigma induces the central bundle and an early stage of
xylem fan formation (*arrow*) immediately below the stigma.
(**C**) Mature gynoecium with germinating pollen grains (*stained
dark blue*). A typical well-developed wide fan of xylem
(*arrow*) descending into two central bundles which were
induced by the high-IAA-producing stigma (*stained blue*). The
discontinuous short veinlets (*arrowheads*) induced by the
ovules do not connect to the gynoecium's central bundles,
because the bundles are well supplied with high-IAA concentration streams descending from the stigma, which prevents
veinlet linkage. Bars = 100 μm. (From Aloni et al. 2006) 169

Fig. 8.5   Free auxin detected by *DR5::GUS* gene expression in transformed *Arabidopsis* during early stages of flower morphogenesis. The micrographs reveal the inhibitory mechanism of the
stamens on nectar-gland activity, which during early flower
development suppress nectar gland activity by the high-IAA

concentrations produced by the stamens. (**A**) Young nectary (*arrow*) without GUS activity, likely inhibited by the high-IAA concentration produced in the anthers. (**B**) Incipient low *GUS* expression at the tip (*arrow*) of a nectary before anthesis (starting simultaneously with nectar secretion). (**C**) Early strong *GUS* expression (*arrow*) in a nectary of a younger flower promoted by four-stamen removal. Both flowers shown in **B** and **C** are from the same inflorescence. (**D**) GUS activity in nectaries (*arrows*) continues after pollination, likely due to the absence of repression by the anthers (that stopped auxin production), and is detectable below a developing silique with differentiating seeds showing strongest *GUS* expression where the embryonic root develops (*arrowheads*), while unfertilized ovules (*small arrows*) lack *GUS* expression. Bars = 100 μm (**B**, **C**), 200 μm (**A**, **D**). (From Aloni et al. 2006) 170

Fig. 8.6   Schematic diagrams showing the gradual changes in sites (*blue spot* locations) and concentrations (*blue symbol* sizes) of free-IAA production (detected by *DR5::GUS* expression) during *Arabidopsis* flower and early fruit development. *Arrows* mark sites of auxin production starting at the tip of floral organs during their development (**A–E**) and at the ovules and developing seeds in the gynoecium (**D**, **E**). The ontogeny of the gynoecium midvein, characterized by its wide fan xylem induced by free IAA descending from the stigma (**D**, **E**), and the short xylem veinlets induced by developing seeds are illustrated by *red lines* (**E**). (**A**) Young floral bud with incipient free-IAA production at the tip of the sepals (the bud is loaded with conjugated auxin). (**B**) Free-IAA production at the sepal tips and massive bioactive-auxin production in the stamens, demonstrating stamen dominance characterized by complete petal suppression. (**C**) Decreased auxin production in the stamens (DR5::GUS activity limited to the anthers) is followed by incipient auxin production in the growing petals and stigma. (**D**) High free-IAA production in the stigma; low-auxin production in the ovules, the nectaries, the petal tips, and stamen-filament tips. (**E**) Residual free-IAA production beneath the stigma, elevated auxin production in developing seeds, and continuous production in nectaries. (From Aloni et al. 2006) 171

Fig. 8.7   Naturally occurring vascular patterns in petals of wild-type *Arabidopsis thaliana* cleared in lactic acid. (**A**) Pattern of closed loops (*arrowhead*). (**B**) Pattern of open branching veins with freely ending veinlets (*arrow*) that lack distal meetings. (**C**) Combined patterns of closed loops (*arrowhead*) and open

branching (*arrow*) with freely ending veins. (**D**) Gradual
development of a closed loop viewed under dark-field illumination, showing that the upper vein (#1) is already mature,
while the lower vein (#2) is still differentiating (*arrow*)
downwards. There are two freely ending veins at the petal's tip.
(**E**) Close-up view on the top of a loop revealing the upper
ends (*arrows*) of the two attached overlapping vessels that
together built the loop. (**F**) Vein containing two vessels induced
by one application of auxin (40 µM NAA spray) to a developing flower. Usually the petal veins contain only one vessel per
vein, which is likely induced by low auxin stimulation.
Bars = 100 µm (**E, F**), 125 µm (**D**), 200 µm (**A–C**) 172

Fig. 8.8   Effects of influx auxin transport inhibitor (2,3,5-triiodobenzoic
            acid, TIBA) and auxin (naphthaleneacetic acid, NAA) spray
            applications (over 14-day period) on vascular differentiation in
            wild-type *Arabidopsis* petals. (**A**) Typical delicate vein system
            in the middle of an intact petal, consisting of two or three
            vessels in the midvein (*arrowhead*) and one in each secondary
            vein (*arrowhead*). (**B**) The auxin transport inhibitor (20 µM
            TIBA) induced a very wide midvein (comprised of 10–12
            vessels), and wide secondary bundles, promoted new bypass-
            ing pathways resulting in additional regenerative bundles
            (*arrows*). (**C**) Repeated alternated-day treatments of 20 µM
            TIBA (on 1 day) and 20 µM NAA (on the next) caused an
            increase in vessel number at the base of the midvein and a
            moderate effect in the secondary veins. Bars = 125 µm. (From
            Aloni 2004)... 173
Fig. 8.9   Experimental modification of fruit development in garden
            strawberry (*Fragaria* × *ananassa*); and bioactive auxin
            accumulation in developing seeds of the DR5::GUS trans-
            formed *Arabidopsis thaliana* (the orientation of the silique is
            correct, with the seed's roots upward towards the stigma). (**A**)
            Selective removal of most of the fertilized ovaries in a straw-
            berry flower (its white sepals remained untouched), resulted in
            the development of two separate miniature "fleshy red fruit"
            induced by the two remaining achenes (*arrowhead*). (**B**)
            Aggregation of five separate "tiny red fruits" each developed
            below its inducing achene (*arrowhead*), demonstrating that the
            developing auxin-producing achene induces the fleshy straw-
            berry fruit tissue below it. (**C**) Longitudinal close-up view of a
            cleared intact strawberry fruit showing the achene (*dark
            brown*) embedded in the surface of the fruit with its supporting
            vascular bundle (*arrow*) that the auxin-synthesizing-achene
            induced, and the remaining basal part of its pistil (*arrowhead*).
            An early developmental stage of the pistil (*arrowhead*) on the

surface of an intact growing white fruit (before it accumulates the red color) is shown in the small inserted photograph. (**D**) Auxin accumulation (due to polar auxin movement in the embryo) at the root side (evident by the green-blue staining of *DR5::GUS* gene expression) inducing the short veinlets during seed development in a silique of *Arabidopsis*. Bars = 200 μm (**C, D**), 500 μm (inserted photo), 10 mm (**A, B**). (**A–C,** R. Aloni unpublished; **D**, from Aloni et al. 2006) 174

Fig. 9.1   Lateral root initiation and development occurs at the xylem pole, shown in cross sections of sweet potato (*Ipomoea batatas*). (**A**) Early stage of two young lateral roots (R) growing through the cortex of a root with five xylem arches (strands). Each lateral root starts from a protoxylem vessel (*arrow*). To mark the primary vessels, they were artificially filled with a red color. (**B**) More developed root region with a wider stele that enabled the development of seven alternating xylem and phloem stands. Note that the lateral root (R) developed at the xylem pole (*arrow*), the lignified vessels were stained red by safranin. Bars = 100 μm. (Courtesy of V. Singh and N. Firon) 178

Fig. 9.2   Model of lateral root initiation (LRi) induced by ethylene and auxin (IAA) and inhibited by cytokinin (CK) at the tip of a young dicotyledonous root. *Arrows* indicate positive LRi regulation; *blunt-ended line* indicates negative regulation. The schematic diagram shows the three outermost cell columns of the vascular cylinder in a radial-longitudinal orientation at the xylem pole, with a differentiating protoxylem vessel (DPV) and the pericycle (Pe). Two pathways of parallel streams of polar IAA (marked with *red arrows*) are shown: in the left side is the high-IAA concentration stream that induces the protoxylem vessel (marked by the *gradual development of secondary wall thickenings*) and in the right side suggested low-IAA concentration stream maintaining the meristematic identity of the pericycle. During vessel differentiation, a local increase of IAA concentration in a differentiating protoxylem vessel element induces low-concentration ethylene (C_2H_4) synthesis. This C_2H_4 signal is released, and in the centrifugal direction (*black arrow*), it locally blocks the polar IAA movement in the adjacent pericycle cells, consequently boosting IAA concentration immediately above the blockage, thus forming a local auxin maximum. This elevated high-IAA concentration induces pericycle cell divisions and LRi. The CK (marked with a *blue line*) arriving from the root cap inhibits LRi in the vicinity of the cap. In *Arabidopsis* the protoxylem vessel is attached to the pericycle. (From Aloni 2013) 181

Fig. 10.1   Polar patterns of xylem regeneration (*arrowheads*) revealing
             the pathways of the inducing polar auxin transport around a
             wound (w). (**A**) In a decapitated young internode of *Cucumis
             sativus* treated with auxin (0.1% IAA in lanolin for 7 days),
             which was applied to the upper side of the internode immedi-
             ately after wounding and removing the leaves and buds above
             it. Showing in a longitudinal view (observed after clearing with
             lactic acid, staining with phloroglucinol, and photographed in
             dark field) that the regenerated vessel elements (re-differenti-
             ated from parenchyma cells) formed above the wound (*arrow*)
             (where IAA was concentrated by the cut) differentiated close
             to the wound, while those below the wound differentiated at
             greater distances from the wound. (**B**) Pattern of limited xylem
             regeneration restricted to veins, around a wound (w) made in
             the midvein (primary vein) of a very young leaf primordium of
             the DR5::GUS-transformed *Arabidopsis thaliana* (harvested
             after 7 days and cleared in lactic acid), revealing that the
             bioactive auxin descending from the leaf's tip is detected
             above the wound (by green-blue staining of *DR5::GUS*
             expression). The *arrows* mark the lower ends of the veins that
             were cut by the wound, which became thicker due to the
             differentiation of additional regenerated vessel elements
             induced in these veins by IAA accumulation. The absence of
             vessels immediately below the wound indicates that the
             wounding was made at the stage of early procambium develop-
             ment. The remote xylem differentiation (*arrowhead*) occurred
             inside veins forming a continuous vein pattern around the
             wound that was determined by the configuration of the
             procambium. Note that there is no xylem regeneration from
             parenchyma cells around the wound in the leaf. Bars = 500 μm
             (**A**), 1 mm (**B**). (By R. Aloni).. 186
Fig. 10.2   (**A**) Diagrams comparing the regeneration of vessels (the heavy
             lines in the "xylem" diagram) with that of the sieve tubes (the
             heavy lines in the "phloem" diagram) from the same *Coleus
             blumei* wounded internode, grown in a greenhouse and
             collected 4 days after wounding. The longitudinal running
             stippled bands represent the pre-existing vascular bundles that
             contained differentiated vessels (in the xylem diagram) or sieve
             tubes (the phloem diagram). Both diagrams are at the same
             magnification (see 1 mm scale bar marked in lower center).
             The different *arrows* show equivalent locations in the two
             preparations, which were separated at the cambium. "W"
             designates the wound, and the small star (lower center of the
             phloem diagram) marks one end of a phloem anastomosis (that
             was not counted as phloem regeneration). The diagrams
             demonstrate that (i) there are more longitudinal phloem strands

(on the right side two phloem-only strands, and on the left side two collateral bundles); (ii) there are more regenerated phloem strands than regenerated xylem strands; (iii) vascular regeneration follows the configuration of the pre-existing strands; (iv) vascular regeneration occurs in discontinuous patterns (*arrows*); (v) the regenerated conduits start differentiation inside the pre-existing phloem strands and bundles and continue regeneration in a fragmental fashion between the bundles; (vi) regeneration between the bundles is polar, occurring first above the wound; (vii) phloem regeneration is faster than the xylem regeneration; and (viii) regenerated vessel elements and wound sieve tube elements around the wound are the connecting parts of continuous new vessels and new sieve tubes along the bundles. (**B**) Time course of complete regenerated sieve tubes and regenerated vessels around a wound in the fifth internode of *Coleus blumei*. Each point is an average of five plants. (Aloni and Jacobs 1977b)............................ 189

Fig. 10.3   Micrographs of typical responses around a wound (w) comparing the ability of very young organs of *Arabidopsis thaliana* to respond and regenerate vascular tissues (harvested after 7 days and cleared in lactic acid). (**A**) Well-developed massive xylem regeneration (*arrows*) was induced around a wound made in the inflorescence. (**B**) Functional vein bypass (*arrowhead*) beside the wound done in the leaf's midvein (*arrow*) connecting the tip veins with the lower part of the midvein. (**C**) No vascular regeneration response could be obtained after wounding a vein (*arrowhead*) in the petal (the petal was photographed in a dark field). Bars = 1 mm. (By R. Aloni)............................... 190

Fig. 10.4   Micrographs of typical regenerated vessels (*arrowheads*) and giant (very wide) regenerated vessels (*arrows*) around a wound (w) in young (**A**) and old (**B**) internodes of *Cucurbita maxima* sampled 7 days after wounding, showing the transition from the typical curved bypass of xylem regeneration from parenchyma cells immediately around the wound (**A**) to xylem regeneration mostly limited to phloem anastomoses further away from the wound (**B**). (**A**) In the third internode from the apical bud, many parenchyma cells have re-differentiated to produce regenerated vessels bypassing the wound and rejoining the damaged bundle. (**B**) In the sixth internode, there are no direct bridges of xylem formed around the wound. The differentiation of the regenerated xylem elements is exclusively associated with phloem anastomoses, leading to the wound being bypassed indirectly via neighboring, undamaged vascular bundles. Note that most of the regenerated xylem in the anastomoses consist of wide regenerated vessels (*arrows*).

Delicate (narrow) regenerated vessels re-differentiated from parenchyma (*arrowheads*) are present (lower left side). Bars = 1 mm. (From Aloni and Barnett 1996)............................. 191

Fig. 11.1 Patterns of vascular differentiation (**A**, **B**) and bioactive auxin distribution in the cambium (**B**). (**A**) Cross section in the vascular tissues of the oak *Quercus calliprinos* (stained with 0.01% aniline blue and observed with epifluorescence microscope) showing a pattern of radial distribution of vessels (*black rounded spots marked by white* V) and their radially located adjacent sieve tubes (*white spots marked by white arrows*) indicating that the conduits were induced by preferably polar IAA movements in specific cambial cells between the conduits. Xylem fibers (F) and phloem fibers (*black arrow*) differentiated in the regions between the conduits likely induced by longitudinal flows of GA in the cambium. (**B**) Cross section in the vascular tissues of the DR5::GUS transformed *Arabidopsis thaliana* presented in the same orientation as the oak (*xylem on left side*), showing free-IAA distribution (marked by *DR5::GUS expression,* which forms the blue spots, marked by *arrows*) located preferably adjacent to the differentiating vessels (V) and absent from the cambium located adjacent of xylem areas without vessels, demonstrating that along the cambium circumference there are preferable longitudinal streams of polar IAA that specifically induce radial patterns of vessels and sieve tubes. Bar = 50 μm (**B**), 200 μm (**A**). (From Aloni 2007).. 201

Fig. 11.2 Naturally occurring patterns of secondary phloem and secondary xylem produced by variant cambium in three liana species: *Perianthomega vellozoi* (**A**), *Fridericia speciose* (**B**), and *Mansoa onohualcoides* (**C**), demonstrating different activity of the cambium along the circumference and consequently formation of more, or less, of each secondary vascular tissue. *Arrows* mark cambial sites that produce mainly phloem, while *arrowheads* mark cambial locations which produced mainly xylem (in **A** and **B**). In some sites the location of the cambium is marked by a *red line* (in **B** and **C**). (**A**) In *P. vellozoi* the cambium alternates, forming wide xylem with thin phloem layers (*arrowheads*) or a cambial region that produced wide phloem layers on the expense of almost no xylem production (*arrow*). (**B**) *F. speciose* shows a cambial variant that produced mainly xylem with gradually enlarging vessels (*arrowheads*) and among them sites with more secondary phloem production (*arrows*) on the expense of xylem formation. (**C**) In *M. onohualcoides*, the cambial variant produced xylem furrowed by phloem arcs/wedges. Starting from four (the lower one is outside the figure frame) equidistant (*white arch*) phloem wedges (near the pith), each of the first-formed wedges is marked by number 1. During cambial growth, more space becomes available on the cambial circumference for a new inductive hormonal stream of a phloem signal that induced a new phloem wedge marked by number 2. Additional cambial growth produced more space on the cambial circumference to form a new hormonal stream that produced a new phloem wedge marked by number 3. The *white arches* mark similar distances on the cambium circumference at the time when a new wedge was formed. Bar = 2 mm (**B**), 3 mm (**A**), 4 mm (**C**). (Courtesy of Marcelo R. Pace)... 202

Fig. 11.3   The nonpolar effect of gibberellin (GA$_3$) on cambium activity and fiber differentiation in young decapitated stems of tobacco (*Nicotiana tabacum*) from which all the leaves were excised. The cross sections shown in the micrographs were done in the middle of internode number 3 (*marked in the experimental scheme by a blue bar; while the hormonal application sites are marked with arrows*), leaving two younger internodes above internode 3 (*as an endogenous source of low IAA production; because GA does not function without an auxin background*). The results were analyzed 3 weeks after treatments. (**A**) Effect of applied lanolin paste (*without gibberellin*), showing the differentiation of relatively large isolated slightly deformed vessels (V) indicating some endogenous auxin production by the two youngest internodes during the experiment (*the deformed thin-wall vessels resulted from shortage of nutrients because all the leaves were removed* – see Sect. 2.9, Fig. 2.32A, B). Note, that in the absence of GA stimulation, there is no cambial activity or any fiber differentiation. (**B**) Effect of 0.8% GA$_3$ in lanolin applied on the youngest inter-node, inducing cambial C activity, secondary fibers (SF) in the xylem, and primary phloem (PF) fibers inside and outside the xylem. (**C**) effect of 0.8% GA$_3$ in lanolin applied on a mature internode, two internodes below internode number 3, showing cambial activity and secondary fiber formation in the above xylem. The results demonstrate that the effect of gibberellin on cambial activity and fiber differentiation is nonpolar; the GA$_3$ promoted cambial activity and fiber differentiation when applied either from above (**B**) or from below (**C**) the studied internode. All micrographs are at the same magnification, bar = 50 μm. (Jonathan Dayan and R. Aloni, unpublished) 205

Fig. 12.1   Photographs showing the effects of a 6-month-old juvenile
            rootstock on 20-year-old adult grafted branch of *Eucalyptus
            globulus*, 4 months after grafting. The graft union is marked
            with a *red ring of a broken line*. (**A**) Showing well-developed
            grafted shoot with adult leaves (*arrow*) only. (**B**) Only juvenile
            (rounded and gray) looking leaves (*arrowhead*) developed
            from a lateral adult bud after the upper branch internodes died.
            (**C**) On a well-developed grafted branch with many adult
            leaves (*arrow*), the lowest adult bud developed into a juvenile
            branch (*arrowhead*). (**D**) Gradual transition of leaves on three
            branches (originated from the grafted adult branch that its
            apical bud was damaged during grafting). Note the two
            original adult leaves of the adult branch (*lower arrow*) and then
            the development of juvenile leaves (*arrowhead*) and the
            gradual transition to adult leaves (*upper arrow*). The scale bars
            are in cm. (From Aloni 2007)   219

Fig. 13.1   Transverse section in a vascular bundle of *Coleus blumei*,
            showing a differentiating vessel with a nucleus and its nucleo-
            lus (*arrow*) in the vessel widening phase that lasts until the
            secondary wall is deposited, which stops cell expansion.
            Cambium, c. Bar = 50 µm   224

Fig. 13.2   Transverse sections of the phloem along the same continuous
            vascular bundle passing through young (**A**) and mature (**B**)
            internodes of an intact stem of *Luffa cylindrica*, stained with
            lacmoid (Aloni 1979), showing the substantial increase of
            sieve elements in the older internode (**B**), with the increasing
            distance from young leaves. Sieve element (s), sieve plate
            (*arrow*), companion cell (*arrowhead*). Bars = 50 µm....................   225

Fig. 13.3   Transverse sections along the same internode, the second
            internode above the cotyledons of *Phaseolus vulgaris* after
            3 weeks of treatment with auxin (0.1% naphthaleneacetic acid
            (NAA) applied in a lanoline paste) renewed every 3 days, on
            the top of the internode after the shoot above it was excised.
            The sections were stained with a mixture of safranin and alcian
            green which gave red color to the lignified xylem cells,
            observed and photographed under a light microscope. (**A**)
            5 mm below the site of NAA application, showing massive
            xylem formation with 7 layers of vessels induced by 7 renewed
            auxin applications, characterized by many narrow vessels in
            high density. (**B**) 40 mm below the site of auxin application,
            showing a substantial decrease in xylem formation, character-
            ized by wide vessels in low density, organized in bundles. Note
            the change of vessel patterns from layers (**A**) to bundles (**B**).
            Bars = 100 µm. (From Aloni and Zimmermann 1983).................   228

Fig. 13.4 Transverse sections along the same internode of *Phaseolus vulgaris*, showing the gradual changes in patterns of the induced vessels with increasing distance from the auxin source, after 3 weeks of treatment with 0.1% NAA applied on the top of the internode after the shoot above it was excised. The auxin was replaced and renewed every 3 days. (**A**) 5 mm below the site of NAA application, (**B**) 10 mm, (**C**) 20 mm, and (**D**) 40 mm below the site of auxin application. The photographs show the gradual increase in vessel diameter and decrease in vessel density with increasing distance from the auxin source, as well as gradual changes of vessel pattern from layers (**A** and **B**) to bundles (**C** and **D**). *Arrow* marks a late-formed secondary vessel. Bars = 100 μm. (From Aloni and Zimmermann 1983) 229

Fig. 13.5 Effects of applied auxin concentration (0.03% NAA, 0.1% NAA, or 1.0% NAA w/w in lanolin, renewed every 3 days) on secondary vessel differentiation in the second internode above the cotyledons of *Phaseolus vulgaris*, observed after 3 weeks of hormonal applications, on the top of the internode after the shoot above it was excised. (**A**) Effect of distance (0.5 and 4.0 cm) from 0.1% NAA application site on the rate of secondary vessel formation, showing intensive vessel differentiation near (0.5 cm) the site of auxin application. (**B**) Effect of 0.03% and 1.0% NAA on the radial diameter of the late-formed secondary vessels, along the studied internode, showing the substantial increase in vessel diameter with increasing distance from the applied auxin. (**C**) Effect of 0.03% and 1.0% NAA on the number of secondary vessels induced along a xylem radius, as affected by distance from auxin source. *Vertical bars* indicate standard errors which are comparable at all points. (From Aloni and Zimmermann 1983) 230

Fig. 14.1 Experimentally induced circular vessels that differentiated above a "butterfly" wound (above an upside down open triangle wound forming a V shape) made in the bark and cambium of the maple *Acer saccharum* (**A**) and the oak *Quercus velutina* (**B**). (**A**) Showing a few circular vessels that differentiated in maple, due to its active cambial activity. Two of the circular vessels are marked by *arrowheads*. (**B**) Close-up view on a circular vessel in the oak, the perforations of the three vessel elements are marked by *arrowheads*. Bars = 100 μm (**B**); 200 μm (**A**). (From Aloni et al. 1997) 238

Fig. 14.2   Naturally occurring circular xylem patterns at the junctions
             of large branch, exposed by bark removal in the oak
             Quercus ithaburensis. (**A**) Late-formed wood patterns
             showing a few large circular vessels. (**B**) Close-up view of
             the same circular vessels (*arrows*). Bars = 50 mm. (From
             Lev-Yadun and Aloni 1990) 239
Fig. 14.3   Photomicrographs of the root-shoot junction, showing longitu-
             dinal thick sections of the root-shoot junction. The tissues are
             all unstained and cleared in lactic acid. All photographs are
             longitudinal views of the junction with the border between the
             root (R) and shoot (S) tissues delineated by a dotted line. (**A**)
             Adventitious root junction in corngrass, which is a corngrass
             mutation of maize (*Zea Mays*, *Cg* mutant) showing the entry of
             two metaxylem vessels from the root (*black arrows*) into the
             shoot. The lower part of a longitudinal vessel of the shoot is
             marked by a *white arrow*. (**B**) Close-up view of a continuous
             metaxylem vessel (*arrows*) in the junction between an adventi-
             tious root and the shoot in corngrass. (**C**) Root-shoot junction
             of a seminal root in winter rye (*Secale cereal* cv. Musketeer)
             showing the upper portion of the central metaxylem vessel,
             which ends at an imperforate wall (*black arrow*), and the small
             tracheids with simple pitting (*white arrows*), which connect the
             vessel of the root to vessels in the shoot. Bars = 50 μm (**C**),
             100 μm (**A, B**). (From Aloni and Griffith 1991)............................ 241
Fig. 15.1   Photomicrographs of longitudinal tangential sections in the
             secondary xylem of *Melia azedarach*, showing the naturally
             occurring increase in the size of vascular rays (*arrows*) with
             increasing distance from the pith (**A, B**) and a substantial
             increase in ray dimensions in response to wounding (**C**). (**A**)
             Narrow and long vascular rays differentiated near the pith. (**B**)
             Wider and larger rays developed away from pith. (**C**) Much
             larger rays differentiated following wounding, likely in
             response to wound-induced ethylene.
             Bars = 100 μm 246
Fig. 15.2   Photomicrographs of cross sections in the secondary xylem of
             Hibiscus cannabinus young stems, in an intact control (**A**) and
             following the effect of ethylene, originating from 1% ethrel
             (2-chloroethylphosphonic acid) in lanolin (**B, C**). The cyto-
             plasm in the ray cells (marked by *arrows*) is dark due to
             accumulation of starch and other reserve materials (toward
             winter). All photographs are at the same orientation (cambium
             in the upper side), from the same experiment and internode
             age. Showing (**A**) Naturally occurring narrow rays in an intact
             stem. (**B**) Substantial increase in ray width induced by ethylene
             (released from 1% ethrel). (**C**) Fusion of rays promoted

by the ethylene stimulation. Bars = 50 μm. (From
Aloni et al. 2000).. 247

Fig. 15.3   Photomicrographs of longitudinal tangential sections in the
secondary xylem of *Suaeda monoica* showing stages in the
development of a polycentric vascular ray, with one (**A**) and
three (**B**) radially elongated "centers" that form a kind of radial
"rays" inside a ray. Each "center" of small parenchyma cells is
marked with an *arrow*. (**A**) Section located away from the
cambium showing an early stage of ray development with one
"center." (**B**) More developed and larger polycentric ray in a
section located close to cambium. The radially continuous
narrow radial "rays" inside the large polycentric ray demonstrate that the three "centers" are the result of individual radial
preferable pathways of a radial signal flow occurring inside the
large polycentric vascular ray. Bars = 75 μm. (From Lev-Yadun
and Aloni 1991) 248

Fig. 16.1   Experim6ental results of *Hibiscus cannabinus* plants grown in
different pot sizes for limiting root development (**A**), and cross
sections (**B–D**) done in their stems (from (**B**) to (**D**) in
Fig. 16.1A) at the same distance from soil surface. (**A**)
Decrease in pot size limited plant development and maximal
stem height; the smallest pot enable limited plant development
and resulted in the shortest plant. (**B**) The secondary vessels (v)
and secondary xylem fibers (xf) were the widest in the tallest
stem. (**C**) Intermediate stem length produced intermediate
diameters in the secondary vessel and xylem fiber. (**D**) The
shortest stem was characterized by the narrowest vessels and
fibers. The latter produce thick secondary walls, as is known to
be induced in fibers by high-auxin concentration (Aloni 1979).
In (**B**)–(**D**): c, cambium; pf, phloem fibers; xf, xylem fibers, v,
secondary vessel. Bars = 100 μm (**B–D**), 50 mm (**A**) (From
Aloni 1988)... 253

Fig. 16.2   Cross sections from the middle of the basal internode of corn
(*Zea mays*) grown under: (**A, C**) favorable conditions in a large
pot, *versus* (**B, D**) stress conditions, due to limited root space
in a small pot, showing the effect of stress on the differentiation of primary xylem vessels and phloem sieve tubes (which
might be influenced also by abscisic acid (ABA) from the root
meri stems, see Sect. 3.5). (**A**) Typical primary vascular
bundles with two or three protoxylem vessels (*arrow*), and
expanding metaxylem vessels (m) that their living cytoplasm
was stained light-blue by lacmoid. (**B**) The stressed plant
developed vascular bundles with many very narrow protoxylem
vessels (*arrows*) and two or four metaxylem (m) vessels,

(**C**) Magnified view of a typical vascular bundle with two functional protoxylem vessels (*arrow*), two expanding metaxylem vessels (before secondary wall deposition), and normal size sieve elements in the phloem (*arrowhead*). (**D**) Magnified bundle affected by stress with many very narrow protoxylem vessels (*arrow*), expanding metaxylem (m) vessels and very narrow sieve tubes (*arrowhead*). Bars = 100 μm (**C**, **D**), 200 μm (**A**, **B**). (By R. Aloni and J. Mattsson, unpublished) 254

Fig. 17.1   Transverse sections showing resin storage structures (*arrows*) in intact mature leaves (**A**) and stems (**B**). (**A**) Showing four rounded resin glands (*white arrows*) from which only the gland in the upper left side was cut in the middle, in lateral scale leaves in unstained secondary shoot of *Thuja plicata*. (**B**) Longitudinal resin ducts (*black arrows*) occurring naturally at the latewood in the trunk of *Pinus halepensis* stained with safranin, which stained the tracheids and rays in red. The photomicrograph shows two complete annual rings. The borders between the years are marked by *arrowheads*. Note that the wider annual wood ring (on the *right side*), which likely differentiated in a rainy year, produced a few resin ducts, clearly more than in the narrow annual wood ring (*left*) produced in a dry year. Bars = 100 μm (**B**); 200 μm (**A**). (**A** is adapted from Foster et al. 2016)... 260

Fig. 19.1   Diagram showing the experimental method for inducing regenerative tracheids from young parenchyma cells in the hypocotyl of *Pinus pinea* seedlings. Half of the hypocotyl is separated from the leaves by a cut. The *arrow* marks the site where a lanolin paste with auxin, gibberellin, or a mixture of auxin and gibberellin is applied to the partly separated half, which remains attached to the root at its base. The *dotted line* inside the isolated half of the hypocotyl marks the sites where regenerative tracheids differentiate. The exposed tissues are separated and protected with parafilm. (Following Kalev and Aloni 1998)... 274

Fig. 19.2   Longitudinal views of tracheids with perforations (*arrows*) induced experimentally by auxin (0.5% 1-naphthaleneacetic acid (NAA) in lanolin w/w) in the hypocotyl of young *Pinus pinea* seedlings (as described in Fig. 19.1). (**A**) A single perforation developed in the tracheid's upper cell wall. (**B**) Two openings developed on the side wall toward the lower side of the perforated tracheid, with a relatively thick rim around the openings. Bar = 50 μm. (From Aloni 2013a).......................... 275

Fig. 19.3   Longitudinal views of experimentally induced regenerative
             tracheids in the hypocotyl of young *Pinus pinea* seedlings
             induced by auxin (0.1% NAA in lanolin) observed in thick
             sections cleared in lactic acid. (**A**) Close view on a short
             tracheid with typical pattern of lignified secondary wall
             thickenings (*arrow*). (**B**) Under low magnification, showing the
             pattern of numerous short tracheids below the site of auxin
             application, produced across the hypocotyl. The auxin induced
             only short tracheids (*arrows*). Bars = 25 μm (**A**), 100 μm (**B**)   276
Fig. 19.4   Longitudinal views of regenerated tracheids in thick sections
             cleared with lactic acid in hypocotyls of *Pinus pinea*, demon-
             strating the effect of gibberellic acid (1.0% GA$_3$ + 0.1% NAA)
             with low auxin (*gibberellin by itself did not induce tracheids*)
             on tracheid elongation. (**A**) Substantial elongation of tracheids
             by intrusive growth of their upper and lower ends (*arrows*).
             Note the swelling of the tracheids' growing points. (**B**)
             Continuous intrusive growth of two regenerative tracheid tips
             (*arrows*) that moved away from each other. Bars = 50 μm (**B**),
             100 μm (**A**). (From Kalev and Aloni 1998 (**A**); From Aloni
             2013a (**B**))   277
Fig. 19.5   Cross sections in *Ephedra campylopoda* stems, characterized
             by a relatively primitive vascular system built of tracheids
             (*white arrowheads*), vessels (*red arrows*), and fibers (*red
             arrowheads*), in the intact stem (**A**), demonstrating the role of
             gibberellin (**B**), and auxin (**C**), applied along 1 month (renewed
             every 3 days) on the differentiation of fibers and vessels,
             respectively. (**A**) The intact stem with vascular system built of
             tracheids, vessels, and fibers. (**B**) Gibberellin (1% GA$_3$ in
             lanolin, with no auxin) induced tracheids in the xylem (with no
             vessels) and many fibers in the phloem. (**C**) Auxin (0.2%
             NAA) induced continuous layers of mainly vessels (with no
             fibers). Bars = 100 μm. (From Aloni 2013a)   278
Fig. 19.6   Schematic diagram illustrating the role of auxin (IAA) and
             gibberellin (GA) in shaping the evolution of vessel elements
             and fibers from the long tracheids of primitive plants. The
             tracheids characterized by bordered pits are induced by a
             mixture of both IAA and GA (**a, b**). During plant evolution,
             GA has become the specific signal for fibers with simple pits
             (**c, d**), and IAA the inducing signal for short vessel elements
             with perforation plates (**e–g**)   279
Fig. 19.7   Cross sections of a diffuse-porous (with tendency to ring
             porosity) wood of *Styrax officinalis* (**A**), in comparison with a
             ring-porous wood of *Robinia pseudoacacia* (**B**). *Asterisks* mark
             earlywood vessels in both photographs. (**A**) In the diffuse-
             porous pattern, there is a gradual decrease in vessel diameter
             during the annual growth season, and the vessels remain

functional (open) for a few years. (**B**) in the ring-porous wood, all the very wide earlywood vessels are functional in the recent season (near the green-stained phloem, at the right side), but tend to become embolized and therefore blocked by the penetration of parenchyma cells to form tyloses (*arrowheads*) in the wood of previous years. The recent year produced wide earlywood vessels followed by a few narrow vessels in the latewood, while the previous years' latewood contained mainly fibers and large groups of parenchyma cells, with almost no vessels. Bar = 500 μm.. 280

Fig. 19.8   Photographs of cross sections comparing typical secondary xylem production and demonstrating xylem conductivity at the base of intact branches of the diffuse-porous maple tree *Acer saccharum* (**A**) and the ring-porous oak tree *Quercus velutina* (**B**) grown at the same site. The red regions near the vessels were stain by safranin, which moved upward with the red-colored water through an intact branch junction and stained the conductive vessels and their surrounding cells. The *arrows* mark actively transporting earlywood vessels. (**A**) Shows that almost all the vessels in the 2-year-old diffuse-porous branch transported water. (**B**) In a 4-year-old ring-porous branch, all the earlywood vessels of the fourth year were active (*right arrow*); some earlywood vessels of the third year remained active (*left arrow*). The *arrowhead* marks the only water transporting latewood vessel of the second season. Both photographs are at the same magnification, Bar = 100 μm. (From Aloni et al. 1997) 281

Fig. 19.9   The effect of auxin (1-naphthaleneacetic acid (NAA) in lanolin) concentration on the width of earlywood vessel differentiation is shown in transverse sections in stems of the ring-porous tree *Melia azedarach*. All photomicrographs were taken from the same experiment; run in Tel Aviv from February 15 to March 15, 1986; and are presented in the same orientation and magnification (Bars = 250 μm). All the sections were taken 50 mm below the apical bud, which was left intact (**A**) or was replaced by a range of auxin concentrations: low, 0.003% NAA (**B**); medium, 0,01% NAA (**C**); or high, 0.1% NAA (**D**). The auxin was applied in the form of a lanolin paste, which was renewed every 3 days. The photomicrographs show a substantial decrease in the diameter of the earlywood vessels (*white arrows*) with increasing auxin concentration (**B–D**). The low auxin concentration induced wide vessels (**B**). The two higher concentrations induced many more xylem cells (along a radius) with narrower vessels (**C**, **D**). The highest auxin concentration tested (0.1% NAA) resulted in very narrow earlywood vessels (**D**). The borderline between the latewood of 1985 (*left*) and the new earlywood of 1986 (*right*) is marked with *white triangles*. The experiment was repeated three times (in 1984, 1985, and 1986) with 5–10 stems per treatment, yielding the same results. (From Aloni 1991) 285

Fig. 20.1   Cross sections showing the haustorium penetration (from the *right side*) of the parasitic plant *Loranthus acacia* into the stem of its host tree *Acacia raddiana*. The *border lines* of the haustorium are marked with arrowheads. The xylem of the host is located in the *upper left side*. (**A**) The actively transporting vessels in both the host and the parasite were stained *red* by safranin transported through these vessels; the dye was applied to the host and moved upward into the parasite leaves (that became red) by transpiration. The 80 μm cross section was not stained after sectioning. (**B**) Magnified view of the same section focusing on a naturally occurring vessel perforation (marked by *arrow*), which forms a continuous open water-transporting lumen from the relatively wide vessel of the host to the narrow vessel of the parasite. h, host vessel; p, parasite vessel. Bars = 100 μm (From Aloni 2015) 295

Fig. 20.2   Tissue differentiation in stem galls induced by the midges *Izeniola obesula* on the host plant *Suaeda monoica*, observed in cleared, hand-cut cross sections. Showing: (**A**) In 1-week-old gall, young larva (L) in its chamber. (**B**) A mature vessel element (*arrow*) and an early stage of vessel element development (*white arrowhead*), starting from the *larva chamber* (LC), that containing fungal mycelia (*black arrowhead*) on which the larva feeds. (**C**) A circular vessel (*white arrowhead*) differentiated near the origin of the larva-induced vessels (*black arrowheads*) descending from the larva chamber towards the stem's vascular bundles. (**D**) In mature gall, showing larva chamber protected by surrounding sclerenchyma (S) and supplied with water by network of novel vessels (*arrowheads*) induced by the larva, which merges into the original stem's vascular *bundle* (*B*). Bars = 25 μm (**A**, **B**), 50 μm (**C**), 100 μm (**D**). (From Dorchin et al. 2002) 300

Fig. 20.3   Effect of the gall-forming aphid *Slavum wertheimae* on secondary xylem differentiation below the gall, in branches of *Pistacia atlantica* trees. (**A**) *P. atlantica* branch before bud break, carrying four coral-like ('cauliflower') galls (*right*), and a control branch with no galls (*left*). (**B**) Close-up view of the coral-like gall (*arrow*). (**C**) and (**D**) are transverse sections from 1-year-old branches of *P. atlantica*. Both sections are at the same magnification. (**C**) is taken from a control branch, free of galls. (**D**) is taken 20 mm below a coral-like gall induced by

S. wertheimae. More xylem is evident immediately below the
gall and is characterized by numerous wide vessels. In the
latewood, there are some very wide vessels (*arrows*).
Bars = 250 μm (**C, D**), 10 mm (**B**), 30 mm (**A**). (From Aloni
et al. 1989) 303

Fig. 20.4   Effect of the gall-forming aphid *Baizongia pistaciae* on
secondary xylem differentiation, shown in transverse sections
made in 3-year old branches, 100 mm below the gall (**B**), and
at an equivalent location in an ungalled branch (**C**) of the same
Pistacia palaestina tree. The earlywood vessels of the current
year are marked with *black arrows* in both branches (**B, C**).
(**A**) *P. palaestina* branch with the typical elongated gall shape
(*arrow*) produced by *B. pistaciae* aphids. (**B**) Substantially
more xylem was induced below the gall (by hundreds of aphids
living inside this gall) with wide vessels (*white arrow*) in the
latewood. (**C**) Normal xylem with narrow latewood vessels
(*white arrow*) differentiated in an adjacent ungalled branch.
Bars = 250 μm (**B, C**), 25 mm (**A**). (From Aloni 1991) 304

Fig. 21.1   Effects of wounding and exposure to *Agrobacterium tumefaciens* on tumor development in sunflower (*Helianthus annuus*)
seedling, 2 weeks after infection. (**A**) Intact control not
exposed to the bacterium. (**B**) Unwounded seedling exposed to
A. tumefaciens did not develop tumor and continue normal
growth. (**C**) Wounded seedlings exposed to *A. tumefaciens*
developed tumors (*arrows*) which substantially retarded plant
growth. The seedling with the crown galls degenerated and
died after another 2–3 weeks.
Bar = 50 mm 310

Fig. 21.2   Photomicrograph of a transverse section in a typically unorganized plant tumor tissue in an *Agrobacterium tumefaciens*-
induced crown gall on sunflower, showing vessel elements
(*arrows*) with lignified secondary cell wall (stained red by
safranin) and parenchyma cells in various orientations and
sizes, due to random cell divisions.
Bar = 100 μm 310

Fig. 21.3   Photomicrographs of vascular tissues in tumors shown in
longitudinal radial sections both stained with lacmoid (according to Aloni 1979). (**A**) Circular patterns (*arrows*) of vessels
and fibers in olive (*Olea europaea*) hardwood gall induced by
the bacterium *Pseudomonas syringae* subsp. *savastanoi*. (**B**)
Typical patterns of vascular differentiation in a 3-week-old
crown gall tumor induced by *A. tumefaciens* on castor bean
(*Ricinus communis*) showing both globular (G) and tree-like
(T) vascular bundles, emerging from a main bundle of the host

(H). Note that the lower side of the host bundle is thicker than its upper side, due to additional vascular tissues induced by the tumor. Bars = 200 μm (**A**); 20 mm (**B**). (From Ullrich et al. 2019 (**A**); and from Aloni et al. 1995 (**B**))..................................... 312

Fig. 21.4   Longitudinal schematic diagram showing the pattern of vascular bundles (*thick lines*), and phloem anastomoses (*dotted lines*) in *Agrobacterium tumefaciens*-induced crown gall on a *Ricinus communis* stem. In addition, three schematic diagrams of transverse sections, illustrating the symmetric structure of a healthy vascular system in the host stem above the tumor (**A**), the pathologic host xylem (*arrow*) with giant rays in a median position (**B**), and the asymmetric xylem differentiation below the tumor (**C**). Numbers 1 and 2 mark the connecting sites of globular vascular bundles to the host vascular system, while 3 and 4 mark the base of tree-like branched vascular bundles, both consisting of xylem and phloem. *Arrowheads* mark some of the phloem anastomoses between the bundles (shown in Fig. 21.5B); F marks regenerative phloem fibers with unique anatomical ramifications (shown in Chap. 2, Fig. 2.25B, C) restricted to the upper and lower basal regions of the tumor. (From Aloni et al. 1995) 313

Fig. 21.5   Photomicrographs of longitudinal views of phloem and xylem in 8-week-old *Agrobacterium tumefaciens*-induced crown galls developed on *Ricinus* cleared in lactic acid and stained with lacmoid (according to Aloni and Sachs 1973). (**A**) Showing a portion of a large vascular bundle build of vessels (*arrowheads*) and sieve tubes, identified by turquoise blue staining of callose on their sieve-plates (*arrow*). (**B**) Phloem anastomoses between the bundles consisting of sieve tubes only (*arrows*). Bar = 100 μm. (From Aloni et al. 1995)....................................... 314

Fig. 21.6   Influence of 8-week-old *Agrobacterium tumefaciens*-induced crown gall on vascular differentiation in a *Ricinus* host stem. Showing transverse sections of healthy xylem formed away from the tumor (**A**), in comparison with pathological xylem close to the tumor (**B**), and very close to the crown gall (**C**) (the location of each section is marked by a circle on the schematic diagram). The vessels (V) and rays (R) had a normal structure away from the tumor (**A**). Narrow vessels and fibers (F) with large unlignified rays differentiated close to the grown gall (**B**). Maximum effect of the crown gall on the host was evident adjacent to the tumor, where vessels were very narrow and the rays were giant and unlignified (**C**). Bars = 100 μm. (From Aloni et al. 1995) 315

Fig. 21.7   Structure and function of xylem in *Agrobacterium tumefaciens*-induced tumor (T) on a stem of the *Ricinus communis* host (H). (**A**) Longitudinal section in an 8-week-old tumor stained with

lacmoid. The xylem of the tumor is connected to the host and
extends nearly up to the tumor periphery. (**B**) Stem cross
sections above and below a 3-week-old tumor stained with
lacmoid. The xylem formed immediately below the tumor
(*arrow*) is twice as large as that above tumor (compare
schematic diagrams A vs C in Fig. 21.4). (**C**) Path of water that
were labelled with the red acid fuchsin, moving from the host
stem into a 1-week-old tumor. Note that the host xylem below
the tumor is thicker than above it. (**D**) Path of water uptake,
labelled with red acid fuchsin, applied from the host stem into
a 3-week-old tumor, extending up to the tumor periphery. The
isolated red patches were found to be connected to the main
water conduits. All the sections shown in the photomicrographs
were viewed under dark field microscopy. Bars = 2 mm. (From
Schurr et al. 1996) 316

Fig. 21.8   Comparison of *A. tumefaciens*-induced crown galls (marked by
yellow arrows) on wild-type tomato (*Lycopersicon esculentum*)
(**A** and **C**) and the ethylene insensitive *Nr* mutant (**B** and **D**)
stems. (**A**) Front view of a 3-week-old tumor developed on a
wild-type plant showing the typical unorganized callus shape
(resulting in enlarged surface) of a young crown gall and the
epinastic response of the leaves (*red arrow*) both above and
below the tumor (typically induced by ethylene). (**B**) Front
view of a 3-week-old tumor developed on the *Nr* mutant,
characterized by a smooth minimal surface and leaves in the
normal orientation (*red arrow*). Note that the lower half of the
gall is protected by epidermis. (**C**) Side view of a 2-month-old
tumor on a wild-type stem with numerous adventitious roots
(white spots marked by *red arrows*) developed both above and
below the crown gall. (**D**) Side view of a 2-month-old tumor on
the *Nr* mutant, showing a fibrous hard gall and a stem almost
free of adventitious roots. All photographs are at the same
magnification, bars = 10 mm. (From Aloni et al. 1998) 318

Fig. 21.9   The effects of 6-week-old *A. tumefaciens*-induced crown galls
on xylem differentiation in tomato host stems, in transverse
sections cleared with lactic acid and stained with lacmoid. The
crown galls were located above the micrographs, and the white
region at the lower part of the photographs is the cleared
parenchymatic pith. The border between the xylem formed
after infection with *A. tumefaciens* (upper part of each micrograph) and the intact xylem developed before the treatments
(lower part) is delineated by *a broken red line*. (**A**) Limited
differentiation of pathologic xylem with very narrow vessels
(*arrows*) and wide rays (R) characterize the wild-type host.

(**B**) Massive xylem with wide vessels (*arrows*) and almost normal rays (R) characterized the host stem formed adjacent to the *tumor* on *the Nr* mutant. Bars = 200 μm. (From Aloni et al. 1998) 319

Fig. 21.10   Typical patterns of vascular tissues in 6-week-old *A. tumefaciens*-induced crown galls on tomato stems observed in thick, longitudinal radial sections cleared with lactic acid and stained with lacmoid. (**A**) Circular vessels surrounded by parenchyma cells in a wild-type stem. (**B**) Circular vessel surrounded by fibers in the *Nr* mutant. Circular vessels (V); fibers (F); parenchyma (P) cells. Bars = 100 μm. (From Aloni et al. 1998) 320

Fig. 21.11   The effects of 4-week-old *A. tumefaciens*-induced crown galls located at the base (*yellow arrow*) of the stems on shoot development and leaf senescence in tomato plants, grown under identical conditions. (**A**) Retarded wild-type shoot (*left*) and a typically taller *Nr* shoot with large leaves (*right*). Note that the older leaves in the wild-type plant started to turn yellow and senesce. (**B**) Moderate water stress caused leaf senescence in the wild-type shoot (*left*), whereas most of the leaves remained green and healthy on the *Nr* shoot (*right*). Bars = 50 mm. (From Aloni et al. 1998) 321

List of Table

Table 5.1   Effect of cytokinin on secondary xylem fiber differentiation
in cultured hypocotyl segments of *Helianthus annuus* after
30 days, demonstrating the need for cytokinin during the first
2 weeks for fiber formation (A, C), while absence of CK in the
early stage prevent fiber differentiation (B, D). Values are
mean ± SE. Sample size was 10 for each treatment. There was
no significant difference in the number of secondary xylem fiber
under treatments A and C. This was also true for the final fresh
weight under any of the four treatments. Kinetin concentration
was 0.2 μg/ml. All culture media contained 0.5 μg/ml IAA
and 0.1 μg/ml GA_3   120

Introduction

This book provides tools for understanding vascular differentiation and plant development by elucidating their regulating hormonal mechanisms. The focus is on the plant vascular system that connects the leaves with the roots and enables long-distance transport of water, nutrition, and hormonal signals between the organs. The book also clarifies the mechanisms how the vascular tissues physically support the plant body, allowing the development of large plants and giant trees. The book explains how continuously moving hormonal signals through the vascular tissues in the whole plant allow the plant to regulate its development, synchronize shoot and root growth, regenerate, continuously respond to changes, and adapt to the environment.

Following my long teaching experience at different universities, the book is oriented to molecular biology students who need the whole organismal background and understanding of the entire vascular system for enabling them to know the system they work on and study plants successfully, to reach the right interpretations and avoid mistakes. It is recommended as a textbook for undergraduate and graduate courses, as well as for active self-learning by advanced research students and scholars. The book is written in a holistic approach, viewing the whole plant as an integrated operating organism, in a simple manner with many supporting original illustrations and photographs, most of them from my research, some of which I prepared especially for the book, to clarify the text and make it attractive to anyone interested in plants, even without early background in plant science. The book is a good source of old and new references, as it covers the field from the initial studies done more than a century ago up to the recent studies with up-to-date explanations, discussing how continuously moving hormonal signals and their precursors induce and control plant vascular differentiation (Fig. 1.1), development, adaptation, evolution, parasitism, gall formation, cancer development, prevention, and therapy, aimed to review the entire field of plant vascular differentiation in a broadened scope and in depth.

New technologies and advanced equipment promote new discoveries. Yet, scientific breakthroughs are conceptual, not merely technological, achievements. A good

© Springer Nature Switzerland AG 2021
R. Aloni, *Vascular Differentiation and Plant Hormones*,
https://doi.org/10.1007/978-3-030-53202-4_1

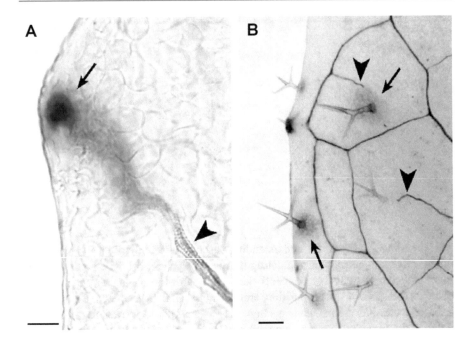

Fig. 1.1 Micrographs of cleared leaf primordia of DR5::GUS transformed *Arabidopsis thaliana* demonstrating how the bioactive auxin hormone, namely, indole-3-acetic acid (IAA), induces vascular veins, visualizing the process by the blue staining of *DR5::GUS* expression during leaf morphogenesis. (**A**) At the lobe of a young leaf primordium, where a hydathode will develop, showing a center of strong expression (*arrow*) marking the synthesis site of high-auxin concentration, from which the auxin starts to flow downward in a diffusible pattern that gradually becomes canalized to a narrow stream which induces the vascular vein (*arrowhead*). (**B**) In a more developed primordium, *GUS* expression is detected at the base of a few auxin-producing trichomes (*arrows*), showing how these trichomes induce their water-supporting vascular veins. Note that two of the trichomes are associated with freely ending veinlets (*arrowheads*) that they induced. Bar = 50 μm (**A**), 250 μm (**B**). (From Aloni et al. 2003)

example in plant sciences is the brilliant study of Bowman, Smyth, and Meyerowitz (1991), who emphasized in their later paper (Bowman et al. 2012) entitled "The ABC model of flower development: then and now" that one of their surprising conclusion is that materials and methods that might have led to their similar work on the control of flower development, and to the same well-known ABC flowering model (see Sect. 8.2), were available 100 years before their experiments, contradicting the belief that progress in biology necessarily comes from improvements in methods, rather than in concepts (Bowman et al. 2012). Therefore, the book exposes the reader to old and new concepts, with critical analysis, evolution of hypotheses, stimulating ideas, and challenging open questions that require additional research work with available and future techniques.

Evolution of ideas and hypotheses is a natural process driven by curiosity and skepticism, which develops for better interpretations, answering questions, and understanding phenomena and results. A scientific hypothesis should provide a comprehensive explanation of the studied phenomenon, provide solid evidence for explanation, and offer means to make predictions about the phenomenon. Only the hypothesis supported and confirmed by solid experimental results survives. A few examples in the field of vascular differentiation will be demonstrated in the book, in order to stimulate imaginative students and promote novel ideas and discoveries.

The progress of molecular research has developed a new language with novel terms, which may be different from the classical organismal terms and expressions, sometimes making it difficult for molecular and organismal scientists to communicate and share their results and ideas. Therefore, molecular scientists might not easily follow organismal studies although their findings might be important for molecular research and interpretation. This book summarizes the organismal knowledge and concepts on vascularization aiming to close the gap by presenting and discussing the organismal knowledge and understanding on the differentiation of plant vascular tissues to molecular students, aiming to transfer the "knowledge torch" to new students and interested scholars. The book exposes students and research scientists to challenging open questions in vascular biology and proposes new directions for further research on vascular differentiation and plant development.

The book discusses the hormonal mechanisms that regulate and control wood (xylem produced by cambium) formation in trees, which enable them to become the oldest and largest organisms on earth. Therefore, we might wonder what the secret to tree longevity is. The oldest continuously living tree is the bristlecone pine (*Pinus longaeva*) that lives in the white mountains of Eastern California, for more than 4850 years (Evert and Eichhorn 2013), while the tallest trees that reach the height of 100 m are the redwoods (*Sequoia sempervirens*) of California (Fig. 1.2A) and *Eucalyptus regnans* trees of Australia (Williams et al. 2019). The book clarifies the hormonal regulation of tree growth and wood development, discussing different aspects of wood formation and function, like the control of cell size along these amazing trees, mechanisms of recovery from bending by reaction wood, protecting mechanisms against insects (e.g., resin) and their associated pathogens, vascular regeneration after injury, biology of stem cells and embryonic tissues (meristems), maturation and rejuvenation, evolution of the vascular cells, hydraulic safety zones in organ junctions, natural grafting between roots of neighboring trees that establishes a cooperative tree community, and more.

Architecturally, unlike the bottom-up design and construction of human-made buildings, trees build their stems and roots by downward flows of supplies (e.g., carbohydrates) produced in their photosynthesizing leaves. As we try to figure out what is the secret to the longevity of trees, we should realize that every year, a tree produces a new outer layer of vascular tissues that envelops the wood produced during previous years. Only a few recent external annual layers of vascular tissues are alive and functioning, while all the old internal wood layers within the tree cease to function gradually and die. The old inside annual wood layers can develop into resistant heartwood by the accumulation of secondary metabolites (Celedon and Bohlmann 2018) or may be destroyed over time (Fig. 1.2B). This means that in a very old tree of a few thousand years, only the newly produced water-transporting sapwood and the nutrient-translocating phloem of the few latest years are alive and

Fig. 1.2 Photographs of coast redwoods (*Sequoia sempervirens*) grown in north California. (**A**) Showing three giant trees of about 100 m height. The two trunks on the right have straight bark patterns, while the left tree has a spiral bark pattern, likely induced by spiral movement of auxin in the phellogen, which is the cork cambium (see Chap. 3, Fig. 3.11A). (**B**) An upward photograph made from the inside of a continuously growing tree, showing a living redwood tree where its inside dry wood was burnt by a strike of lightning, while the outer moisturized wood layers remain alive, and the tree is covered by its own many green active leaves and growing branches, demonstrating that only the outer vascular layers are the living tissues of a tree and the dead core of a tree can be burnt and the tree continues to grow. *Arrow* marks some annual wood rings that were not burnt; T, top of tree damaged by the lightning

functioning, which are continuously produced by the lateral embryonic tissue – the cambium, while the inner solid old wood skeleton that builds the heartwood is composed of thousands annual layers of dead wood containing the old tissues that were alive many years ago. The book clarifies the biology of the living vascular tissue of trees and reveals how they are regulated by different, sometimes opposing, hormonal mechanisms.

The book widens the scope in specific topics and relates also to additional developmental aspects of shoot and root communication (see Chap. 6), as well as plant responses to phosphate and nitrate concentrations in the soil because they modify the biosynthesis of two important hormones that shape shoot developmental architecture and vascular differentiation (see Chap. 4). These issues are important for understanding synchronized organismal development, plant architecture and competition, the control of organ initiation, inhibition, and growth of lateral organ. Additionally, the book discusses gall formation in response to insects and bacteria, parasitism, and environmental adaptation and evolution, which are expected to interest the readers and provide better understanding of plants and their behavior.

Special attention is given to tumor development and prevention, which has an enormous impact on crop yield and human cancer therapy (see Chap. 21). These unorganized tumorous tissues provide a wider perspective on normal plant development and better understanding of vascular differentiation.

A living plant cell with a nucleus can potentially produce different hormones. However, the synthesis of a hormonal signal is determined by the location of this cell within the plant body and might be influenced by age and external environmental conditions. Thus, cells in young growing leaves would tend to produce auxin, and those in the root cap produce cytokinin (Aloni et al. 2006). Experimental evidence shows position-dependent inductive signaling in plant tissues (van der Berg et al. 1995) where cell activity is synchronized with the microenvironment of the cell, namely, cells will behave and produce signals according to their location in the plant body and that the positional control is most important in the determination of cell fate and behavior. Similarly, stem cell behavior is determined by their niche (Zhou et al. 2015) but not vice versa (Singh and Bhalla 2006), namely, that stem cells in embryonic tissues (meristems) are not the source of patterning information and that they are regulated by already differentiated tissues around them.

There are excellent books for studying general and specific topics of plant anatomy (Easu 1965; Fahn 1990; Evert 2006) and likewise superb books for studying broad spectrums of plant physiology, biochemistry, molecular biology of plants (Taiz and Zeiger 2006; Jones et al. 2012; Taiz et al. 2014, 2018; Buchanan et al. 2015), and plant biology (Evert and Eichhorn 2013). However, students in my courses and colleagues have convinced me that there is a need for an up-to-date specific textbook that focuses on the differentiation of vascular tissues, which combines their structure, development, and hormonal regulation in a holistic approach. The present book was written accordingly for students at other universities as well as scholars interested in plant sciences and those who practically work with plants in the fields of biology, agriculture, forestry, and ecology as well as medical scholars who study human and animal vascular tissues and cancer (Ullrich et al. 2019).

References

Aloni, R, Aloni E, Langhans M, Ullrich CI (2006) Role of cytokinin and auxin in shaping root architecture: regulating vascular differentiation, lateral root initiation, root apical dominance and root gravitropism. Ann Bot 97: 883–893.

Aloni R, Schwalm K, Langhans M, Ullrich CI (2003) Gradual shifts in sites of free-auxin production during leaf-primordium development and their role in vascular differentiation and leaf morphogenesis in *Arabidopsis*. Planta 216: 841–853.

Bowman, JL, Smyth DR, Meyerowitz EM (1991) Genetic interactions among floral homeotic genes. Development 112: 1–20.

Bowman JL, Smyth DR, Meyerowitz EM (2012) The ABC model of flower development: then and now. Development 139: 4095–4098.

Buchanan BB, Gruissem W, Jones RL (2015) *Biochemistry and Molecular Biology of Plants*, 2nd edn. Amer Soc Plant Biol and Wiley & Sons, New York.

Celedon JM, Bohlmann J (2018) An extended model of heartwood secondary metabolism informed by functional genomics. Tree Physiol 38: 311–319.

Easu K (1965) *Vascular Differentiation in Plants*. Holt, Rinhart & Winston, New York.

Evert RF (2006) *Esau's Plant Anatomy: Meristems, Cells, and Tissues of the Plant Body: Their Structure, Function, and Development*. Wiley & Sons, Hoboken, New Jersey.

Evert RF, Eichhorn SE (2013) *Raven Biology of Plants*, 8th edn. WH Freeman/Palgrave Macmillan, New York.

Fahn A (1990) *Plant anatomy*, 4th edn. Pergamon, Oxford.

Jones RL, Ougham H, Thomas H, Waaland S (2012) *The Molecular Life of Plants*. Amer Soc Plant Biol and Wiley-Blackwell, New York.

Singh MB, Bhalla PL (2006) Plant stem cells carve their own niche. Trends Plant Sci 11: 241–246.

Taiz L, Zeiger E (2006) *Plant Physiology*, 4th edn. Sinauer, Sunderland, MA.

Taiz L, Zeiger E, Møller IM, Murphy A (2014) *Plant Physiology and Development*. 6th edn. Sinauer, Sunderland, MA.

Taiz L, Zeiger E, Møller IM, Murphy A (2018) *Fundamentals of Plant Physiology*. Sinauer, Sunderland, MA.

Ullrich CI, Aloni R, Saeed MEM, Ullrich W, Efferth T (2019) Comparison between tumors in plants and human beings: Mechanisms of tumor development and therapy with secondary plant metabolites. Phytomedicine 64: 153081.

van den Berg C, Willemsen V, Hage W, Weisbeek P, Scheres B (1995) Cell fate in the *Arabidopsis* root meristem determined by directional signalling. Nature 378: 62–65.

Williams CB, Anfodillo T, Crivellaro A, Lazzarin M, Dawson TE, Koch GW (2019) Axial variation of xylem conduits in the earth's tallest trees. Trees 33: 1299–1311.

Zhou Y, Liu X, Engstrom EM, Nimchuk ZL, Pruneda-Paz JL, Tarr PT, Yan A, Kay SA, Meyerowitz EM (2015) Control of plant stem cell function by conserved interacting transcriptional regulators. Nature 517: 377–380.

Structure, Development, and Patterns of Primary, Secondary, and Regenerative Vascular Tissues

2

To discuss vascular differentiation, there is a need to introduce their basic structures, development, and patterns. Accordingly, major important developmental patterns that will be needed for discussion in the following chapters will be presented. For more detailed and general anatomical information, the reader is directed to comprehensive leading anatomy books (Esau 1965a, 1965b; Fahn 1990; Evert 2006).

2.1 Primary and Secondary Vascular Meristems

The longitudinal growth of plants occurs in their growing points, the apices of both the shoots and roots (Figs. 2.1 and 2.2). In these growing points, apical embryonic tissues, namely, the **apical meristems**, which originated from **stem cells** of the embryo, divide and generate new cells. The produced cells continue to divide and differentiate to mature tissues that build the plant body. The meristems remain active and produce new cells from their stem cells as long as the plant lives. However, they can enter into a dormancy stage during extreme environmental periods, e.g., during defoliation in fall and winter. The **shoot apical meristem** (SAM) is very gentle and protected by folded leaf primordia and young leaves that may develop trichromes and cuticle that form a bud, while the **root apical meristem** (RAM) is protected by the root cap, where the first layer of root cap cells may be covered by an electron-opaque cell wall modification resembling a plant cuticle (Berhin et al. 2019) and mucilage secretion.

The stem cells located at the tip of the apical meristem are surrounded by a group of embryonic cells which form the **stem cell niche** (SCN), creating a microenvironment which provides intercellular signals for stem cell regulation. The maintenance of stem cells is dependent upon reciprocal signaling between the stem cells and the specialized tissue of their niche. When stem cells and their SCN are removed experimentally, e.g., by cutting the youngest tissue at the root tip, a new tip will

© Springer Nature Switzerland AG 2021
R. Aloni, *Vascular Differentiation and Plant Hormones*,
https://doi.org/10.1007/978-3-030-53202-4_2

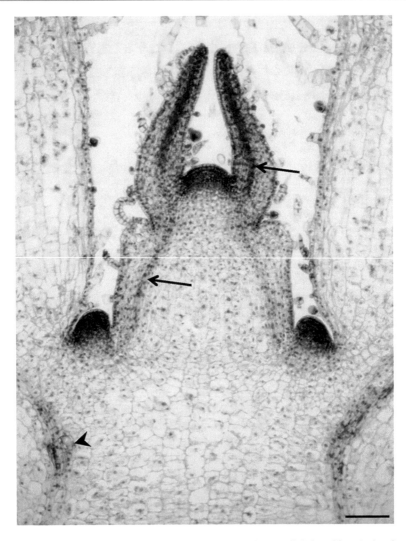

Fig. 2.1 Longitudinal view of the shoot apical meristem (SAM) of *Coleus blumei*, showing the primary vascular meristem, the procambium (*arrows*), descending from the leaf primordia tips. Primary phloem and later also primary xylem differentiate at the node (*arrowhead*). Scale Bar = 200 μm

regenerate. New stem cell niche and new stem cells will develop from the remaining young parenchymatic cells above the cut, demonstrating that the hormonal signals arriving from the shoot to the root tip determine the position, initiation, and maintenance of the newly formed regenerated stem cells. The most embryonic cells divide at the lowest frequency, which gives them an advantage of lowering the frequency of mutations, thus keeping the original DNA with no or almost no modifications. At

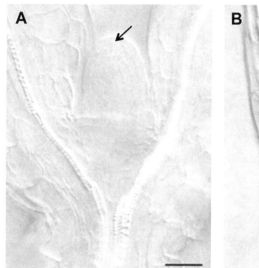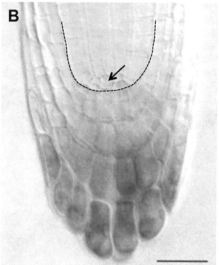

Fig. 2.2 Longitudinal views of both the SAM and the root apical meristem (RAM) in a 3-day-old seedling of free-cytokinin-responsive ARR5::GUS transformant of *Arabidopsis thaliana*, comparing the plant apices regarding the distribution of the active root hormone cytokinin in both cleared plant's poles (see Aloni et al. 2004, 2005). (**A**) The SAM of a 3-day-old seedling is cytokinin-free. (**B**) The RAM is protected by the root cap that produces high quantities of free cytokinin (detected by blue *ARR5::GUS expression*). The *arrows* mark the quiescent center (QC) at both apices, where the frequency of cell divisions is the lowest. The dotted line marks the border of the vascular cylinder with the cortex and root cap. Bars = 25 μm. (Aloni et al. 2004)

the root tip, the slowly dividing embryonic cells form the **quiescent center** (QC) (marked by arrow in Fig. 2.2B).

The youngest provascular meristematic tissue, namely, the **procambium**, develops into a vein pattern along the center of the youngest leaf primordium, forming the early stages of the midvein (arrows in Fig. 2.1). The vascular tissues that differentiate from the procambium are defined as **primary vascular tissues**.

At nodes, where the auxin signal descending from young leaves merges, the local increased signal concentration promotes the differentiation of the first phloem elements, which is followed by xylem elements (arrowhead in Fig. 2.1). As will be discussed in Chaps. 3 and 5, the young leaf hormonal signal is auxin (see Chap. 3), which first induces the procambium, then the phloem, and later also the xylem. The auxin signal which descends from the tips of young leaf primordia downward toward the root tip induces a continuous pattern of the vascular system from leaves to root tips. Nodes tend to remain more embryonic than internodes, keeping their ability to regenerate adventitious roots for a longer period.

All the procambium initials are elongated cells and have a similar structure. Naturally occurring transverse cell divisions in the long procambial stem cells produce short **ray initials** (Fig. 2.3). From the ray's initiation site, the vascular meristem is termed a secondary meristem, namely, the **cambium**. Therefore, the

Fig. 2.3 Longitudinal
tangential view of the
cambium cells in *Populus*
maximowiczii, showing the
short ray initials (r) that
build the vascular rays and
the long fusiform initials
(f). Bar = 50 μm, Courtesy
of R. Funada and
Y. Murakami

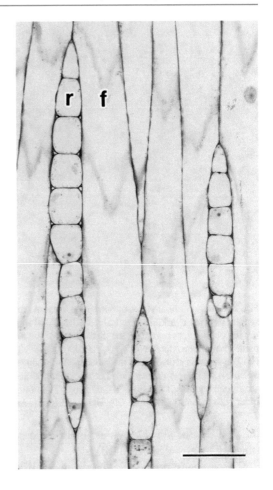

procambium and the cambium are a continuous vascular meristem along the differentiating vascular bundles (Larson 1994). As the transition from procambium to cambium is the appearance of the short ray stem cells. Therefore, the cambium consists of two types of stem cells: the long, axially oriented **fusiform initials** and the short-rounded ray initials (Fig. 2.3). As the main signal promoting ray differentiation is the hormone ethylene (for further information, see Chap. 15), I suggest that the transition from procambium to cambium is induced by a local increase in ethylene production in the xylem of maturing regions along the plant axis.

Vascular cambium develops in gymnosperms and dicots. Monocots usually do not produce cambium. The cambium, like the procambium, produces phloem cells toward the outside and xylem cells toward the inside of shoots and roots.

The cambium may also develop from parenchyma cells between vascular bundles to form the **interfascicular cambium**, to create a continuous ring containing both the fascicular cambium and interfascicular cambium. Following injury, a cambium may differentiate from parenchyma cells of a callus tissue that covers the

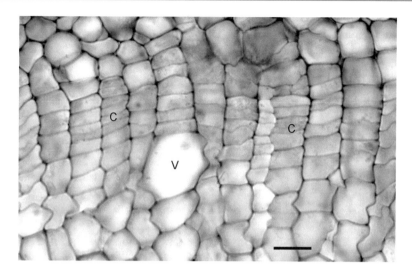

Fig. 2.4 Cambial zone pattern in a cross section of *Nicotiana tabacum*, showing similar undifferentiated cambial derivatives on both sides of the cambial initials (c). An early stage of vessel (v) expansion is observed on the xylem side. Bar = 50 µm

wound, which usually originate from parenchyma cells of the vascular rays. Usually more cell layers are produced toward the xylem side. When the bark is separated from the wood at the cambium, the active meristematic cells usually remain on the phloem side, enabling molecular extractions or direct experimental treatments on the cambium. The cambium is a single layer of stem cells. Because the neighboring cell layers produced by the cambium have a similar undifferentiated structural appearance, these several layers form a **cambial zone** (Fig. 2.4), which might become very wide during the active growth season.

The vascular tissues that differentiate from the cambium are defined as **secondary vascular tissues**. Therefore, the cells produced from the cambium build the secondary plant body, which increase stem and root width. The cambium of trees produces the secondary xylem, which is the wood. The secondary xylem of conifer trees that contains tracheids is called **softwood**, whereas the secondary xylem of angiosperms which is characterized by vessels and fibers is termed **hardwood**, due to the stiff lignified secondary cell walls of the fibers.

Molecular studies have identified that the *WUSCHEL-related HOMEOBOX (WOX)* gene family function in the initiation and maintenance of various meristematic stem cells and are conserved throughout the plant kingdom (Haecker et al. 2004; Alvarez et al. 2018). The *WOX4* expression in the *Arabidopsis thaliana* procambial cells defines the vascular stem cell niche. *WOX4* transcripts are detected in the procambium of both *Arabidopsis* (Hirakawa et al. 2010; Ji et al. 2010) and tomato (Ji et al. 2010) and in the cambium of *Arabidopsis* (Suer et al. 2011; Zhang et al. 2019; Shi et al. 2019), showing that *WOX4* is an essential factor that makes both vascular meristems responsive to the regulation of the polar auxin transport along the plant. *WOX4*-like genes are specifically expressed in the cambial region of

Populus trees during active vegetative growth, but not after growth cessation and dormancy (Kucukoglu et al. 2017). WUS/WOX family proteins are key factors in the specification and maintenance of stem cells within all meristems studied (Ge et al. 2016; Segatto et al. 2016; Alvarez et al. 2018).

WUSCHEL-RELATED HOMEOBOX 5 (WOX5), which is specifically expressed in the root quiescent center (QC), defines QC identity and functions interchangeably with WUSCHEL (WUS) in the control of shoot and root stem cell niches. WUS, a homeodomain transcription factor expressed in the *Arabidopsis* shoot apical meristem (SAM), is a key regulatory factor controlling SAM stem cell populations and is thought to establish the shoot stem cell niche through a feedback circuit involving the CLAVATA3 (CLV3) peptide signaling pathway (Miyashima et al. 2013; Fletcher 2018).

Procambial cell fate is selected by the polar transport of auxin, as visualized by the activation of auxin response and the accumulation of the PIN-FORMED1 (PIN1) auxin efflux transporter protein at the basal membrane of the differentiating vascular cells (Mattsson et al. 2003; Scarpella et al. 2006; Scheres and Xu 2006; Wisniewska et al. 2006; Wenzel et al. 2007). PIN1 subcellular localization indicates that auxin is directed into developing vascular vein subsequent to PIN1 accumulation (see Chap. 3).

The polar auxin flow activates the expression of the *HD-ZIPIII* transcription factor *HOMEOBOX GENE8* (*AtHB8*) which defines the pre-procambial cells and is considered a committed step toward procambial cell fate (Baima et al. 1995, 2001; Kang and Dengler 2002; Sawchuk et al. 2007; Biedroń and Banasiak 2018). AtHB8 restricts pre-procambial cell specification to narrow files (Donner et al. 2009). Among auxin signaling factors, the auxin-dependent transcription factor *MONOPTEROS/AUXIN RESPONSE FACTOR 5 (MP/ARF5)* plays a major role in translating auxin accumulation into the establishment of procambium identity (Hardtke and Berleth 1998; Jouannet et al. 2015). The *ATHB8* expression at the pre-procambial stage is directly and positively controlled by the auxin-response transcription factor *MP (ARF5)* through an auxin-response element in the *ATHB8* promoter (Donner et al. 2009; Biedroń and Banasiak 2018).

2.2   The Two Conducting Tissues of Vascular Plants

2.2.1   Longitudinal Vascular Bundles of Phloem and Xylem

Vascular plants develop two types of nutrient and water-conducting vascular tissues that together build vascular strands or bundles (Fig. 2.5). The most common bundles in seed plants are the **collateral vascular bundles** in which the phloem is produced toward the outside and the xylem is located in the inside part of the strands (Figs. 2.6 and 2.7). In **bicollateral vascular bundles**, the phloem differentiates also inside the xylem (Fig. 2.8A). Additional bundle types can be found in plant species, the amphivasal vascular bundles in which the xylem surrounds the phloem and the amphicribral vascular bundles in which the phloem surrounds the xylem (Esau

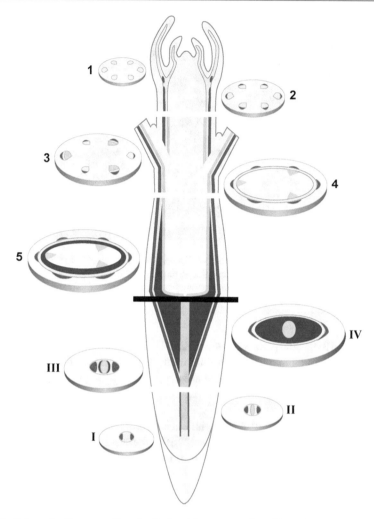

Fig. 2.5 Schematic diagrams of longitudinal and its cross sections illustrating the developmental patterns of the meristematic (pro)cambium (yellow), the phloem (red), primary xylem (light blue), and secondary xylem (dark blue) in a dicotyledonous plant. Besides the longitudinal drawing, the cross sections 1–5 show vascular development along the stem, while cross sections I–IV illustrate vascular development in the root. Note that the phloem starts to differentiate before the xylem along the stem and root. In the stem, the vascular bundles differentiate at the periphery, and wide pith develops in the center. In the root, the vascular tissues occupy the center and are surrounded by a wide cortex. These patterns are primarily induced by polar auxin streams, originating in leaves, which merge into the root

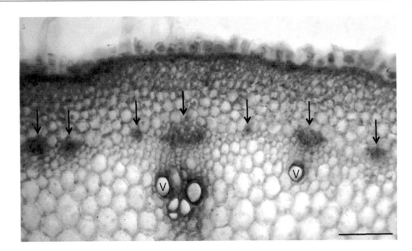

Fig. 2.6 Cross section in a young internode of *Coleus blumei*, showing phloem (*arrows*) and collateral (when vessels develop) vascular bundles at the stem periphery. Vessels (v) differentiate in the more developed bundles, while the narrow bundles are built only of phloem. During internode development, additional vessels will differentiate in the larger bundles. Bar = 200 μm. (From Aloni 1988)

1965a; Fahn 1990; Scarpella and Meijer 2004). The first conducting tissue to differentiate in young nodes and later in internodes is the phloem (Esau 1965b) (Fig. 2.5), and strands of only phloem are common in young internodes (Fig. 2.6). Xylem differentiation may follow the path of phloem differentiation during shoot development.

Long-distance transport of photosynthetic assimilates and other signaling molecules occur through living **sieve elements** (SEs) that lose their nuclei during maturation (Esau 1969). The plasmodesmata in their cell walls that connect adjoining SEs become modified to enlarged sieve plate pores. In angiosperms, the SEs are accompanied by nucleate **companion cells** (CCs) (Figs. 2.8B and 2.32C). The SE requires permanent support by CC, and the exchange of macromolecules between the CC and SE is indispensable for the survival of the sieve elements (van Bel et al. 2002), which allow the mature enucleate SEs to remain functional for long periods (up to about hundred years in a palm tree, R. Aloni personal observation) in longitudinal translocation of nutrients and signaling molecules. Both the SE and the CC originate from the same mother cell and are surrounded by parenchyma cells (Fig. 2.8B) and fibers. Sugars, amino acids, hormones, and other signaling molecules are translocated through the sieve tubes from the producing leaves to developing young tissues at the plant tips (Dayan et al. 2012; Ham and Lucas 2017) and to storage organs (Ma et al. 2015). Translocation in the phloem is bidirectional, up and down, from source to sink, and the flow is driven by a pressure gradient (Knoblauch et al. 2016). The sieve tubes are characterized by specific P proteins (phloem proteins) whose function is not completely understood. When they are arranged in longitudinal filaments along an intact sieve tube, they might form a path for fast

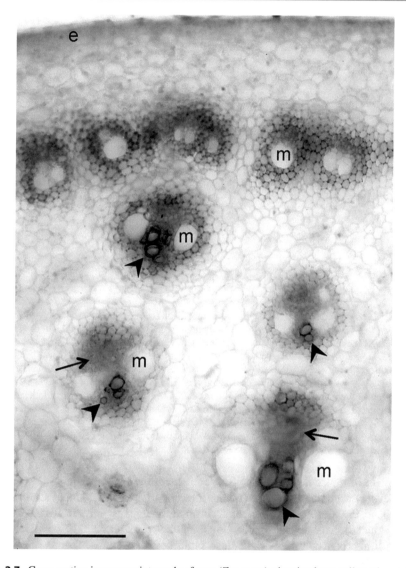

Fig. 2.7 Cross section in a young internode of corn (*Zea mays*), showing large collateral vascular bundles in the center of the stem (lower side) with phloem (*arrows*), functioning narrow protoxylem vessels (*arrowheads*), and expanding metaxylem vessels (m). The bundles at the periphery near the epidermis (e) descended from an upper internode, where they were located in the center of the stem. Because they are away from their inducing leaf, they receive low-concentration polar auxin stimulation; therefore, these bundles are small, and they do not produce protoxylem vessels; their low-auxin stimulation is enough to produce one or two slow differentiating metaxylem vessels. Bar = 200 µm

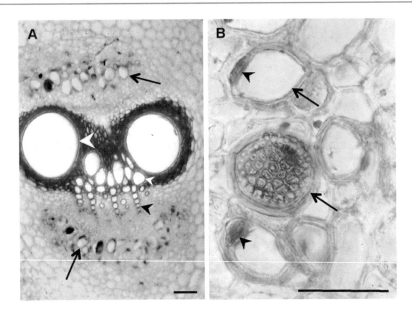

Fig. 2.8 Cross sections in well-developed internodes of *Luffa cylindrical* stained with lacmoid according to Aloni (1979). (**A**) Showing a bicollateral vascular bundle, with phloem (*arrows*) on both sides of the xylem. In the xylem, the protoxylem vessels are marked with *black arrowhead*, and the first produced secondary vessels are marked with *small white arrowhead*, while the giant vessels (*large arrowhead*) are the last ones to differentiate by very low-auxin stimulations descending from far away leaves, which therefore induce very slow differentiation for a few weeks. (Auxin application to such an internode prevents the formation of these giant vessels and induces additional regular size secondary vessels; the experimental evidence is not presented, R. Aloni unpublished). (**B**) Magnified view of the phloem, showing sieve elements (*arrows*), and the middle one with a sieve plate covered with callose (stained sky blue by lacmoid). The companion cells (*arrowheads*) have dense cytoplasm, which was stained blue. Bars = 100 μm

movement of molecules along the SEs. In response to phloem injury, they tend to immediately accumulate as slime bodies attached to the nonfunctional sieve plates.

Around the sieve plate pores and plasmodesmata, the polysaccharide **callose** (β-1,3-glucan) is usually deposited (Figs. 2.8, 2.9, 2.12 and 2.32F) (van Bel et al. 2002). It is produced rapidly by callose synthases in response to wounding which plugs the SEs (Fig. 2.10B, C) preventing leaking of nutrients on the wound surface that might attract pathogens. At early stages of SE differentiation, the sieve plates lack callose, and it gradually accumulates during time and along the season (Aloni and Peterson 1991). In spring, the sealed winter-dormant sieve plate pores resume active translocation, after the callose is degraded by β-1,3-glucanases during the removal of dormancy callose (Aloni et al. 1991). The hormone auxin promotes dormancy callose removal from the sieve plates, which resumes phloem activity in the beginning of the growth season (Aloni et al. 1991; Aloni and Peterson 1997).

The cambium, which normally develops between the primary phloem and primary xylem, produces secondary phloem cells outside toward the organ periphery and secondary xylem cells inside, toward the pith. However, a relatively large

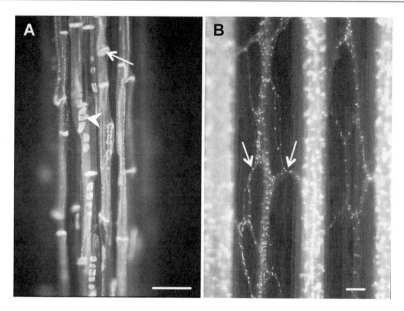

Fig. 2.9 Longitudinal views of the phloem (which was removed from the xylem at the cambial zone and is observed from the cambium side) in intact stems of *Dahlia pinnata*, stained with aniline blue and observed under epifluorescence microscope. (**A**) Showing sieve tubes in a bundle with simple (*arrow*) and compound sieve plates (*arrowhead*) with callose (stained yellow). Callose is also detected in the plasmodesmata scattered on the longitudinal walls of the sieve elements (observed as many small yellow points). (**B**) Patterns of phloem anastomoses (*arrows*) are evident between the longitudinal vascular bundles. Bars = 100 μm

number of dicotyledonous species, in about 55 families (Fahn 1985), exhibit anomalous secondary vascular development, where **interxylary phloem** differentiates inside the secondary xylem (Fig. 2.11) (Fahn 1985; Carlquist 2013). Phloem strands become included in the secondary xylem as a result of anomalous cambium activity, of either numerous successive cambia activities (Fahn 1985) or internal development of vascular cambium in axial xylem parenchyma (Rajput et al. 2018). The formation of interxylary phloem creates safe transport pathways for translocation of photosynthetic assimilates from leaves to roots, which are protected by the massive secondary xylem; this differentiation pattern enables the stem to survive a serious damage that might occur to the regular peripheral phloem near the organ surface.

Xylem of higher plants is characterized by both tracheids and vessels, which are dead conduits at maturity. Through these tracheary elements, water, minerals, and hormones are transported via hollow elements with no cell contents. The driving force for the upward water transport from soil to the transpiring leaves is a tensional gradient of **negative xylem pressure** induced by the leaves, which evaporate water to the atmosphere mainly through their stomata (Tyree and Zimmermann 2002; Lucas et al. 2013; Zhang et al. 2018).

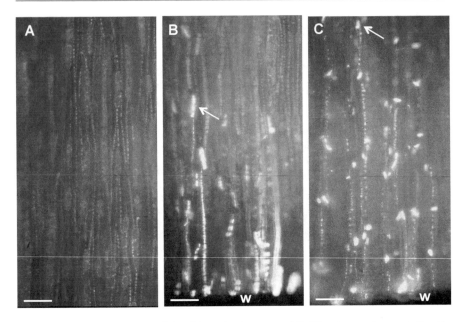

Fig. 2.10 Longitudinal views of the phloem in the stems of *Dahlia pinnata*, demonstrating the effect of wounding on callose accumulation in the same studied location, showing (**A**) the zero-time control before wounding, (**B**) 8 minutes after injury, and (**C**) 15 min after injury. (To view the phloem from the cambium side, 50-mm-long portion of the xylem was gently removed at the cambium, while a wide strip of continuous phloem was kept intact in 0.01% aniline blue, in 0.05 M K-Na phosphate buffer, pH 7.2, and observed under epifluorescence microscope.) (**A**) No callose can be detected in the intact phloem (during spring, in a 6-week-old stem) before making the horizontal wound. (**B**) 8 min after wounding, large quantities of callose accumulation are evident immediately above the wound (w) plugging the damaged sieve tubes. (**C**) 15 min after wounding, callose accumulation is observed further away at a greater distance from the wound. *Arrows* mark callose on remote sieve plates. Bar = 100 μm

The xylem conduits develop hard lignified secondary cell walls that prevent conduit collapse under pressure of surrounding parenchyma cells (Figs. 2.12, 2.13 and 2.23). The secondary wall thickenings develop in various patterns which are correlated with tissue elongation. During early stages of rapid elongation, tracheids and vessels develop annular rings and continuous helical or spiral thickenings (Figs. 2.12, 2.13 and 2.23). When the growth is moderate, reticulated secondary walls are formed, which limit cell elongation. Usually, when there is no tissue elongation, the conduits develop an extensive secondary thickening, and only the pits remain with primary wall, forming pitted walls (Fig. 2.14A).

Water is transported between adjacent tracheary elements either via vessel perforations or through the primary wall in pits at the lower and upper ends of vessels (Fig. 2.14A) and through the primary cell walls of the bordered pits of tracheids (Fig. 2.15A, B). As the secondary xylem conduits produce thick lignified secondary cell walls, their walls have pit cavities allowing water transport via primary cell walls. The pits allow water to pass between xylem conduits, but limit the spread of

Fig. 2.11 Interxylary phloem strands embedded in the secondary xylem of a *Salvadora persica* stem stained with aniline blue and observed under epifluorescence microscope. (**A**) Showing the anatomy of a well-developed stem with the external phloem outside the xylem (*arrowhead*) and interxylary phloem strands (*arrow*) of different sizes embedded inside the secondary xylem, forming safer pathways for photosynthate translocation than through the external phloem. (**B**) in a young internode, narrow interxylary phloem strands built of sieve tubes (the *arrow* marks a sieve plate with callose accumulation) surrounded by parenchyma cells (*black color*). The xylem is mainly composed of fibers (F) and vessels (V). (**C**) A wide (tangentially flattened) interxylary phloem strand in an older internode, resulting from the merging of narrow phloem strands (the *arrow* marks phloem parenchyma inside the strand). Bars = 50 µm (**A**), 100 µm (**B**, **C**)

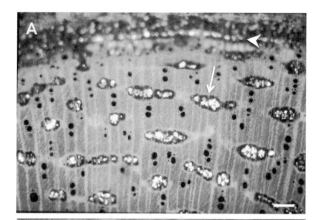

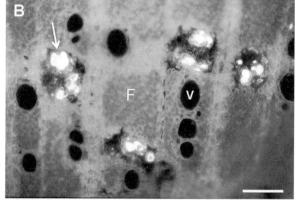

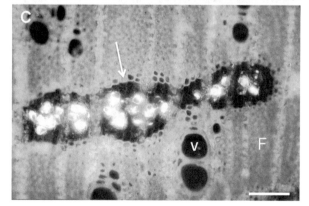

embolism by preventing entrance of an air bubble to the water-transporting conduits (Tyree and Zimmermann 2002). Tracheids and some of the pitted vessels produce **bordered pits** (Figs. 2.15 and 2.16) (Evert 2006; Choat et al. 2008), which have specialized cavity architecture that both maximizes the primary wall surface area available for water transport between conduits and minimizes the interruption to

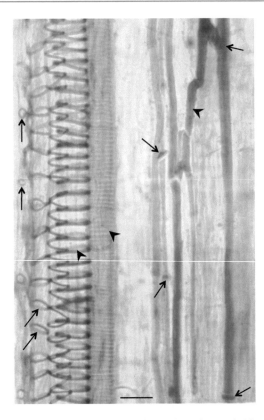

Fig. 2.12 Longitudinal three-dimensional view of both the xylem and phloem in a thick section of cucumber (*Cucumis sativus*) stem cleared with lactic acid and stained with lacmoid according to Aloni and Sachs (1973). Showing in the xylem, annular rings (*arrows*) and spiral secondary wall thickenings (*arrowheads*) inside vessels. Due to early internode elongation, the annular rings of the first formed narrow protoxylem vessel (at the left side) were stretched away from each other (*arrows*). Note that the distances between the annual rings and late formed spiral secondary wall thickenings become smaller (from left to right), as the internode slows its elongation. In the phloem, sieve tubes can easily be followed due to the blue staining of their P protein and the sky blue staining of their sieve plates (*arrows*). The first induced sieve tube has very long protophloem elements (on the right side). Late formed sieve elements are shorter (close to xylem). A phloem anastomosis (*arrowhead*) differentiated inside the bundle. Bar = 100 *μm*

secondary wall continuity by the narrow **pit aperture** (opening in the secondary wall), thus providing maximum physical support by the thick secondary wall. In the middle of a bordered pit, the primary cell wall produces a central primary thickening (**torus**) and a relatively porous region of primary cell wall (**margo**) in the pit cavity around the torus. The bordered pits act as safety valves between adjacent conduits to prevent embolism. In case of cavitation and embolism, the torus is sucked by the negative xylem pressure in the functioning tracheids toward the water-transporting tracheid (Fig. 2.16B). Thus the torus becomes attached to the minimized pit aperture in the secondary cell wall and prevents entrance of air to the water-transporting tracheid (Delzon et al. 2010; Tyree and Zimmermann 2002).

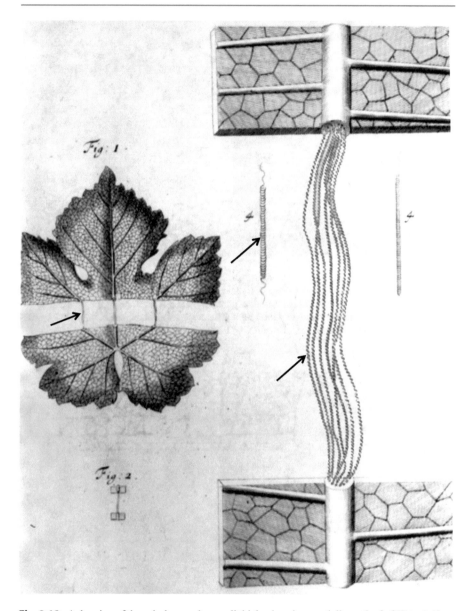

Fig. 2.13 A drawing of the spiral secondary wall thickenings in a partially cut leaf of *Vitis vinifera* published in the pioneering *Anatomy of Plants* book, written by Nehemiah Grew (1682), indicating that the spiral secondary wall thickenings (*arrows*) along a vessel might be continuous along a few vessel elements

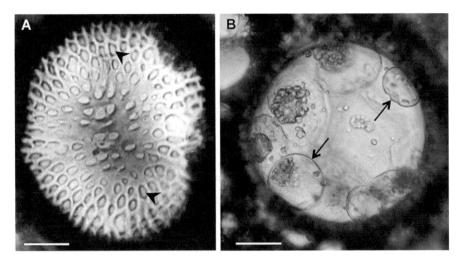

Fig. 2.14 (**A**) Experimentally induced vessel ending in the upper side of a vessel of *Luffa cylindrica*, characterized by a pitted secondary wall pattern. The *arrowheads* mark pits (From Indig and Aloni 1989). (**B**) Early stages of penetration of parenchyma cells (*arrows*) into an air-filled nonfunctional vessel in hops (*Humulus lupulus*) stem. These tyloses, which have balloon-like appearance, tend to expand and prevent invasion and movements of pathogens along the vessel. Bars = 25 μm

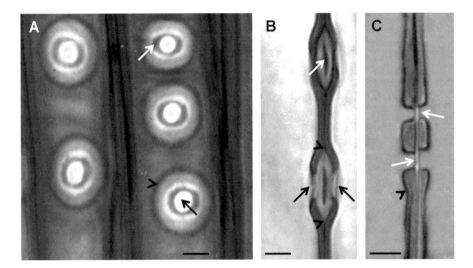

Fig. 2.15 Longitudinal views of basic pit structures in vascular elements. (**A**) Showing the radial and (**B**) The tangential views of bordered pits in earlywood tracheids of *Cupressus sempervirens*. In (**A**) and (**B**), the *black arrows* mark the minimized pit aperture (opening) in the secondary walls of the tracheids, the *white arrows* mark the torus, and the *arrowheads* mark the margin of the relatively porous region of primary wall (margo) inside the pit cavity through which water is transported between adjacent tracheids (see schematically drawn in Fig. 2.16). (**C**) Showing simple pit between adjacent primary phloem fibers in *Coleus blumei*. Their primary cell wall is marked by an *arrow* and the secondary wall by *arrowhead*. Bars = 5 μm

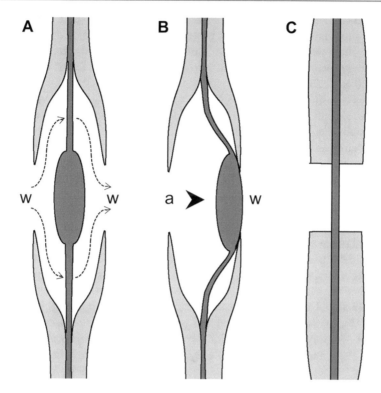

Fig. 2.16 Schematic diagrams showing how bordered pit function. (**A**) Pathway of water (w) flow through the primary wall margo around the torus of intact transporting tracheids. (**B**) Air (a) entrance from an embolized tracheid to the water (w)-transporting tracheid is prevented by the torus that seals the opening (pit aperture) at the functioning tracheid side, due to the negative xylem pressure in the water-transporting tracheids. (**C**) Simple pit enables water movement through the primary cell wall of the pit cavity between two secondary walled fibers

A vessel is a long tube built of vessel elements with endings walls in the lower and upper ends (Fig. 2.14A). Along a vessel, transport of water between the vessel elements occurs through perforations. In "primitive" early plants, the early vessels were characterized by scalariform perforations with numerous closely spaced bars (Esau 1965a; Fahn 1990; Evert 2006). Fossil records analysis promoted (Wheeler and Baas 2019) the suggestion that during evolution, tropical conditions have accelerated xylem evolution toward greater hydraulic efficiency and the development of simple perforations (see Sect. 19.1 and Fig. 19.6). Transport of water from vessel to vessel occurs through pits along their vessel elements or their end cell wall (Fig. 2.14A). It should be noted that vessels and not vessel elements are the physiologically operating units of water transport regarding cavitation and embolism (Zimmermann 1983; Zhang et al. 2018). Embolism of a large vessel is usually followed by the outgrowths of the surrounding parenchyma cells into the vessel (Fig. 2.14B and Fig. 19.7B in Chap. 19), a phenomenon known as **tyloses** (Zimmermann 1983), which forms blockages to penetration and movement of fungi and bacteria into and along the air-filled nonfunctional vessel. There are species that gum may plug the air-filled vessel to prevent possible penetration damage. Tyloses and gum plugs develop naturally in hardwood trees when the functional water-transporting sapwood gradually turns into a stable heartwood that is resistant to rot (De Micco et al. 2016).

There is a positive correlation between **conductive efficiency** and **vulnerability** to water stress and freezing inducing embolism. Wide and long vessels that are efficient conduits are more vulnerable to cavitation and embolism induced by freezing and water stresses than narrow vessels and tracheids. The widest earlywood vessels of ring-porous trees (see in Sect. 19.3) operate for a relatively short duration, usually for only one growth season, and become nonfunctional at the end of the season. On the other hand, tracheids and narrow vessels are safe conduits that function for long periods of a few years, but are less efficient in water transport (Tyree and Sperry 1989; Tyree and Ewers 1991; Baas et al. 2004). In *Acer pseudoplatanus* leaves, embolism events along the midvein occurred first in the wider vessels at the leaf's basipetal part and afterward at the leaf tip where vessels are narrower. The decrease of vessel diameter from leaf base to tip influences the hydraulic safety of leaves, since the wider vessels at the leaf base embolize first (Lechthaler et al. 2019).

The functional lifespans of vessels, which are induced by auxin (Sachs 1981; Aloni 2015), gradually differentiate and produce secondary wall thickenings for protection against pressure from their surrounding tissue. Their nucleus and cytoplasm undergo autolysis to hydraulically function as nonliving hollow conduits, which form water-transporting vessel networks along the xylem. Vessels become nonfunctional due to the formation and penetration of gas emboli (Tyree and Zimmermann 2002). In some species, as lianas of tropical forests, vessel functionality of embolized conduits may be restored through water refilling by positive predawn xylem pressures produced by the root (Ewers et al. 1997: Tyree and Zimmermann 2002). Blockages, such as tyloses, or gum, cause permanent losses in hydraulic functional capacity (Jacobsen et al. 2018).

The vessels are accompanied by xylem parenchyma cells and fibers with thick and hard lignified secondary cell walls. The latter provide the mechanical strength for supporting the plant body (Figs. 2.8A, 2.20, 2.31 and 2.32D).

2.2.2 Phloem Anastomoses

The xylem is usually composed of longitudinal vessels inside bundles extending along the plant body, whereas in the stem, the phloem also builds networks of anastomoses made of sieve tubes and companion cells between the longitudinal vascular bundles, thus forming a three-dimensional phloem network (Figs. 2.9B, 2.12, 2.17 and 2.18B). Phloem anastomoses are also common between the inside and outside sieve tubes of bicollateral vascular bundles. These phloem anastomoses are common in many plant species and could be numerous (Aloni and Sachs 1973). A few thousands anastomoses per internode were recorded in *Dahlia pinnata* (Aloni and Peterson 1990) and in *Cucurbita* species (Aloni and Barnett 1996), thus building a

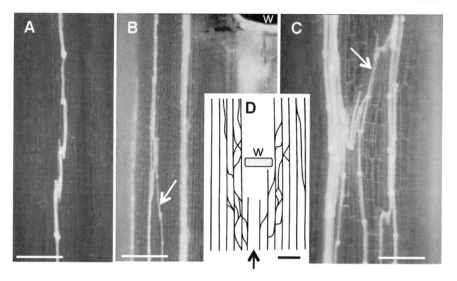

Fig. 2.17 Longitudinal views of transporting phloem in the stems of *Dahlia pinnata*, showing the pattern of functioning sieve tubes transporting the fluorescent dye, fluorescein, in living phloem tissue observed under an epifluorescence microscope. (**A**) In an intact stem, the fluorescein (yellow stain) is translocated along a longitudinal sieve tube, but not in anastomoses. (**B**) Wound (w) promoted fluorescein translocation in a phloem anastomosis (*arrow*). (**C**) Magnified view of fluorescein-filled sieve tubes (*arrow*) in phloem anastomoses beside a wound. Note that there is a local increase of fluorescein concentration at the sieve plates. (**D**) Diagram showing the transport of fluorescein in the sieve tubes around the wound (w) in a *Dalia* stem. The fluorescein was applied to the internode below the wounded one shown in the diagram. The diagram demonstrates that wounding activated the phloem anastomoses, which serve as an emergency system providing alternative pathways for assimilates around the stem. Bars = 125 μm (**C**), 250 μm (**A, B, D**). (From Aloni and Peterson 1990)

dense network of phloem. The number and density of phloem anastomoses increase under summer conditions and decrease in internodes developed during winter (Aloni and Sachs 1973). The phloem anastomoses are possible channels for lateral and upward assimilates distribution from mature leaves to the upper parts of the stem. Phloem anastomoses are a type of normally occurring regenerative phloem, as they develop between the bundles during internode elongation originating from young parenchyma cells following cell divisions (Fig. 2.18A). Very young internodes do not have any phloem anastomoses. Early stages of the regenerative sieve element differentiation can be detected during internode elongation.

The role of phloem anastomoses in translocation was studied experimentally in intact and wounded internodes of *Dahlia pinnata* (Fig. 2.17). Translocation was visualized with fluorescein, a fluorescent dye capable of movement in the sieve tubes. Translocation was analyzed in large areas of living phloem tissue which were partially peeled from the xylem at the cambium zone. Under normal conditions, fluorescein was observed in sieve tubes of the longitudinal phloem strands but very rarely in the sieve tubes of the anastomoses. However, when a few longitudinal

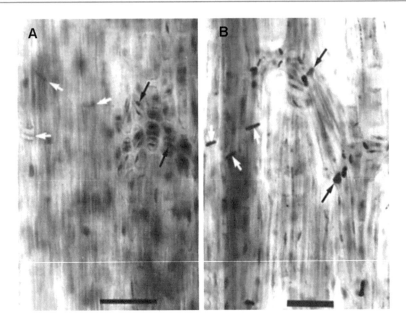

Fig. 2.18 Phloem anastomoses originate from parenchyma cells between longitudinal vascular bundles in the stem of *Dahlia pinnata*, showing longitudinal views of phloem cleared with lactic acid and stained with lacmoid. (**A**) Early stage of anastomosis differentiation in a very young internode, showing the pattern of parenchyma cell divisions (*black arrows*) from them, the sieve elements of the anastomosis develop between longitudinal bundles. (**B**) Developed anastomosis built of a few sieve tubes with callose (*black arrows*) on their sieve plates in a mature internode. *White arrows* mark sieve plates in the longitudinal bundles. Bars = 1 mm. (From Aloni and Barnett 1996)

strands were cut, fluorescein was translocated through the anastomoses located around the wound, as was evident in the anastomoses within 24 h (Fig. 2.17). Therefore, it was suggested that the phloem anastomoses in mature internodes of *Dahlia* serve mainly as an emergency system which enable a fast response to damage by providing alternative pathways for assimilates around the stem (Aloni and Peterson 1990). A possible regulatory mechanism was then proposed, based on differences in resistance to flow in lateral sieve tubes versus the longitudinal ones, namely, that phloem anastomoses do not function under normal conditions because resistance to flow through them is greater than through the longitudinal sieve tubes. The sieve elements of the anastomoses originate from parenchyma cells (Fig. 2.18A) and therefore they are short and the flow has to pass through many more sieve plates in the anastomoses, which reduces flow speed and gives priority to faster flow through the long longitudinal sieve elements.

Phloem anastomoses in *Cucurbita* species (Aloni and Barnett 1996) serve as preferable lateral pathways for auxin after injury resulting in differentiation of very wide regenerative vessels around a wound in these anastomoses (see Chap. 10, Fig. 10.4B).

2.3   Patterns of Primary Vascular Differentiation in Roots and Shoots

Although the primary vascular system of flowering plants is continuous from the shoot organs to the root tips (Fig. 2.5), the vascular patterns in the shoot form three-dimensional complex patterns that are significantly different from the relatively simple ones of the roots. This is due to the many sources of auxin that originate in developing shoot organs (Jacobs 1952, Sachs 1981; Aloni et al. 2003, 2006b), while the root tip acts mainly as a sink for the auxin descending from the shoot (Sachs 1968; Aloni et al. 2006a).

1. **In roots**, the primary vascular tissues are characterized by a **radial pattern** of alternating arches (strands) of phloem and xylem (Figs. 2.19 and 2.20) surrounded by the **pericycle** and around it the **endodermis** and embedded in a relatively wide cortex. There is a linear correlation between the number of protoxylem strands and the circumference of the vascular cylinder (Sachs 1981; Fahn 1990). This linear correlation indicates that the number of protoxylem strands is determined by the free available space at the circumference of the vascular cylinder. An increase in the circumference surface enables more free space for additional separate streams of shoot inducing signals. Accordingly, high-auxin concentration versus low-auxin concentration streams result in an increased number of protoxylem strands versus protophloem strands, respectively (see Chap. 5 for the regulation of phloem and xylem differentiation). For example, the primary root of a monocot seedling (important only in the early stages of seedling development) has a narrow circumference and develops only a few protoxylem strands, like in a typical dicot root. During monocot development and growth, they produce, usually from their nodes, many adventitious roots (Bellini et al. 2014). Because the adventitious roots are induced by much higher-auxin concentrations originating in larger leaves, they develop wide vascular cylinder that enables many more discrete auxin streams at the circumference of the vascular cylinder, resulting in many strands of protoxylem and protophloem in their typical **polyarch monocot adventitious root** (Fig. 2.20). Pattern analysis of phloem and xylem strands in adventitious monocot roots has revealed modifications (e.g., differentiation of phloem and xylem strands on the same radius) in the typical radial pattern of roots (R. Aloni, unpublished data), indicating that each phloem strand, or xylem strand, is induced by a separate independent auxin stream.

In *Arabidopsis thaliana*, which has a narrow vascular cylinder, usually two protoxylem strands differentiate (a diarch root). However, by manipulating the width of the vascular cylinder, the increasing circumference increased the number of xylem poles and formed three protoxylem strands (a triarch root) (Mellor et al. 2019). In vascular cylinders of sweet potato (*Ipomoea batatas*) adventitious roots, usually four or five xylem poles are produced (Fig. 2.19); in naturally occurring wider vascular cylinders, six to ten xylem poles can be produced in *I. batatas* (Fig. 9.1, in Chap. 9) (Ma et al. 2015).

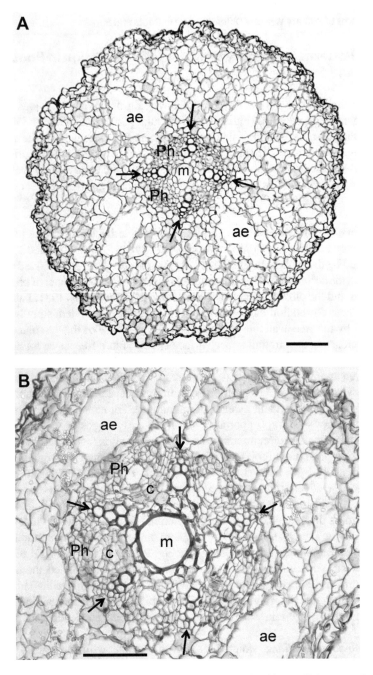

Fig. 2.19 Primary vascular differentiation in roots is characterized by a radial pattern of alternating strands of xylem and phloem, showing cross sections of developmental stages of adventitious dicotyledonous roots of sweet potato (*Ipomoea batatas*) grown in a wet soil, therefore developing large air spaces of aerenchyma (ae) in the cortex. (**A**) Early stage of a young root with four xylem (*arrows*) and phloem (ph) strands and early expansion of a metaxylem (m) vessel. (**B**) More developed adventitious root with a wider stele that enables the development of five alternating xylem and phloem stands with an already functional, wide lignified (stained red) metaxylem vessel in the center. Early stage of cambium (C) development appears between the primary phloem and primary xylem. Bars = 100 µm. (Courtesy of V. Singh and N. Firon)

In dicot roots, characterized by a few phloem and xylem radial archs, one or a few wide metaxylem vessels slowly differentiate in the center of the vascular cylinder (Fig. 2.19A, B), whereas in monocots the mechanism that induces the numerous phloem and xylem archs, by elevated hormonal stimulation, retards vascular differentiation in the center of the vascular cylinder, resulting in a pith of parenchyma cells (Fig. 2.20). The primary xylem in the root matures centripetally, and therefore, the root is characterized by an exarch xylem, i.e., the metaxylem is situated inside the protoxylem. It has been suggested (Aloni 2004) that auxin flow in the pericycle influences the width of primary vessels in the root vascular cylinder. Accordingly, in primary roots, although the metaxylem and protoxylem vessels begin differentiation almost simultaneously, auxin flow in the pericycle enhances the rate of vessel differentiation in the neighboring protoxylem elements, which therefore deposit their secondary walls earlier, resulting in narrow protoxylem vessels. By contrast, the inside metaxylem vessels, which differentiate away from the pericycle, are induced by lower auxin concentration, have more time to expand before secondary wall deposition, and therefore become wide metaxylem vessels (Fig. 2.19A, B). This

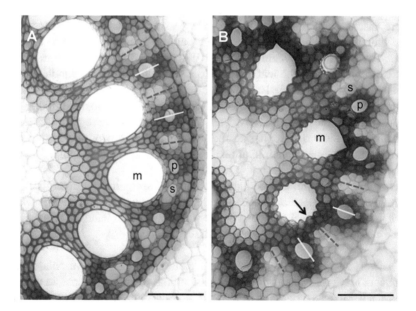

Fig. 2.20 Cross sections showing radial patterns in monocot adventitious roots of corn (*Zea mays*), with many alternating protoxylem (*p*, or *yellow lines*) archs and protophloem (*s*, or *broken red lines*) strands at the periphery of the vascular cylinder, around wide metaxylem (m) vessels and parenchyma pith in the center. (**A**) Large and rounded metaxylem vessels differentiated in a well-developed root grown under favorite water irrigation and root space. (**B**) Deformed metaxylem vessels due to thin secondary wall formation in a stressed-developed root grown under limited root space and irrigation, demonstrating that environmental stress decreases secondary wall deposition, resulting in vessels with thin secondary cell walls, which become deformed under pressure of surrounding cells. The deformed vessels may substantially reduce the efficiency of water transport to leaves. Bars = 100 μm. (R. Aloni and J. Mattsson, unpublished)

explanation follows ideas suggested by Aloni and Zimmermann (1983). (For more information on the control of vessel width, see Chap. 13.)

Genetic and molecular analyses together with mathematical approaches yield continuously improving explanations and developmental models of how vascular patterns are induced and controlled at the molecular level in roots and possibly in the entire plant (Scarpella and Meijer 2004; Scarpella and Helariutta 2010; Bishopp et al. 2011; Bellini et al. 2014; De Rybel et al. 2016; Heo et al. 2017; Ohtani et al. 2017; Ruonala et al. 2017; Mellor et al. 2017, 2019; Augstein and Carlsbecker 2018; Smetana et al. 2019; Miyashima et al. 2019; Agustí and Blázquez 2020). Additional experimental investigations are expected to further improve comprehensive molecular understanding of vascular differentiation in plants.

2. **In stems**, there are two main patterns of primary vascular strands: the dicot pattern of collateral or bicollateral bundles around parenchymatic pith (Fig. 2.6) and the monocot pattern characterized by collateral bundles embedded in pith parenchyma (Fig. 2.7). In monocots, the bundles descending from a leaf enter inside the node toward the center of the stem and then gradually continue down along the stem toward the epidermis. Therefore, the central bundles are larger with more vascular cells including protoxylem vessels because they are close to the leaf-inducing signals, and gradually the bundles become smaller as they continue downward to the stem periphery. With increasing distance from the leaf, the concentrations of the leaf's signals decrease; therefore, the bundle structure is also gradually modified as the concentrations of the leaf signals decline (Fig. 2.7). Accordingly, only the upper portion of a bundle located in the stem's center forms protoxylem vessels, while in the lower bundle region near the epidermis, only slow differentiating metaxylem vessels are produced due to the gradual decline in auxin stimulation along the differentiating bundle.

Dicot plants, like conifers, produce cambium and secondary vascular tissues that become their main vascular tissues during development (Larson 1994), whereas most of the monocot species depend on the primary vascular bundles, as they do not produce cambium or secondary vascular tissues.

3. **In leaves**, there are two main typical patterns of primary vascular strands. Monocot leaves are characterized by longitudinal venation patterns, in which the vascular strands are parallel to the leaf axis and may be connected by minor veins, often near the leaf's tip, whereas most dicot leaves have reticulated vein patterns, where veins and veinlets of different sizes and orders show branching patterns induced by auxin signaling (Scarpella and Meijer 2004; Fambrini and Pugliesi 2013; Biedroń and Banasiak 2018) (for the control of vascular differentiation in leave, see Chap. 7).

The flattening of leaves to form broad blades is an important evolutionary adaptation of land plants for light harvesting that maximizes photosynthesis. Leaves are characterized by their adaxial/abaxial (upper/lower) polarity. At the molecular level, the core of polarity establishment in *Arabidopsis* leaves occurs via mutual repression interactions between adaxial/abaxial factors after primordium emergence. The *REVOLUTA (REV)* gene is expressed at the upper side and determines the adaxial domain in leaves, while the *KANADI (KAN)* genes are expressed at the lower side and promote the abaxial leaf domain (Eshed et al. 2001). The REV/KAN1 module directly and antagonistically regulates the expression of several genes involved in controlling leaf polarity (Eshed et al. 2001; Merelo et al. 2017), as well as spatial auxin signaling that together control the flattening of leaves in *Arabidopsis* (Guan et al. 2017).

2.4 Discontinuous Vascular Patterns During Bundle Maturation

Discontinuous xylem and phloem patterns occur normally during early stages of vascular bundle differentiation in wild-type plants (Aloni 2004). A vessel initiates and matures in discontinuous basipetal and acropetal patterns, which are easily observed during leaf and flower development (Figs. 2.21 and 2.22). Along the stem, a long vessel may normally start vessel element formation and maturation first at the nodes and only later continue the differentiation process along the intervening internodes between these nodes (R. Aloni, unpublished observations). The direction of xylem maturation does not provide enough information to determine the source of auxin production or transport direction of the inducing signal. Although xylem and phloem regeneration around a wound is induced by the auxin signal arriving from the leaves above the injury, the regeneration of a vessel or a sieve tube around the wound is characterized by downward, upward, and discontinuous fragments during the process (Aloni and Jacobs 1977a, b; Aloni and Barnett 1996), finally forming a continuous and functional regenerative vessel or sieve tube (see details in Chap. 10). This typical regeneration patterns emphasize that a polar signal flow of auxin from leaves to roots can normally induce vascular maturation in opposite directions and discontinuous vascular patterns.

In leaves, the naturally occurring discontinuous vessel patterns (Fig. 2.21) follow the configuration of the vascular meristem. Loops of procambium develop downward (see Chap. 7, Fig. 7.1A, B) in a very young leaf primordium. In *Arabidopsis*, the first procambial loop is produced at the leaf's tip. Additional loops develop basipetally toward the leaf base. The differentiation of procambium in the downward direction is likely induced by incipient low auxin stimulation at the tip hydathode. The late formed basal loops also receive auxin supply from developing lateral hydathods at the lobes (Scarpella et al. 2006; Wenzel et al. 2007). This basic procambium framework of midvein and loops determines the pattern of both xylem and phloem differentiation, which follow the procambium pattern (Fig. 2.21) during leaf morphogenesis.

Analogously, normal discontinuous patterns of primary xylem differentiation also characterize the early stages of vascular differentiation during the development of young flower organ (Fig. 2.22) (Aloni et al. 2006b). Vessel differentiation starts

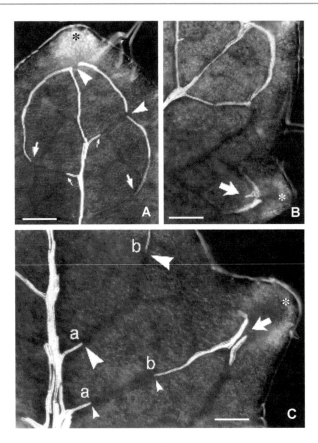

Fig. 2.21 Naturally occurring discontinuous vessel patterns in cleared intact wild-type *Arabidopsis* leaf primordia observed under Nomarski illumination. Under this illumination, the procambium is observed as dark lines, and vessels have a white appearance. (**A**) Discontinuous xylem (*arrowheads*) at the leaf tip, showing basipetal (*arrows*) and acropetal (*small arrows*) vessel differentiation. (**B**) Typical discontinuous xylem initiation (*arrow*) in a lobe, promoted by the high-auxin concentration of a developing hydathode (asterisk). (**C**) Typical early stages of lateral vascular bundle maturation, showing two discontinuous vessels (*each vessel is marked by the same size arrowheads*) progressing acropetally (a) and basipetally (b) along the procambium. Bars = 200 μm. (From Aloni 2004)

from the site of high-auxin production at the anthers (Fig. 2.22A) and immediately below the auxin-producing stigma (Fig. 2.22B). Hence, it is critical to realize that normal vascular differentiation in shoot organ primordia of wild-type plants starts from the site of auxin production at the tips of the organs and progresses basipetally in discontinuous patterns, which should not be confused with possible patterns of discontinuities found in mature organs of defective mutants (e.g., Przemeck et al. 1996; Carland et al. 1999; Deyholos et al. 2000; Koizumi et al. 2000).

On the other hand, acropetal vascular development may progress from the base of shoot organs upward. Such acropetal pattern of xylem and phloem differentiation

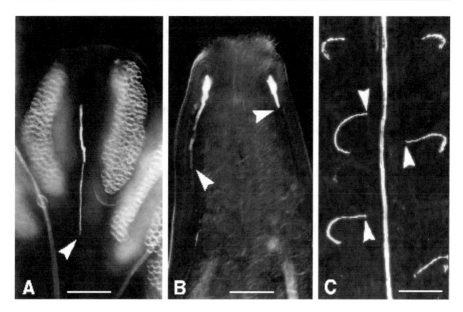

Fig. 2.22 Naturally occurring discontinuous vessel patterns in cleared intact wild-type *Arabidopsis* flower primordia photographed under Nomarski illumination (giving the xylem a white appearance). (**A**) and (**B**), showing early discontinuous xylem patterns immediately beneath the tips of developing flower organs (at the sites of high-auxin concentrations). (**A**) In the young stamen primordium, the vein differentiates at the anther (*arrowhead*) and progresses basipetally. (**B**) Two sites of xylem initiation at the tip of a very young gynoecium (*arrowheads*), starting beneath the stigma and progressing downward. (**C**) In mature gynoecium, the short veinlets (*arrowheads*) originated from the ovules, but could not connect to the central bundle of the gynoecium (due to higher-auxin streams descending from the stigma through the central gynoecium bundle; for explanation, see Fig. 3.7C). Bars = 200 μm. (From Aloni 2004)

characterizes the midvein and the base of the secondary bundles. It is suggested that the acropetal differentiation patterns result from the local buildup of auxin concentration above sites of short cells that locally slow auxin movement or at merging junctions. Thus, above locations which slow auxin movements (e.g., at the base of leaf petiole, the short cells, from them the abscission zone will develop, see also for gibberellin flow in Fig. 2.27), a local auxin accumulation occurs and will induce rapid maturation of vessel elements in the acropetal direction in young leaf primordium. In junctions between the midvein and the secondary bundles (Fig. 2.21C), the streams of auxin descending from the marginal hydathods merge at the midvein junction with the auxin flow descending from the tip hydathode, resulting in a local increase of auxin concentration at the junction. Thus, the merging junction site elevates auxin concentration locally, promoting fast acropetal vessel maturation at the base of the secondary vessel (Fig. 2.21C), which can normally be observed during early stages of leaf-primordium ontogeny.

At the end of the bundle-maturation process, the vessel fragments which develop in the basipetal and acropetal directions (Fig. 2.21) will join into one functional vessel. This gradual and fragmental vessel maturation, progressing in opposite directions, demonstrates that a vessel is composed of a population of differentiating vessel elements that may have different rates of maturation. Those procambial initials exposed to high hormonal stimulation, either near the hydathode (marked with arrows in Fig. 2.21B, C) or at the midvein-stem junction, will differentiate faster than the intervening procambial initials.

To avoid any possible confusion between naturally occurring and defective-mutant discontinuities, the latter should be analyzed in mature organs (at the time that wild-type organs do not show vascular discontinuities). Xylem discontinuities in organs of defective mutants (e.g., Przemeck et al. 1996; Deyholos et al. 2000; Koizumi et al. 2000) can be regarded as early developmental stages that have become fixed during differentiation owing to a genetically induced failure to complete the maturation process or due to other possible interruptions to vascular differentiation. Similar discontinuities in mature phloem have been observed also in defective mutants (Przemeck et al. 1996; Carland et al. 1999). Such discontinuities have been correlated with analogous discontinuities in the xylem (Carland et al. 1999).

2.5   Parenchyma Cells in the Vascular System

The parenchyma cells are simple-looking cells that fill the spaces between and around the vascular elements. The parenchyma cells in the vascular tissues are usually living cells that enable transport between cells via their plasmodesmata. The axial parenchyma cells form continuous **symplastic pathways** along the plant axis. While in the radial direction, parenchyma cells connect the phloem with the xylem and vice versa through the vascular rays (Sokołowska 2013; Spicer 2014), in angiosperm wood (secondary xylem), the vessel-associated parenchyma cells function in radial uptake of water and solutes from the apoplastic transport route occurring through the vessels into the symplastic parenchyma routes; thus, the apoplastic route and the symplastic one function collectively and continuously.

In a large survey of 2332 woody angiosperm species (Morris et al. 2018), the relationship between the proportion and spatial distribution of axial parenchyma to vessel size was analyzed. The analysis reveals that mean vessel diameter shows a positive correlation with axial parenchyma proportion and arrangement. Species with wide vessels tend to have more axial parenchyma tissue, and most of the parenchyma cells are packed around vessels, whereas species with small diameter vessels show a reduced amount of axial parenchyma that is not directly connected to vessels. This finding provides indication that in large-vesselled species, the axial parenchyma participates in radial uptake of water and solutes from the long-distance water-transporting vessels.

Tracer uptake experiments from water-conducting vessels to living vessel-associated xylem parenchyma cells revealed that the radial solute uptake is an endocytosis process, showing that the vessel-associated parenchyma cells are actively transporting solute molecules into their cytoplasm by engulfing them with the plasma membrane (Słupianek et al. 2019).

Interestingly, vessel elements may be dependent on neighboring parenchyma cells for their **lignification**. Evidence from *Zinnia* tracheary-element differentiation in cell cultures (Pesquet et al. 2013) and in *Arabidopsis* plants (Smith et al. 2013) indicates that living xylem parenchyma cells contribute lignin precursors (monolignols) for the lignification of their neighboring vessel elements in a *post-mortem* lignification process (Pesquet et al. 2013; Smith et al. 2013; Ménard and Pesquet 2015).

In the phloem, the long-distance rapid symplastic transport occurs via the sieve tubes, and the relatively short-distance and slow transport occurs through neighboring parenchyma cell. Essentially, radial parenchyma cells interconnect the phloem and the xylem systems via the vascular rays (see Chap. 15) building a functional vascular parenchyma complex in the entire plant.

The parenchyma cells are active in nutrient transport and accumulation. Thus, for example, toward winter dormancy, sugars are translocated centripetally from the sieve tubes via the vascular rays into the pith that becomes loaded with starch (see Fig. 2.31). During early spring the vascular rays transport sugars centrifugally, from the pith to the cambium and differentiating vascular tissues.

Parenchyma cells have an important role in wound repair. At the exposed injured surface, parenchyma cells start to divide, usually parallel to the surface, and form a callus that covers the wound for healing the exposed tissue. In young tissues and callus formed around a wound, regenerative vascular elements re-differentiate from parenchyma cells allowing recovery and repair of the damaged vascular system by forming new bypass around the injury (see Chap. 10). In wood injuries, the callus formed on the wound surface originates mainly from parenchyma cells of the vascular rays.

Various glands and secreting resin ducts in the vascular tissues contain an inner layer of parenchyma cells that synthesize and secrete into the gland's lumen (for more information, see Chap. 17).

It was already mentioned above that the vessel-associated parenchyma cells around large vessels with wide pits may penetrate through the primary wall of the pits into nonfunctional vessel after cavitation to form tyloses (Figs. 2.14B and 19.7B), which plug a nonfunctional embolized vessel and protect the plant from a fungal and bacterial invasion through the air-filled vessel.

Specialized parenchyma cells, characterized by wall ingrowths that grow into the cell and increase the surface of the plasma membrane, which likely increase transport activities, are called "transfer cells." They can be found in both the xylem and the phloem mainly at the nodes of many plant species from different families. In dicot leave, transfer cells can differentiate in minor veins and hydathodes. They are likely involved in short-distance transport of solutes, but there is limited information about their initiation, differentiation, and function (Gunning and Pate 1969; Pate and Gunning et al. 1972).

2.6   Usual, Uncommon, and Regenerative Tracheids and Fibers

Tracheids are both the water-conducting and supporting vascular elements typically building the softwood of gymnosperms. Tracheids may differentiate also in both dicots and monocots. Along the plant axis and leaves, tracheids originating from procambium and cambium are elongated cells (Fig. 2.23A). In the wood of conifers, secondary tracheid elongation is relatively limited because the tracheids are densely packed. However, in the monocot dragon tree (*Dracaena draco*), the cambium produces secondary tracheids in bundles surrounded by parenchyma. This relatively soft surrounding parenchyma tissue permits considerable intrusive elongation during secondary tracheid development and formation of small protrusions on the tracheid surface (Jura-Morawiec 2017).

Short tracheids are common in conifer leaves and are called **transfusion tracheids** (Frank 1864; Mohl 1871). Frank (1864) emphasized that transfusion tracheids are missing in the stems of conifer trees, which produce only long primary and secondary tracheids. Mohl (1871) gave these relatively short tracheids and their associated plasma-rich (transfusion) parenchyma cells the name Transfusionsgewebe

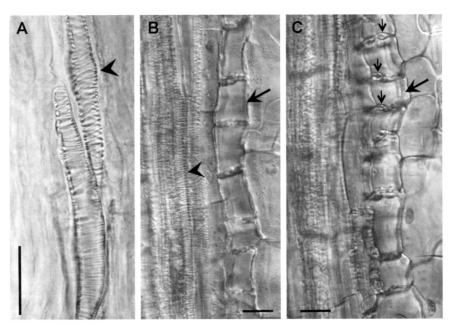

Fig. 2.23 Longitudinal views of cleared juvenile needle-like leaves of the conifer *Thuja plicata*. (**A**) Showing primary tracheids (*arrowhead*) (originated from procambium) in an early stage of bundle differentiation in a young leaf primordium (before any transfusion tracheid development). (**B**) and (**C**), showing short tracheids (*arrows*), which form a vessel-like pattern in well-developed leaves. Water moves from one transfusion tracheid to the other which occurs via bordered pits (*small vertical arrows*). Bars = 100 μm (**A**), 25 μm (**B**, **C**). (From Aloni et al. 2013)

(transfusion tissue). On the basis of the development of transfusion tracheids in cotyledons of a few gymnosperm species, Carter (1911) suggested that they originate from parenchyma cells. Therefore, the parenchyma originating transfusion tracheids may be considered as naturally occurring regenerative tracheids (Aloni et al. 2013). Regenerative tracheids can be experimentally induced in young hypocotyls of *Pinus pinea* seedlings from parenchyma cells (Kalev and Aloni 1998).

The transfusion tracheids in *Thuja plicata* differentiate after all the leaf's procambial initials have already differentiated to long primary tracheids (Fig. 2.23A) and sieve cells. Because the leaf's tip continues to produce auxin, it induces the short transfusion tracheids from parenchyma cells in the vascular bundle, more transfusion tracheids near the tip and less at the leaf base, due to a decreasing gradient of auxin concentration. The short transfusion tracheids may form a vessel-like pattern (Fig. 2.23B, C), but they do not have perforation plates, and water moves from one transfusion tracheid to the other via bordered pits. These short tracheids also differentiate at the tip of the bundle (Fig. 2.24A, B) near the leaf's tip. Wounding and auxin application to very young leaves promote vascular regeneration and the formation of short regenerative tracheids around the injury (Aloni et al. 2013).

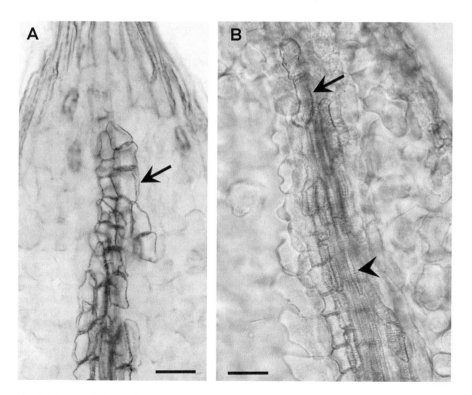

Fig. 2.24 Longitudinal views of cleared juvenile needle-like leaves of *T. plicata*, showing short transfusion tracheids (*arrows*) and long primary tracheids (*arrowheads*) at the tip of the leaf's vascular bundle. Bars = 100 μm (**A**), 50 μm (**B**). (From Aloni et al. 2013)

Fibers are long vascular cells characterized by thick secondary walls that are usually lignified. Their main function is supporting the plant body. Commonly long fibers along stems and roots originate from procambium or cambium. The elongation of xylem fiber (termed libriform fibers) in the secondary xylem is limited because of their high densely packed arrangements which almost prevent intrusive growth. However, in primary bundles surrounded by parenchyma, the primary phloem fibers may show remarkable intrusive elongation (Fig. 2.25A).

Maturation of the xylem elements involves extensive deposition of secondary cell wall material and autolytic process resulting in cell death. Although there are species in which mature fibers remain alive and can actively accumulate starch (Fahn 1990), in many species the xylem fibers are dead at maturity. In *Populus*

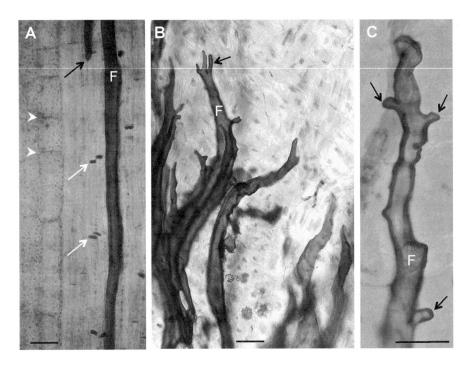

Fig. 2.25 Longitudinal views of phloem fibers (F) in tissues cleared and stained with lacmoid (according to Aloni and Sachs 1973). (**A**) The central part of a primary phloem fiber and the lower end (*black arrow*) of a nearby fiber, after substantial intrusive elongation in an intact internode of *Coleus blumei*, in comparison to sieve elements (the callose on their sieve plates was stained sky blue, marked by *white arrows*), and surrounding parenchyma cells that remained short (upper and lower ends marked by *arrowheads*). (**B**) The upper portions of typical tumorous regenerative phloem fibers with several ramifications (*arrow*) grown intrusively in various direction in the lower region of a 6-week-old *Ricinus* crown gall (tumor tissue experimentally induced by *Agrobacterium tumefaciens*, Aloni et al. 1995), due to interruptions to their growth caused by neighboring dividing and expanding parenchyma cells. (**C**) A magnified view a tumorous regenerative phloem fiber in a *Ricinus* crown gall with three ramifications (*arrows*) developed after the upper tip stop growth due to interruption caused by neighboring parenchyma. Bars = 50 μm. (**B** and **C** are from Aloni et al. 1995)

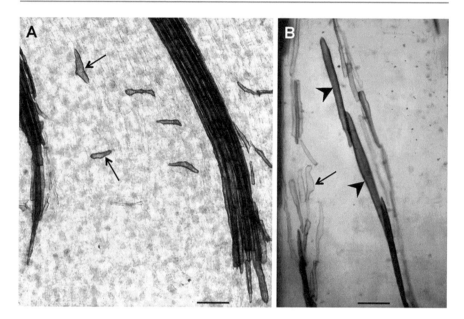

Fig. 2.26 Regenerative phloem fibers induced above wounds in young *Coleus blumei* internodes (cleared and stained with lacmoid). (**A**) Showing a few short isolated fibers (*arrows*) originated from parenchyma cells between the cut bundles. (**B**) 60-day-old long regenerative phloem fibers after intrusive growth (*arrowheads*) and shorter ones that were likely developed later during the regeneration process (*arrow*). Bars = 100 μm. (From Aloni 1976)

stems, the xylem fibers undergo a **programmed cell death** (PCD), which differs from the process found in tracheary elements (TE) of *Zinnia* grown in culture (in TE the vacuole plays a central role; see Fukuda 2000), showing that in fibers there is a gradual degradative process occurring in both the nucleus and cytoplasm concurrently with a phase of active cell wall deposition (Courtois-Moreau et al. 2009).

Wounding young internodes may induce short regenerative fibers from parenchyma cells between the severed vascular bundles (Fig. 2.26A, B). Regenerative phloem and xylem fibers, which originate from parenchyma cells, may develop in the periphery of well-developed *Agrobacterium*-induced tumors, called crown galls, on *Ricinus* stems (Fig. 2.25B, C). These unique fibers in tumor tissue show ramifications due to interruptions to fiber elongation caused by the fast development and expansion of neighboring parenchyma cells. These **tumorous fibers** are characterized by branching patterns showing growing fingers or hook-like shapes with "growing points" (Fig. 2.25B, C). The tumorous fibers elongate intrusively in various directions among the *Agrobacterium*-induced parenchyma cells (Aloni et al. 1995) (see Chap. 21).

2.7   Vascular Differentiation in Organ Junctions

Along the shoot, the junctions between stem and leaves, branches, or adventitious roots are important physiological locations characterized by the development of short or narrow cells, tracheids between vessels, or circular vessels, which may locally slow hormonal movement and increase concentration of inductive signals arriving from above the junctions, at the base of organs, or at stem nodes where adventitious roots develop; likewise, the junctions can regulate upward water flow from stem to lateral organs. The junctions may cause a local increase in signal concentration because the signal has to move through more short cell membranes which slow signal flow (Fig. 2.27) (Dayan et al. 2012). Thus, the short cells (from them the abscission zone will develop before leaf abscission) at the base of a leaf primordium may cause a local increase in auxin concentration above the leaf-stem junction resulting in acropetal vessel maturation in the midvein (see Chap. 7).

Tracheids differentiate between the vessels of the leaf and stem vessels in the palm *Rhapis excelsa*, prevent and protect the primary large metaxylem vessels of the stem from embolism that originate in drying leaves (Zimmermann and Tomlinson 1965; Zimmermann and Sperry 1983; Sperry 1985). This leaf-stem **hydraulic** "safety zone" in palms is important for trunk survival, as palms do not produce cambium and therefore the trunk cannot replace an embolized vessel.

Similarly, in cereal species, like barley (*Hordeum vulgare*) (Luxová 1986; Aloni and Griffith 1991), wheat (*Triticum aestivum*), and rye (*Secale cereale*) (Aloni and Griffith 1991), junctions between the stem and adventitious roots may result in differentiation of short tracheids between the root vessels and shoot vessels forming a "safety zone" in their root-shoot junctions (see Chap. 14; Fig. 14.3C); this vessel segmentation protects the vessels of the roots from embolism originating in the shoot.

In rice (*Oryza sativa*), there is a substantial decline in vessel diameter at the nodes that regulates hydraulic conductance to leaves, acting like a "nodal switch" when the soil water content is severely restricted, which protects rice from drought stress (Chen et al. 2018).

Circular vessels (Sachs and Cohen 1982) are commonly produced in dormant buds and branch junctions of trees (Aloni and Wolf 1984; Kurczyńska and Hejnowicz 1991), and their size and frequency increase continuously with age and branch width (Lev-Yadun and Aloni 1990). These circular vessels do not function in water transport and may interrupt water flow to branches, giving priority in water supply to the main stem's apical bud at the expense of the lateral branches. The mechanism of circular vessels differentiation is discussed in Chap. 14.

2.8   Reaction Wood and Gelatinous Fibers

A terrestrial environmental adaptation of large arborescent plants is their response to gravity, or reaction to wind; branches and leaning trunks adapt their orientation by producing a **reaction wood** in an asymmetric pattern, a result of eccentric growth

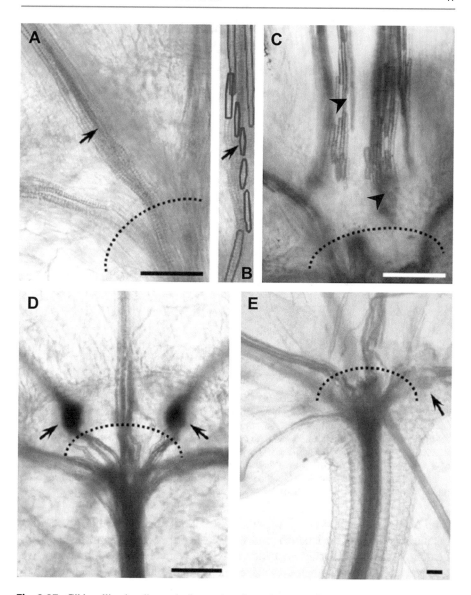

Fig. 2.27 Gibberellin signaling at leaf-stem junctions, demonstrating the structure and function of the leaf-stem junction. The figures show longitudinal images of cleared *Arabidopsis* plants with rosette leaves. Dotted lines mark petiole-stem junction line. (**A**)–(**C**), the leaf-stem junction (where the abscission zone will develop) is characterized by short cells. Short tracheary elements (marked with blue in **B**) differentiated between the long vessels (marked with red in **B**) of the petiole and stem. An identical vessel element is marked by short-thick arrows in (**A**) and (**B**). (**B**) The cell walls of some of the short vessel elements between the long ones are outlined in blue and red, respectively. (**C**) Longitudinal section through the base of the petiole, showing short vessel elements (*arrowhead*) with high *ProGA2ox2:GUS* expression, due to high local gibberellin concentration. (**D**) Accumulation of *ProGA2ox2:GUS* expression is observed just above the petiole-stem junctions. *Arrows* point to site of GUS accumulation. (**E**) Paclobutrazol (a gibberellin inhibitor) treatment abolishes the *GUS* expression. Bars = 55 μm. (From Dayan et al. 2012)

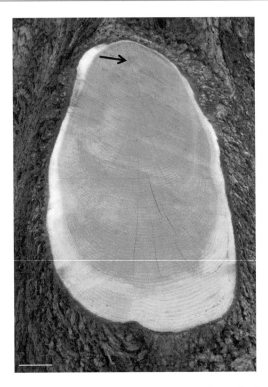

Fig. 2.28 A large cross section revealing the anatomy at the base of a large branch after its removal from a conifer tree, showing the asymmetric pattern of massive compression wood produced in the lower side of the branch (below the *arrow*, which marks the original central pith of the cut branch), demonstrating minimum growth of the opposite wood above the *arrow*, and the massive compression wood produced in the lower side, which supports the weight of the heavy branch. Note that the lower side of the branch develops most of the active living compression sapwood, distinguished by its lighter color (while the dead heartwood in the center has a dark brown color). Bar = 50 mm. (R. Aloni, unpublished)

(Fig. 2.28). The reaction wood supports the shoot by resisting bending and correcting the orientation of the inclined trunk up to a vertical position (Barnett et al. 2014). Leaning stems and branches are exposed to both compressive strain in their lower side and tensile strain in their upper side. Interestingly, during evolution, trees have responded to the gravity stresses by completely opposite mechanisms, either by producing **compression wood** in their lower side or by forming **tension wood** in their upper side. In the latter xylem, **gelatinous fibers** (G-fibers) produce the tensile strain that upright the bended shoot organs.

In climbing plants, the coiling of tendrils and the twining of vines depends on the development of G-fibers in the twining organs, enabling the climbing stem to become attach to other plants or other supporting objects.

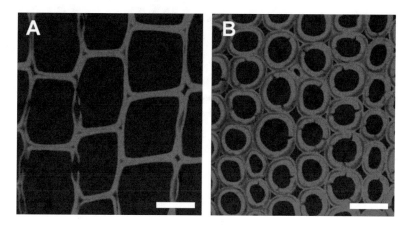

Fig. 2.29 Confocal laser scanning microscopic images of cross sections of tracheids in *Abies sachalinensis*, after staining with safranin. (**A**) Normal wood; (**B**) Compression wood. The compression wood differentiated in the lower side of the inclined stem. Bar = 25 μm. (Courtesy of R. Funada, Y. Sano, Y. Yamagishi, and S. Nakaba)

2.8.1   Compression Wood

The reaction wood in gymnosperms is the compression wood that is produced at the lower side of inclined stem and branches (Fig. 2.28). While in normal wood the tracheids have a square appearance (Fig. 2.29A), the compression wood is characterized by rounded tracheids with thick secondary walls (Fig. 2.29B). The compression tracheids have low cellulose (38%) content in increased microfibril angles and high lignin (36%) content, which allow them to support heavy weights (Timell 1986; Donaldson and Singh 2016), while the wood in the opposite side of the stem, called **opposite wood** (in the upper side), is relatively inhibited and has normal chemistry and anatomy (Fig. 2.29A).

2.8.2   Tension Wood

Conversely, in angiosperms, most species develop tension wood at the upper side of leaning stems, where maximum tension is produced. The tension wood is characterized by the formation of gelatinous fibers (G-fibers) with a thick inner gelatinous layer (Fig. 2.30) characterized by high cellulose (54%) content, arranged in crystalline cellulose fibrils orientated longitudinally, includes the "molecular muscle" xyloglucan (Mellerowicz et al. 2008) and low lignin (14%) content (Wardrop 1964; Clair et al. 2006b; Du and Yamamoto 2007; Gritsch et al. 2015; Donaldson and Singh 2016).

The G-fibers usually appear as continuous layers or a continuous region along the upper side of leaning stems and branches (Fig. 2.30). The gelatinous layer (G-layer) induces longitudinal shrinkage which acts as a "muscle" that brings the

Fig. 2.30 Light micrograph of a cross section from the upper side of inclined stem of *Betula platyphylla*, showing gelatinous fibers, which differentiated after stem inclination. The gelatinous fibers are characterized by the development of the inner gelatinous layer, stain blue (*arrow*) by safranin and Astra blue according to Nugroho et al. (2012). The continuous region that developed the G-fibers is marked by a line on the right side of the micrograph. Bar = 100 μm. (Courtesy of R. Funada)

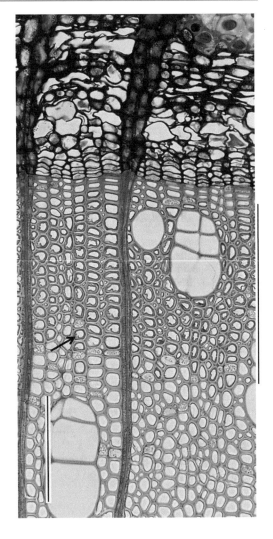

stem to vertical position. The cellulose microfibrils in the G-layer are oriented parallel to the longitudinal axis of the fibers. Their swelling produces the shrinkage, which brings the stem to upright position (for hypotheses on the mechanism how gelatinous fibers function, see Clair et al. 2006a, 2011; Donaldson and Singh 2016).

Reaction woods have an impact on the quality of wood for industry. Reaction woods are generally considered natural defects of woods, causing severe twisting and warping of boards. On the other hand, compression wood contains air spaces among the compression tracheids (Fig. 2.29B), which makes it easy to separate them for pulp than from normal wood. Likewise, tension wood is attractive for pulping and bioethanol production because of its low lignin content (Sawada et al. 2018).

The hormonal mechanisms that regulate the two types of reaction wood formation are discussed in Chap. 18.

2.8.3   Gelatinous Fibers in Climbing Shoots

Climbing plants are "structural parasites" that use other plants or nonliving objects to support themselves, rather than investing in massive wood production to support the weight of their shoots. The climbing habit has evolved multiple times during the evolution of angiosperms and is common in many plant families. Climbing plants like woody vines, called lianas, are very common in tropical regions where they climb on forest trees. Some of them produce beautiful anatomy of vascular patterns (Fig. 11.2) which will be discussed in Chap. 11. Many of the climbing plants produce gelatinous fibers which enable them to coil on other plants (Bowling and Vaughn 2009).

In climbing stems, G-fibers can differentiate as individual cells, as small fiber groups, or in continuous fiber patterns around the entire climbing stem inside the phloem and xylem (Fig. 2.31), rather than only on one side (like the upper xylem side of leaning trunks of angiosperm trees). Usually, the G-layer in fibers of climbing stems is not observed during the rapid elongation of young internodes and tendrils, in the early rapid circumnutating shoot growth. The G-layers develop at later stages, during which they induce longitudinal shrinkage, which cause the coiling and twining of climbing stems and tendrils before their position becomes fixed (Bowling and Vaughn 2009). The G-fibers are common in many tendrils or twining stems, and all tendrils perform helical growth during contact-induced coiling, indicating that such ability is not correlated with their ontogenetic genetic origin or phylogenetic history (Sousa-Baena et al. 2018).

2.9   Vascular Differentiation Under Metabolic and Environmental Stresses

Induced metabolic mutations are tools for studying the influence of deficiencies on plant development and on vascular differentiation. The term "irregular xylem (irx)" was first used to describe the phenotype of three *Arabidopsis* mutants (*irx1, irx2, and irx3*) that had collapsed and deformed vessels (Turner & Somerville 1997; Hao and Mohnen 2014). These irregular xylem (*irx*) mutations were identified by screening plants from a mutagenized population by microscopic examination of stem sections. Analysis of their cell walls revealed a decrease in their cellulose content, suggesting that the irx cell walls are not resistant to compressive forces and therefore their vessels become deformed.

Likewise, reducing the level of the monosaccharide fructose, by lowering the expression of *LeFRK2*, producing the major fructose-phosphorylating enzyme in tomato (*Solanum lycopersicon*) plants, depresses the level of carbohydrates. Transgenic tomato plants with antisense suppression of *LeFRK2* (German et al. 2003) are smaller than wild-type plants and show reduced xylem and phloem differentiation (Fig. 2.32). The suppression of *LeFRK2* results in a significant reduction in the size of vascular cells and slows fiber maturation. The vessels in stems of *LeFRK2*-antisense plants are narrower than in wild-type (WT) plants and have thinner secondary cell walls. Although the cambium produces rounded secondary vessels, these vessels become deformed during early stages of xylem maturation due to

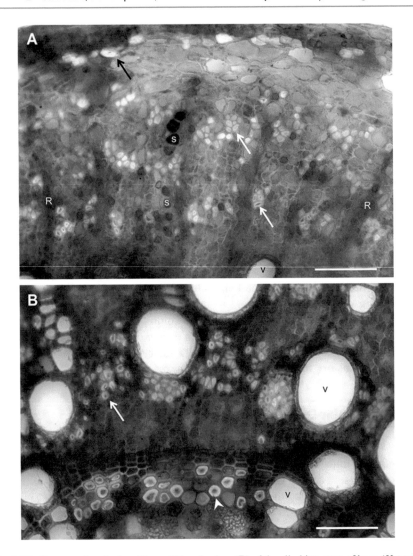

Fig. 2.31 Cross sections in the phloem (**A**) and xylem (**B**) of the climbing stem of hops (*Humulus lupulus*) stain with lacmoid according to Aloni and Sachs (1973), showing gelatinous fibers in both vascular tissues. Their lignified walls stained blue and their inside G-layer remained white. (**A**) In the phloem, wide primary phloem G-fibers with cellulose and almost no lignin are the first to differentiate (*black arrow*) during internode development. They are followed by the formation of many narrow secondary phloem G-fibers (*white arrows*) among the secondary sieve tubes (s). (**B**) In the xylem, the large secondary xyle fibers (*arrowhead*) differentiate near the pith during internode elongation, but during the early stage of internode elongation, they do not produce the G-layer. They are followed by the differentiation of many narrow secondary G-fibers (*white arrow*) among wide vessels (v). During later internode developmental stage, at the time of longitudinal stem shrinkage, which causes the coiling and twining of the climbing stem, only then the wide xylem fibers (*arrowhead*) produce their G-layer and the stem becomes fixed. Note that the rays (R) have a dark appearance due to winter starch accumulation. Starch is also observed in the pith. Bars = 100 μm. (R. Aloni, K. Sims, J. Mattsson, unpublished)

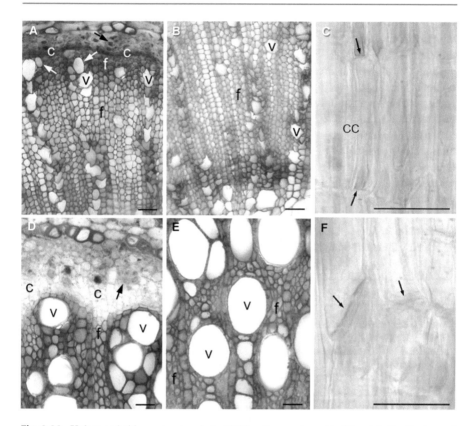

Fig. 2.32 Xylem and phloem structure in *LeFRK2*-antisense stems (**A–C**) metabolically stressed due to reduced fructose production, compared to wild-type (**D–F**) tomato stems. (**A**) Cross section from the middle of a stem showing that suppression of *LeFRK2* reduces cambium (C), vessel (V), and sieve element (*black arrow*) width and slows fiber differentiation, resulting in a few layers of differentiating fibers (*white f*) with dense cytoplasm. Note that the secondary vessels near the cambium differentiate with a circular shape (*white arrows*) and become deformed (V) away from the cambium due pressure of neighboring cells on their thin secondary wall. Mature fibers (*black f*) are narrow and have thin cell walls. (**B**) Xylem from the base of the stem characterized by deformed vessels and narrow fibers (f) both with thin cell walls. (**C**) Longitudinal section from the base of the antisense stem showing short and narrow sieve elements with almost no callose on the sieve plates (*arrows*) and a companion cell (CC). (**D**) Cross section from the middle of a wild-type stem characterized by wide cambium, large vessels and sieve tubes (*arrow*), and mature fibers (f). (**E**) Widest vessels and fibers, both with thick secondary walls, from the stem base. (**F**) longitudinal section from the base of a wild-type stem showing normally wide sieve elements with callose deposited on their sieve plates (*arrows*). *Scale bars* = 50 μm. (From Damari-Weissler et al. 2009)

compressive pressure inflicted by their surrounding cells. Water conductance is then reduced in stems, roots, and leaves, showing that *LeFRK2* influences xylem development throughout the entire vascular system. Suppression of *LeFRK2* reduced the length and width of the sieve elements, as well as their callose deposition (Fig. 2.32C) (Damari-Weissler et al. 2009).

The effect of *LeFRK2* suppression demonstrates how carbohydrate stress reduces the thickness of secondary walls resulting in narrow and **deformed vessels**. Similar effect on secondary wall formation in vessels of wild-type plants may be induced by various environmental stresses. When plants are grown under limiting environmental conditions, e.g., limiting root space, the root and shoot remain small. Restricted root space limits plant development, which diminishes root development and vessel differentiation. In corn (*Zea mays*) roots grown in small limiting pots, their wide metaxylem vessels produce thin secondary walls that may become deformed due to compressive pressure of the neighboring cells (Fig. 2.20B; R. Aloni and J. Mattsson, unpublished). Deformed cell walls reduce vessel diameter and increase resistance to water flow, consequently decreasing water transport to the leaves, which limits shoot development.

Environmental stresses might also substantially influence the number and size of the vascular conduits, which might not become deformed. Such changes are considered as an environmental adaptation of the vascular tissues to stress (see Chap. 16).

References and Recommended Reading[1]

*Agustí J, Blázquez MA (2020) Plant vascular development: mechanisms and environmental regulation. Cell Mol Life Sci 77: 3711–3728.

Aloni R (1976) Regeneration of phloem fibres around a wound: a new experimental system for studying the physiology of fibre differentiation. Ann Bot 40: 395–397.

Aloni R (1979) Role of auxin and gibberellin in differentiation of primary phloem fibers. Plant Physiol 63: 609–614.

Aloni R (1988) Vascular differentiation within the plant. In: *Vascular Differentiation and Plant Growth Regulators*, LW Roberts, PB Gahan, R Aloni. Springer-Verlag, Berlin, Heidelberg, New York, pp. 39–62.

Aloni R, Schwalm K, Langhans M, Ullrich CI (2003) Gradual shifts in sites of free-auxin production during leaf-primordium development and their role in vascular differentiation and leaf morphogenesis in Arabidopsis. Planta 216: 841–853.

Aloni R (2004) The induction of vascular tissue by auxin. In: *Plant Hormones: Biosynthesis, Signal Transduction, Action!* PJ Davies (ed). Kluwer Academic Publishers, Dordrecht, Boston, London, pp. 471–492.

*Aloni R (2015) Ecophysiological implications of vascular differentiation and plant evolution. Trees 29: 1–16.

Aloni, R, Aloni E, Langhans M, Ullrich CI (2006a) Role of cytokinin and auxin in shaping root architecture: regulating vascular differentiation, lateral root initiation, root apical dominance and root gravitropism. Ann Bot 97: 883–893.

Aloni R, Aloni E, Langhans M, Ullrich CI (2006b) Role of auxin in regulating *Arabidopsis* flower development. Planta 223: 315–328.

Aloni R, Barnett JR (1996) The development of phloem anastomoses between vascular bundles and their role in xylem regeneration after wounding in *Cucurbita* and *Dahlia*. Planta 198: 595–603.

*Aloni R, Foster A, Mattsson J (2013) Transfusion tracheids in the conifer leaves of *Thuja plicata* (Cupressaceae) are derived from parenchyma and their differentiation is induced by auxin. Am J Bot 100: 1949–1956.

---

[1] Papers of particular interest for suggested reading have been highlighted (with *).

Aloni R, Griffith M (1991) Functional xylem anatomy in root-shoot junctions of six cereal species. Planta 184: 123–129.

Aloni R, Jacobs WP (1977a) Polarity of tracheary regeneration in young internodes of *Coleus* (Labiatae). Am J Bot 64: 395–403.

Aloni R, Jacobs WP (1977b) The time course of sieve tube and vessel regeneration and their relation to phloem anastomoses in mature internodes of *Coleus*. Am J Bot 64: 615–621.

Aloni R, Langhans M, Aloni E, Dreieicher E, Ullrich CI (2005) Root-synthesized cytokinin in *Arabidopsis* is distributed in the shoot by the transpiration stream. J Exp Bot 56: 1535–1544.

Aloni R, Langhans M, Aloni E, Ullrich CI (2004) Role of cytokinin in the regulation of root gravitropism. Planta 220: 177–182.

Aloni R, Peterson CA (1990) The functional significance of phloem anastomoses in stems of *Dahlia pinnata* Cav. Planta 182: 583–590.

Aloni R, Peterson CA (1991) Seasonal changes in callose levels and fluorescein translocation in the phloem of *Vitis vinifera* L. IAWA Bull ns 12: 223–234.

Aloni R, Peterson CA (1997) Auxin promotes dormancy callose removal from the phloem of *Magnolia kobus* and callose accumulation and earlywood vessel differentiation in *Quercus robur*. J Plant Res 110: 37–44.

Aloni R, Pradel KS, Ullrich CI (1995) The three-dimensional structure of vascular tissues in *Agrobacterium tumefaciens*-induced crown galls and in the host stems of *Ricinus communis*. Planta 196: 597–605.

Aloni R, Raviv A, Peterson CA (1991) The role of auxin in the removal of dormancy callose and resumption of phloem activity in *Vitis vinifera*. Can J Bot 69: 1825–1832.

Aloni R, Sachs T (1973) The three-dimensional structure of primary phloem systems. Planta 113: 343–353.

Aloni R, Wolf A (1984) Suppressed buds embedded in the bark across the bole and the occurrence of their circular vessels in *Ficus religiosa*. Am J Bot 71: 1060–1066.

Aloni R, Zimmermann MH (1983) The control of vessel size and density along the plant axis - a new hypothesis. Differentiation 24: 203–208.

Alvarez JM, Bueno N, Cañas RA, Avila C, Cánovas FM, Ordás RJ (2018) Analysis of the *WUSCHEL-RELATED HOMEOBOX* gene family in *Pinus pinaster*: New insights into the gene family evolution. Plant Physiol Biochem 123: 304–318.

Augstein F, Carlsbecker A (2018) Getting to the roots: a developmental genetic view of root anatomy and function from Arabidopsis to Lycophytes. Front Plant Sci 9: 1410.

*Baas P, Ewers FW, Davis SD, Wheeler EA (2004) Evolution of xylem physiology, In: *The Evolution of Plant Physiology*. AR Hemsley, I Poole (eds). Linnean Society Symposium Series, no. 21, Academic Press, Amsterdam, pp 273–295.

Baima S, Nobili F, Sessa G, Lucchetti S, Ruberti I, Morelli G (1995) *The expression of the Athb-8 homeobox gene is restricted to provascular cells in Arabidopsis thaliana*. Development 121: 4171–4182.

Baima, S., Possenti, M., Matteucci, A., Wisman, E., Altamura, M. M., Ruberti, I. and Morelli, G (2001) *The Arabidopsis ATHB-8 HD-zip protein acts as a differentiation-promoting transcription factor of the vascular meristems*. Plant Physiol 126: 643–655.

Barnett JR, Gril J, Saranpää P (2014) Introduction. In: *The Biology of Reaction Wood*. B Gardiner, J Barnett, P Saranpää, J Gril (Eds). Springer-Verlag, Berlin.

Bellini C, Pacurar DI, Perrone I (2014) Adventitious roots and lateral roots: similarities and differences. Annu Rev Plant Biol 65: 639–666.

Berhin A, de Bellis D, Franke RB, Buono RA, Nowack MK, Nawrath C (2019) The root cap cuticle: a cell wall structure for seedling establishment and lateral root formation. Cell 176: 1367–1378.

Biedroń M, Banasiak A (2018) Auxin-mediated regulation of vascular patterning in Arabidopsis thaliana leaves. Plant Cell Rep 37: 1215–1229.

Bishopp A, Help H, El-Showk S, Weijers D, Scheres B, Friml J, Benková E, Mähönen AP, Helariutta Y (2011) A mutually inhibitory interaction between auxin and cytokinin specifies vascular pattern in roots. Curr Biol 21: 917–926.

Bowling AJ, Vaughn KC (2009) Gelatinous fibers are widespread in coiling tendrils and twining vines. Am J Bot 96: 719–727.

Carland FM, Berg BL, FitzGerald JN, Jinamornphongs S, Nelson T, Keith B (1999) Genetic regulation of vascular tissue patterning in *Arabidopsis*. Plant Cell 11: 2123–37.

*Carlquist S (2013) Interxylary phloem: diversity and functions. Brittonia 65: 477–495.

Carter MG (1911). A reconsideration of the origin of "transfusion tissue". Ann Bot 25: 975–982.

Chen T, Feng B, Fu W, Zhang C, Tao L, Fu G (2018) Nodes protect against drought stress in rice (*Oryza sativa*) by mediating hydraulic conductance. Environ Exp Bot 155: 411–419.

Choat B, Cobb AR, Jansen S (2008) Structure and function of bordered pits: new discoveries and impacts on whole-plant hydraulic function. New Phytol 177: 608–625.

Clair B, Alméras T, Pilate G, Jullien D, Sugiyama J, Riekel C (2011) Maturation stress generation in poplar tension wood studied by synchrotron radiation microdiffraction. Plant Physiol 155: 562–570.

Clair B, Alméras T, Yamamoto H, Okuyama T, Sugiyama J (2006a) Mechanical behavior of cellulose microfibrils in tension wood, in relation with maturation stress generation. Biophys J 91: 1128–1135.

Clair B, Ruelle J, Beauchêne J, Prévost MF, Fournier M (2006b) Tension wood and opposite wood in 21 tropical rain forest species. IAWA J 27: 329–338.

Courtois-Moreau CL, Pesquet E, Sjödin A, Muñiz L, Bollhöner B, Kaneda M, Samuels L, Jansson S, Tuominen H (2009) A unique program for cell death in xylem fibers of *Populus* stem. Plant J 58: 260–74.

Damari-Weissler H, Rachamilevitch S, Aloni R, German MA, Cohen S, Zwieniecki MA, Michele Holbrook N, Granot D (2009) LeFRK2 is required for phloem and xylem differentiation and the transport of both sugar and water. Planta 230: 795–805.

Dayan J, Voronin N, Gong F, Sun T-p, Hedden P, Fromm H, Aloni R (2012) Leaf-induced gibberellin signaling is essential for internode elongation, cambial activity, and fiber differentiation in tobacco stems. Plant Cell 24: 66–79.

De Micco V, Balzano A, Wheeler E, Baas P (2016) Tyloses and gums: a review of structure, function and occurrence of vessel occlusions. IAWA J. 37: 186–205.

*De Rybel B, Mähönen AP, Helariutta Y, Weijers D (2016) Plant vascular development: from early specification to differentiation. Nat Rev. Mol Cell Bio 17: 30–40.

Delzon S, Douthe C, Sala A, Cochard H (2010) Mechanism of water-stress induced cavitation in conifers: bordered pit structure and function support the hypothesis of seal capillary-seeding. Plant Cell Environ 33: 2101–2111.

Deyholos MK, Cordner G, Beebe D, Sieburth LE (2000) The SCARFACE gene is required for cotyledon and leaf vein patterning. Development 127: 3205–3213.

*Donaldson LA, Singh AP (2016) Reaction wood. In: *Secondary Xylem Biology, Origins, Functions, and Applications,* Kim YS, Funada R, Singh AP (eds). Elsevier/Academic press, Amsterdam, Boston, pp. 100–109.

Donner TJ, Sherr I, Scarpella E (2009) Regulation of preprocambial cell state acquisition by auxin signaling in *Arabidopsis* leaves. Development 136: 3235–3246.

Du S, Yamamoto F (2007) An overview of the biology of reaction wood formation. J Integr Plant Biol 49: 131–143.

Esau K (1965) *Plant Anatomy*. 2nd ed. John Wiley, New York.

Easu K (1965) *Vascular Differentiation in Plants*. Holt, Rinhart & Winston, New York.

Esau K (1969) *The Phloem*. In: *Encyclopedia of Plant Anatomy*, Gebrüder Borntraeger, Berlin.

Eshed Y, Baum SF, Perea JV, Bowman JL (2001) Establishment of polarity in lateral organs of plants. Curr Biol 11: 1251–1260.

Evert RF (2006) *Esau's Plant Anatomy: Meristems, Cells, and Tissues of the Plant Body: Their Structure, Function, and Development*. John Wiley & Sons, New Jersey.

Ewers FW, Cochard H, Tyree MT (1997) A survey of root pressures in vines of a tropical lowland forest. Oecologia 110: 191–196.

Fahn A (1985) The development of the secondary body in plants with interxylary phloem. In: *Xylorama*. LJ Kučera (ed). Birkhäuser, Basel, pp. 58–67.

Fahn A (1990) *Plant anatomy*, 4th edn. Pergamon, Oxford.

Fambrini M, Pugliesi C (2013) Usual and unusual development of the dicot leaf: involvement of transcription factors and hormones. Plant Cell Rep 32: 899–922.

Fletcher JC (2018) The CLV-WUS stem cell signaling pathway: a roadmap to crop yield optimization. Plants (Basel) 7: 87.

Frank AB (1864) Ein Beitrag zur Kenntniss der Gefässbündel. Bot Zeitung 22: 165–174.

Fukuda H (2000) Programmed cell death of tracheary elements as a paradigm in plants. Plant Mol Biol 44: 245–253.

Ge Y, Liu J, Zeng M, He J, Qin P, Huang H, Xu L (2016) Identification of WOX family genes in Selaginella kraussiana for studies on stem cells and regeneration in Lycophytes. Front Plant Sci 7: 93.

German MA, Dai N, Matsevitz T, Hanael R, Petreikov M, Bernstein N, Ioffe M, Shahak Y, Schaffer AA, Granot D (2003) Suppression of fructokinase encoded by *LeFRK2* in tomato stem inhibits growth and causes wilting of young leaves. Plant J 34: 837–846.

Grew N (1682) *The Anatomy of Plants with an Idea of a Philosophical History of Plants*. Rawlins, London.

Gritsch C, Wan Y, Mitchell RA, Shewry PR, Hanley SJ, Karp A (2015) G-fibre cell wall development in willow stems during tension wood induction. J Exp Bot 66: 6447–6459.

Guan C, Wu B, Yu T, Wang Q, Krogan NT, Liu X, Jiao Y (2017) Spatial Auxin Signaling Controls Leaf Flattening in Arabidopsis. Curr Biol 27: 2940–2950.

Gunning BES, Pate JS (1969) Transfer cells. Plant cells with wall ingrowths, specialized in relation to short distance transport of solutes—their occurrence, structure, and development. Protoplasma 68: 107–133.

Haecker A, Gross-Hardt R, Geiges B, Sarkar A, Breuninger H, Herrmann M, Laux T (2004) Expression dynamics of WOX genes mark cell fate decisions during early embryonic patterning in *Arabidopsis thaliana*. Development 131: 657–668.

*Ham BK, Lucas WJ (2017) Phloem-mobile RNAs as systemic signaling agents. Annu Rev Plant Biol 68: 173–195.

Hao Z, Mohnen D (2014) A review of xylan and lignin biosynthesis: foundation for studying Arabidopsis irregular xylem mutants with pleiotropic phenotypes. Crit Rev Biochem Mol Biol 49: 212–241.

Hardtke CS, Berleth T (1998) The *Arabidopsis* gene *MONOPTEROS* encodes a transcription factor mediating embryo axis formation and vascular development. EMBO J 17: 1405–1411.

Heo JO, Blob B, Helariutta Y (2017) Differentiation of conductive cells: a matter of life and death. Curr Opin Plant Biol 35: 23–29.

*Hirakawa Y, Kondo Y, Fukuda H (2010) TDIF peptide signaling regulates vascular stem cell proliferation via the WOX4 homeobox gene in *Arabidopsis*. Plant Cell 22: 2618–2629.

Indig FE, Aloni R (1989) An experimental method for studying the differentiation of vessel endings. Ann Bot 64: 589–592.

Jacobs WP (1952) The role of auxin in differentiation of xylem around a wound. Amer J Bot 39: 301–309.

Jacobsen AL, Valdovinos-Ayala J, Pratt RB (2018) Functional lifespans of xylem vessels: Development, hydraulic function, and post-function of vessels in several species of woody plants. Am J Bot 105: 142–150.

*Ji J, Strable J, Shimizu R, Koenig D, Sinha N, Scanlon MJ (2010) WOX4 promotes procambial development. Plant Physiol 152: 1346–1356.

Jouannet V, Brackmann K, Greb T (2015) (Pro)cambium formation and proliferation: two sides of the same coin? Curr Opin Plant Biol 23: 54–60.

Jura-Morawiec J (2017) Atypical origin, structure and arrangement of secondary tracheary elements in the stem of the monocotyledonous dragon tree, *Dracaena draco*. Planta. 245: 93–99.

Kalev N, Aloni R (1998) Role of auxin and gibberellin in regenerative differentiation of tracheids in *Pinus pinea* seedlings. New Phytol 138: 461–468.

Kang J, Dengler N (2002) Cell cycling frequency and expression of the homeobox gene ATHB-8 during leaf vein development in *Arabidopsis*. Planta 216: 212–219.

Koizumi K, Sugiyama M, Fukuda H (2000) A series of novel mutants of *Arabidopsis thaliana* that are defective in the formation of continuous vascular network: calling the auxin signal flow canalization hypothesis into question. Development 127: 3197–3204.

Knoblauch M, Knoblauch J, Mullendore DL, Savage JA, Babst BA, Beecher SD, Dodgen AC, Jensen KH, Holbrook NM (2016) Testing the Münch hypothesis of long distance phloem transport in plants. eLife 5: e15341.

Kucukoglu M, Nilsson J, Zheng B, Chaabouni S, Nilsson O (2017) WUSCHEL-RELATED HOMEOBOX4 (WOX4)-like genes regulate cambial cell division activity and secondary growth in *Populus* trees. New Phytol 215: 642–657.

Kurczyńska EU, Hejnowicz Z (1991) Differentiation of circular vessels in isolated segments of *Fraxinus excelsior*. Physiol Plant 83: 275–280.

Larson PR (1994) *The vascular cambium: Development and Structure*. Springer Verlag, Berlin, Heidelberg.

Lechthaler S, Colangeli P, Gazzabin M, Anfodillo T (2019) Axial anatomy of the leaf midrib provides new insights into the hydraulic architecture and cavitation patterns of *Acer pseudoplatanus* leaves. J Exp Bot 70: 6195–6201.

Lev-Yadun S, Aloni R (1990) Vascular differentiation in branch junction: circular patterns and functional significance. Trees 4: 49–54.

Lucas WJ, Groover A, Lichtenberger R, Furuta K, Yadav SR, Helariutta Y, He XQ, Fukuda H, Kang J, Brady SM, Patrick JW, Sperry J, Yoshida A, López-Millán AF,Grusak MA, Kachroo P (2013) The plant vascular system: evolution, development and functions. J Integr Plant Biol 55: 294–388.

Luxová M (1986) The hydraulic safety zone at the base of barley roots. Planta 169: 465–470.

Ma J, Aloni R, Villordon A, Labonte D, Kfir Y, Zemach H, Schwartz A, Althan L, Firon N (2015) Adventitious root primordia formation and development in stem nodes of 'Georgia Jet' sweetpotato, *Ipomoea batatas*. Am J Bot 102: 1040–1049.

Mattsson J, Ckurshumova W, Berleth T (2003). Auxin signaling in *Arabidopsis* leaf vascular development. Plant Physiol. 131: 1327–1339.

Mellerowicz EJ, Immerzeel P, Hayashi T (2008) Xyloglucan: the molecular muscle of trees. Ann Bot 102: 659–665.

*Mellor N, Adibi M, El-Showk S, De Rybel B, King J, Mähönen AP, Weijers D, Bishopp A (2017) Theoretical approaches to understanding root vascular patterning: a consensus between recent models. J Exp Bot 68: 5–16.

Mellor N, Vaughan-Hirsch J, Kümpers BMC, Help-Rinta-Rahko H, Miyashima S, Mähönen AP, Campilho A, King JR, Bishopp A (2019) A core mechanism for specifying root vascular patterning can replicate the anatomical variation seen in diverse plant species. Development 146: dev172411.

*Ménard D, Pesquet E (2015) Cellular interactions during tracheary elements formation and function. Curr Opin Plant Biol 23:109–115.

Merelo P, Paredes EB, Heisler MG, Wenkel S (2017) The shady side of leaf development: the role of the REVOLUTA/KANADI1 module in leaf patterning and auxin-mediated growth promotion. Curr Opin Plant Biol 35: 111–116.

Miyashima S, Roszak P, Sevilem I, Toyokura K, Blob B, Heo JO, Mellor N, Help-Rinta-Rahko H, Otero S, Smet W, Boekschoten M, Hooiveld G, Hashimoto K, Smetana O, Siligato R, Wallner ES, Mähönen AP, Kondo Y, Melnyk CW, Greb T, Nakajima K, Sozzani R, Bishopp A, De Rybel B, Helariutta Y (2019) Mobile PEAR transcription factors integrate positional cues to prime cambial growth. Nature 565: 490–494.

Miyashima S, Sebastian J, Lee JY, Helariutta Y (2013) Stem cell function during plant vascular development. EMBO J 32: 178–193.

Mohl HV (1871) Morphologische Betrachtung der Blätter von *Sciadopitvs*. Bot Zeitung 29: 1–14, 17–23.

Morris H, Gillingham MAF, Plavcová L, Gleason SM, Olson ME, Coomes DA, Fichtler E, Klepsch MM, Martínez-Cabrera HI, McGlinn DJ, Wheeler EA, Zheng J, Ziemińska K, Jansen S (2018)

Vessel diameter is related to amount and spatial arrangement of axial parenchyma in woody angiosperms. Plant Cell Environ 41: 245–260.

Nugroho WD, Yamagishi Y, Nakaba S, Fukuhara S, Begum S, Marsoem SN, Ko JH, Jin HO, Funada R (2012) Gibberellin is required for the formation of tension wood and stem gravitropism in *Acacia mangium* seedlings. Ann Bot 110: 887–895.

*Ohtani M, Akiyoshi N, Takenaka Y, Sano R, Demura T (2017) Evolution of plant conducting cells: perspectives from key regulators of vascular cell differentiation. J Exp Bot 68: 17–26.

Pate JF, Gunning BES (1972) Transfer cells. Annu Rev Plant Physiol 23: 173–196.

*Pesquet E, Zhang B, Gorzsás A, Puhakainen T, Serk H, Escamez S, Barbier O, Gerber L, Courtois-Moreau C, Alatalo E, Paulin L, Kangasjärvi J, Sundberg B, Goffner D, Tuominen H (2013) Non-cell-autonomous postmortem lignification of tracheary elements in Zinnia elegans. Plant Cell 25: 1314–1328.

Przemeck GK, Mattsson J, Hardtke CS, Sung ZR, Berleth T (1996) Studies on the role of the *Arabidopsis* gene MONOPTEROS in vascular development and plant cell axialization. Planta 200: 229–237.

Rajput KS, Gonda AD, Lekhak MM, Yadav SR (2018) Structure and ontogeny of intraxylary secondary xylem and phloem development by the internal vascular cambium in Campsis radicans (L.) Seem. (Bignoniaceae). J Plant Growth Reg 37: 755–767.

Ruonala R, Ko D, Helariutta Y (2017) Genetic networks in plant vascular development. Annu Rev Genet 51: 335–359.

Sachs T (1968) The role of the root in the induction of xylem differentiation in pea. Ann Bot 32: 391–399.

*Sachs T (1981) The control of patterned differentiation of vascular tissues. Adv Bot Res 9: 151–262.

Sachs T, Cohen D (1982) Circular vessels and the control of vascular differentiation in plants. Differentiation 21: 22–26.

Sawada D, Kalluri UC, O'Neill H, Urban V, Langan P, Davison B, Pingali SV (2018) Tension wood structure and morphology conducive for better enzymatic digestion. Biotechno Biofules 11: 44.

Sawchuk MG, Head P, Donner T, Scarpella, E (2007) Time-lapse imaging of *Arabidopsis* leaf development shows dynamic patterns of procambium formation. New Phytol 176: 560–571.

*Scarpella E, Helariutta Y (2010) Vascular pattern formation in plants. Curr Top Dev Biol 91: 221–265.

Scarpella E, Marcos D, Friml J, Berleth T (2006). Control of leaf vascular patterning by polar auxin transport. Genes Dev 20:1015–1027.

*Scarpella E, Meijer AH (2004) Pattern formation in the vascular system of monocot and dicot plant species. New Phytol 164: 209–242.

Scheres B, Xu J (2006) Polar auxin transport and patterning: grow with the flow. Genes Dev 20: 922–924.

Segatto AL, Thompson CE, Freitas LB (2016) Molecular evolution analysis of WUSCHEL-related homeobox transcription factor family reveals functional divergence among clades in the homeobox region. Dev Genes Evol 226: 259–268.

Shi D, Lebovka I, López-Salmerón V, Sanchez P, Greb T (2019) Bifacial cambium stem cells generate xylem and phloem during radial plant growth. Development 146: dev171355.

*Słupianek A, Kasprowicz-Maluśki A, Myśkow E, Turzańska M, Sokołowska K (2019) Endocytosis acts as transport pathway in wood. New Phytol 222: 1846–1861.

Sokołowska K (2013) Symplasmic transport in wood: The importance of living xylem cells. In: *Symplasmic Transport in Vascular Plants*, K Sokołowska and P Sowiński (eds). Springer Science and Business Media New York, pp. 101–132.

Smetana O, Mäkilä R, Lyu M, Amiryousefi A, Sánchez Rodríguez F, Wu MF, Solé-Gil A, Leal Gavarrón M, Siligato R, Miyashima S, Roszak P, Blomster T, Reed JW, Broholm S, Mähönen AP (2019) High levels of auxin signalling define the stem-cell organizer of the vascular cambium. Nature 565: 485–89.

*Smith RA, Schuetz M, Roach M, Mansfield SD, Ellis B, Samuels L (2013) Neighboring paren-
  chyma cells contribute to *Arabidopsis* xylem lignification, while lignification of interfascicular
  fibers is cell autonomous. Plant Cell, 25: 3988–3999.
Sousa-Baena MS, Lohmann LG, Hernandes-Lopes J, Sinha NR (2018) The molecular control of
  tendril development in angiosperms. New Phytol 218: 944–958.
Sperry JS (1985) Xylem embolism in the palm *Rhapis excelsa*. IAWA Bull ns 6: 283–292.
Spicer R (2014) Symplastic networks in secondary vascular tissues: parenchyma distribution and
  activity supporting long-distance transport. J Exp Bot 65: 1829–1848.
*Suer S, Agusti J, Sanchez P, Schwarz M, Greb T (2011) WOX4 imparts auxin responsiveness to
  cambium cells in *Arabidopsis*. Plant Cell 23: 3247–3259.
Timell TE (1986) Compression Wood in Gymnosperms, Vol 2. Springer-Verlag, Heidelberg,
  pp. 983–1262.
Turner SR, Somerville CR (1997) Collapsed xylem phenotype of Arabidopsis identifies mutants
  deficient in cellulose deposition in the secondary cell wall. Plant Cell 9: 689–701.
*Tyree MT, Ewers FW (1991) The hydraulic architecture of trees and other woody plants. New
  Phytol 119: 345–360.
Tyree NT, Sperry JS (1989) Vulnerability of xylem to cavitation and embolism. Ann Rev Plant
  Physiol 40: 19–36.
Tyree MT, Zimmermann MH (2002) Xylem Structure and the Ascent of Sap, 2nd ed.
  Springer, Berlin.
van Bel AJE, Ehlers K, Knoblauch M (2002) Sieve elements caught in the act. Trends Plant Sci
  7: 126–132.
Wardrop AB (1964) Reaction anatomy of arborescent angiosperms. In: *The Formation of Wood in
  Forest Trees,* MH Zimmermann (ed). Academic Press, New York.
*Wenzel CL, Schuetz M, Yu Q, Mattsson J (2007). Dynamics of *MONOPTEROS* and *PIN-
  FORMED1* expression during leaf vein pattern formation in *Arabidopsis thaliana*. Plant J 49:
  387–398.
Wheeler E, Baas P (2019) Wood evolution: Baileyan trends and functional traits in the fossil
  record. IAWA J 40: 488–529.
Wisniewska J, Xu J, Seifertová D, Brewer PB, Ruzicka K, Blilou I, Rouquié D, Benková E,
  Scheres B, Friml J (2006) Polar PIN localization directs auxin flow in plants. Science 312: 883.
Zhang J, Eswaran G, Alonso-Serra J, Kucukoglu M, Xiang J, Yang W, Elo A, Nieminen K, Damén
  T, Joung JG, Yun JY, Lee JH, Ragni L, Barbier de Reuille P, Ahnert SE, Lee JY, Mähönen AP,
  Helariutta Y (2019) Transcriptional regulatory framework for vascular cambium development
  in *Arabidopsis* roots. Nat Plants. 5: 1033–1042.
Zhang W, Feng F, Tyree MT (2018) Seasonality of cavitation and frost fatigue in *Acer mono*
  Maxim. Plant Cell Environ 41: 1278–1286.
Zimmermann MH (1983) *Xylem Structure and the Ascent of Sap*. Springer-Verlag, Berlin.
Zimmermann MH, Sperry JS (1983) Anatomy of the palm *Rhapis excelsa*. IX. Xylem structure of
  the leaf insertion. J Arnold Arbor 64: 599–609.
Zimmermann MH, Tomlinson PB (1965) Anatomy of the palm *Rhapis excelsa*. I. Mature vegeta-
  tive axis. J Arnold Arbor 46: 160–180.

The Hormonal Signals that Regulate Plant Vascular Differentiation

The major signaling molecules that regulate vascular differentiation and plant development are the plant hormones, also called **phytohormones** (Went and Thimann 1937). The word hormone is derived from Greek, meaning to set in motion, excite, and stimulate. Plant hormones regulate gene expression, cellular activity, cell and tissue polarity, and growth and differentiation of tissues, organs, and the whole plant. The hormones can be produced in any living plant cell at extremely low concentrations. They may act locally or at a distance from the producing cells. Synthesized man-made chemicals with similar effects on plants are called plant growth regulators, which are mainly used in agriculture.

Since plants are sessile organisms, namely, immobile, they use continuously moving signaling molecules, mainly hormones, to maintain synchronized growth and response to changing environmental cues. Very few phytohormonal signals enable regulation and adaptation in remarkably simple mechanisms. The developmental process could be carried out either by a single developmental signal or by very limited number of signals. The use of one or two signals is an economical way for carrying out major integrating roles. Vascular developmental processes occur gradually and sometimes may occur independently, or be repeated a few times in different plant groups, during evolution through competition and selection (see Chap. 19).

In addition to the phytohormones, today there is enormous detailed information regarding signaling molecules, such as RNAs, transcription factors, proteins, and small signaling peptides, which drive growth and development in plants. In the future, more signaling molecules and new mechanisms will be discovered in plants (Regnault et al. 2015; De Coninck and De Smet 2016; Ham and Lucas 2017), but there will always be a basic requirement for general, updated concepts and holistic understanding where these molecules are produced, transported, and function. This book aimed to provide the background and holistic understanding how vascular differentiation and plants development are influenced and controlled by signaling molecules, focusing on the biology of the plant hormones.

© Springer Nature Switzerland AG 2021
R. Aloni, *Vascular Differentiation and Plant Hormones*,
https://doi.org/10.1007/978-3-030-53202-4_3

The four primary phytohormonal signals that control vascular differentiation are auxin, gibberellin, cytokinin, and ethylene (Fig. 3.1). Additional hormonal signals are discussed below, which are involved in specific responses to the environment, various stresses, wounding, regulation of specific cell differentiation, and plant development. The role of the hormonal signals and their molecular mechanisms in vascular differentiation were extensively reviewed in recent years by Caño-Delgado et al. (2010), Scarpella and Helariutta (2010), Lucas et al. (2013), Aloni (2013a, 2015), Furuta et al. (2014), Zhang et al. (2014a), De Rybel et al. (2016), Scarpella (2017), Hellmann et al. (2018), Fukuda and Ohashi-Ito (2019), and Agustí and Blázquez (2020).

All these four primary hormones (Fig. 3.1) are moving signals that are transported in specific pathways through the primary (originate from procambium) and secondary (originate from cambium) vascular tissues (Aloni 2010, 2013a, 2015). In addition, hormonal movement through young parenchyma cells can induce naturally occurring regenerative differentiation of transfusion tracheids which are typical to conifer leaves (Aloni et al. 2013), naturally occurring regenerative vascular tissues at lateral root junctions (Aloni and Plotkin 1985), naturally occurring regenerative differentiation of phloem anastomoses (Aloni and Barnett 1996), naturally occurring xylem regeneration around a new lateral root (Aloni and Baum 1991), and vascular regeneration around a wound (Jacobs 1952, 1984; Sachs 1981; Aloni 2010).

Auxin is the young leaf signal (Jacobs 1952; Sachs 1981; Aloni et al. 2003), gibberellin is the mature leaf signal (Dayan et al. 2012), cytokinin is the root cup signal

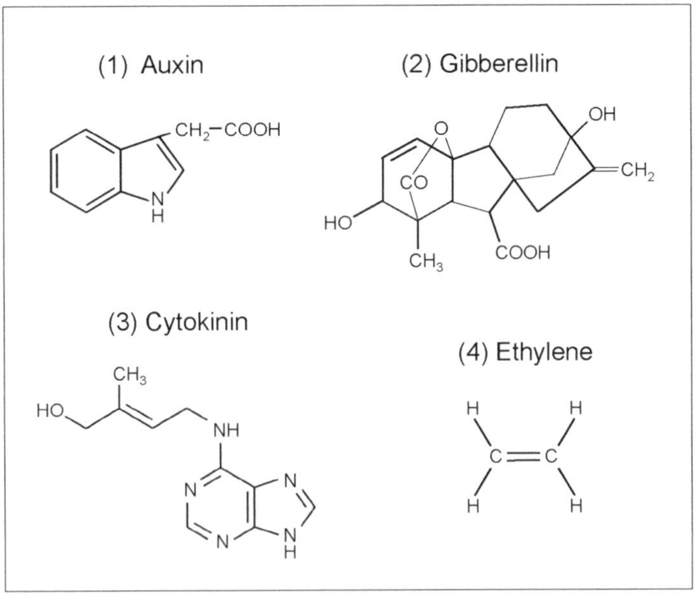

Fig. 3.1 The molecular structure of the four primary hormonal signals that control vascular differentiation in plants: auxin, gibberellin, cytokinin, and ethylene

(Aloni et al. 2004, 2005), and ethylene is regularly produced locally in maturing vessels (Pesquet and Tuominen 2011; Bollhöner et al. 2012) and in response to wounding and stress (Aloni 2013b). The continuous flow of these hormonal signals enables the plant to continuously respond to changing environmental cues, as will be clarified in this book.

The first three signals are mainly induced by different plant organs and thus informing the stem cells of the vascular cambium on the physiological strength and quantity of the producing organs and their developmental stage. The vascular tissues are induced and regulated accordingly, and the produced vascular elements reflect the developmental phase and amount of the plant organs. Thus, for example, during early spring when there are mainly young leaves on the stem of a hardwood tree, the auxin they produce is the main signal flowing through the cambium which, therefore, produces mainly sieve tubes and vessels, while during late summer when there are mainly mature leaves building large foliage biomass, their produced gibberellin becomes the dominant signal resulting in the formation of numerous fibers building stronger wood, which supports the enlarged shoot. Leaf development and biomass are regulated by environmental conditions (i.e., photoperiod, water availability, temperature, and nutrients), which control the production of wood and the type, quantity, and patterns of its differentiating vascular cell.

Understanding the role of each hormonal signal is the key to understand how these moving signals design plant development, structure, and vascular tissue differentiation under different environmental conditions.

3.1 Auxin (IAA) Is the Young Leaf Signal

Charles Darwin's observations on the phototropism of grass coleoptiles led him to hypothesize the existence of a signal that was transported downward from the tip of the coleoptile to the bending region (Darwin 1880). The Dutch biologist Frits Warmolt Went was the first to isolate the signal by diffusion from coleoptile tips into agar blocks, which stimulated decapitated coleoptile growth and bending. This substance was originally named Wuchsstoff by Went and was later renamed to auxin (Went 1935). Kenneth V. Thimann determined the auxin chemical structure as indole-3-acetic acid (Fig. 3.1) (Thimann 1935).

Indole-3-acetic acid (IAA) is the most common naturally occurring auxin in plants. IAA is the major shoot signal which regulates all aspects of vascular differentiation in plants (Jacobs 1979; Aloni 2001). Auxin is produced in plants primarily by the two-step tryptophan-dependent auxin biosynthesis pathway catalyzed by the tryptophan aminotransferase of *Arabidopsis* (TAA) family of transaminases and the YUCCA (YUC) family of flavin monooxygenases. The pathway is conserved throughout the plant kingdom (Zhao 2018).

The pioneering study of Williams P. Jacobs demonstrated that the auxin produced in young leaves is the limiting and controlling signal of xylem regeneration around a wound (Jacobs 1952) (see Chap. 10). Auxin is synthesized in young growing leaves, developing flowers, seeds, and fruits (Aloni et al. 2003, 2006b; Baylis et al. 2013; Robert et al. 2015) (see Chaps. 7 and 8) as well as in roots by the TAA/ YUC pathway, when shoot-derived auxin is not sufficient to support root growth (Zhao 2018). The polar transport of IAA from the tips of young shoot organs downward via the procambium, cambium, and parenchyma cells to the root tips induces and controls vascular differentiation (Sachs 1981; Aloni 1987; Mattsson et al. 1999; Aloni et al. 2006a; Scarpella and Helariutta 2010; Scarpella 2017). The continuity of the vascular tissues along the plant axis is a result of the continuous, steady polar flow of IAA from leaves to roots (Sachs 1981; Berleth et al. 2000; Aloni 2010; Scarpella and Helariutta 2010).

Auxin is a weak organic acid (IAA-H $\Leftrightarrow$ IAA$^-$ + H$^+$) consisting of an indole ring coupled to a side chain with a terminal carboxyl group (Fig. 3.1). The carboxyl group is protonated at low pH (IAA-H), making the molecule less polar, and in this form, it can diffuse across cell membranes. Whereas, the auxin in its unprotonated form (IAA$^-$) is negatively charged and is too polar to diffuse through a membrane. The pH in the different cellular and tissue compartments changes from being $\sim$5.0–5.5 in vacuoles, in the apoplastic fluids of the cell wall and spaces outside the cell, whereas $\sim$7.0 in the cytoplasmic matrix (cytosol). IAA-H in the apoplast (outside the plasma membrane) and in vacuoles can diffuse via cellular membranes, whereas the IAA$^-$ is trapped within the cell membrane and cannot outflow from the cytosol without the aid of specific **protein transporters**, namely, **efflux and influx carriers**. The PIN-FORMED (PIN) protein family includes essential carriers involved in IAA efflux out of cells, whereas the AUXIN-RESISTANCE/LIKE-AUX (AUX/LAX) protein family includes major auxin influx carriers that function in IAA transport into the cell. The polar location of the efflux and influx carriers at the plasma membrane directs the transport of auxin from cell to cell and determines auxin flow direction and cell polarity (Gälweiler et al. 1998; Swarup et al. 2001; Zazímalová et al. 2010; Swarup and Péret 2012; Ljung 2013; Adamowski and Friml 2015).

Interestingly, the auxin produced in young leaves is also the hormonal signal that induces the initiation of new leaf primordia at the shoot tip (Reinhardt et al. 2003; Benková et al. 2003) (Fig. 3.2) (see Sect. 7.1). Although young leaves produce auxin that most of it moves downward, toward the root tip, in high conductance polar auxin transport streams (Bennett et al. 2016), relatively small amounts of the young-leaf-produced auxin are transported upward to the shoot apical meristem (SAM) via the epidermis, and continuously toward the shoot tip through the protoderm and tunica (the outermost meristematic cell layer). The upward movement (Figs. 3.2, 3.3A) is directed by an asymmetrically upward membrane distribution of the PIN1 protein (Fig. 3.3B), which directs an upward dynamic auxin flow gradient, creating an orderly pattern of auxin maxima at the shoot tip, which induces the initiation of new leaf primordia (Reinhardt et al. 2003; Benková et al. 2003). During early stages of leaf initiation, the leaf acts as a sink for auxin, which is accumulated as conjugated auxin (Figs. 3.4 and 3.5A). Gradually, during leaf development and elongation, the young leaf becomes a major source of IAA production.

Auxin is the plant hormone involved in an extremely broad diversity of biological mechanisms acting as a general coordinator of vascular differentiation and plant development (Leyser 2018). These range from basic cellular and tissue processes,

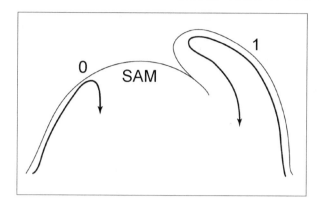

Fig. 3.2 Schematic diagram illustrating the upward transport of auxin from young leaves to the shoot apical meristem (SAM). Auxin moves via the epidermis, and the outermost meristematic cell layer, creating auxin maxima at the shoot tip, which induce the initiation of new leaf primordia. 0, shows the site of auxin maximum inducing a new leaf primordium. 1, shows auxin transport in the youngest growing leaf primordium, upward in the differentiating epidermis and then downward in the center of the leaf, where the signal induces the midvein

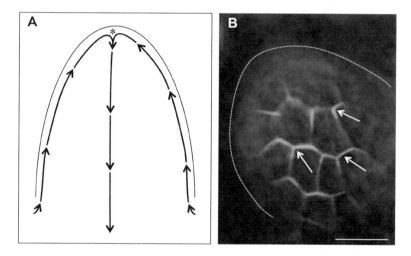

Fig. 3.3 (**A**) Schematic diagram illustrating the upward transport of auxin in the epidermis, from the base of the leaf primordium to its tip, establishing auxin maximum (*asterisk*) at the tip, from which auxin continues downward and induces the midvein in the center of the growing leaf. (**B**) Upward PIN1 orientation during early development of *Arabidopsis thaliana* rosette-leaf primordium, directing polar auxin transport from leaf-base to tip, showing tip-directed polarity of PIN1 in the epidermis (*arrows*), detected by PIN1:GFP under GFP fluorescence and transmitted light illumination. Scale bar = 10 μm. Micrograph **B** was reproduced from Scarpella et al. 2006, *Genes & Development* 20: 1015–1027, with permission of Cold Spring Harbor Laboratory Press and the authors

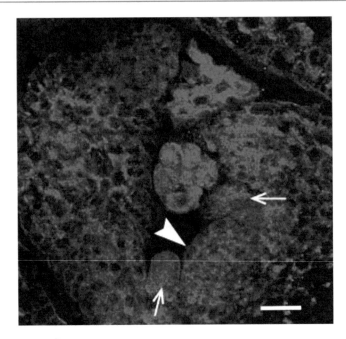

Fig. 3.4 Concentrated conjugated auxin accumulation in the youngest rosette-leaf primordium (*arrowhead*) and in its stipules (*arrows*) of a DR5::GUS transformed *Arabidopsis*. Showing auxin immunolocalization (stained red with mouse monoclonal red Alexa Fluor 568 fluorescence and rabbit polyclonal antibodies) viewed with a confocal laser-scanning microscope (CLSM). The accumulation of conjugated auxin in the youngest leaf primordium demonstrates that young leaf primordia are sink for auxin arriving from adjacent young leaves (see Sect. 7.1). This youngest leaf does not show free auxin (with DR5::GUS staining) at this early developmental stage. Bar = 20 μm. (From Aloni et al. 2003)

such as cell polarity, cell growth and differentiation, tissue patterning and organ formation. Although the initial observations of Charles Darwin during his study of coleoptile bending due to phototropism (Darwin 1880) and early steps in the long history of auxin research (Holland et al. 2009) that started more than 100 years ago, we are still far from a comprehensive understanding of the mechanisms how auxin regulates and controls such a wide range of responses. Therefore one would expect more new future discoveries in this field. The title of the article "Auxin: simply complicated" (Sauer et al. 2013) might give us the idea why auxin gets and will receive so much attention in this chapter and along the entire book.

3.1.1 Free and Conjugated Auxin

The percentage of IAA, the bioactive hormone, namely, the **free auxin,** is usually very low in a given tissue and might range from 1% to 5% of the total amount of auxin (Ljung et al. 2002; Ljung 2013) and up to 25% depending on the tissue and the plant species studied (Ludwig-Müller 2011). The synthesis of auxin conjugates

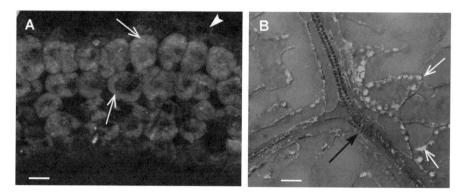

Fig. 3.5 Total auxin (mainly conjugated auxin) label in young *Arabidopsis* rosette-leaves stained with green Alexa Fluor 488 fluorescence and rabbit polyclonal antibodies, viewed with CLSM. (**A**) Longitudinal section in a very young leaf primordium showing strong labelling in its compact cytoplasm (*arrows*) of densely packed mesophyll cells, but nearly no label in the epidermal cells (*arrowhead*). (**B**) Paradermal section in a more developed leaf, showing in its lamina strong auxin label in the chloroplasts (*white arrows*) and weaker labeling in the thin cytoplasmic layer in the elongated cells around the vascular bundles, containing vessels (*black arrow*), emphasizing the important role of chloroplasts in auxin metabolism. Bar = 20 μm. (From Aloni et al. 2003)

and auxin conjugate hydrolases was detected in early stages of plant evolution, since it has been already found in moss and fern species (Ludwig-Müller 2011). Most of the auxin in cells is covalently bound to carbohydrates, amino acids, or peptides, and these conjugated molecules are inactive as hormonal signals and serve as a reservoir from which the free auxin can be released. Free auxin (detected by *DR5::GUS* expression; see Fig. 1.1) is not detected in the apical bud and the youngest leaf primordia (Aloni et al. 2003) or youngest flower primordia (Aloni et al. 2006b) (see Chap. 8, Fig. 8.3A), which are all loaded with **conjugated auxin**, detected by antibodies (Figs. 3.4, 3.5 and 8.2) (Aloni et al. 2003, 2006b). The conjugated auxin, which is a **bound auxin**, is accumulated in the shoot apex, youngest leaves, and youngest flowers likely due to the local upward polar auxin flow from the adjacent young leaves (Reinhardt et al. 2003; Benková et al. 2003; Scarpella et al. 2006). The youngest leaf primordia likely start as sinks for IAA and become sources of free auxin during later stages of leaf growth and development (Aloni et al. 2003; Aloni 2010).

This developmental auxin pattern stimulated me (Aloni 2007, 2013a) to put forward the following concept, proposing that due to the importance and requirement for continuous IAA supply from the shoot organs to the rest of the plant, the shoot establishes a large reservoir of conjugated auxin molecules at the shoot apex and youngest shoot organs (Figs. 3.4, 3.5 and 8.2) from which the IAA will later be hydrolyzed continuously in young growing leaves. This is a powerful strategy of plants to preserve a bound auxin pool which prevents situations of IAA shortage. Thus, the massive bound auxin pool in the young shoot organs guarantees the constant supply of free auxin, which maintains cell polarity and plant's polar orientation, regulates organ development, tissue formation, and vascular differentiation throughout the whole plant (Aloni 2001).

*** **A technical note** - The localization of **free auxin** (the bioactive hormone) can be visualized in leaf primordia (Aloni et al. 2003) by the expression of *GUS* fused to the highly active synthetic auxin-response element (AuxRE), referred to as *DR5* (Ulmasov et al. 1997). The *DR5::GUS* expression pattern was demonstrated to reflect free-auxin concentrations in shoots and roots of *Arabidopsis* transformants between 10^{-8} and 10^{-4} *mol* (Ulmasov et al. 1997; Sabatini et al. 1999). A positive correlation was also found between *GUS* expression in *Arabidopsis* flowers and exogenously applied auxin (Aloni et al. 2006b). The comparison of quantified GUS activity (Sabatini et al. 1999) with auxin concentration gradients measured by mass spectrometry (Marchant et al. 2002) in *Arabidopsis* roots clearly showed a positive correlation of free auxin and *GUS* expression in *DR5::GUS* transformed *Arabidopsis*. Therefore, analysis of *DR5::GUS* gene expression in transformed plants visualizes the sites of intense free-IAA concentrations (Sabatini et al. 1999; Friml et al. 2002; Marchant et al. 2002; Aloni et al. 2003, 2006b; Wenzel et al. 2007) (e.g., see Figs. 1.1, 7.2, 7.6, 8.4 and 8.5).

The **total auxin** distribution of both conjugated auxin and free auxin (IAA) can be elucidated in plant tissues by immunolocalization with specific polyclonal antibodies, as was shown with polyclonal and monoclonal antibodies in *Arabidopsis* leaves (Aloni et al. 2003, 2006b) (see Figs. 3.4, 3.5 and 8.2).

3.1.2 Sachs' Canalization Hypothesis

On the basis of elegant experimental evidence, Tsvi Sachs explained the orderly pattern of vascular tissues from leaves to roots by the *canalization hypothesis* (Sachs 1981). According to this hypothesis, IAA flow which starts by diffusion induces a polar auxin transport process which promotes IAA movement and leads to the canalization of the auxin flow along a narrow file of cells. These cells become more polarized and more efficient transporters of auxin (Fig. 3.6). As the auxin flow becomes diverted into the canalized file, lateral neighboring cells would receive decreasing amounts of the hormone. The continuous polar transport of IAA through the canalized cells induces a further complex sequence of events which terminates in the formation of a vessel, or a sieve tube (Sachs 1981, 1991, 2000). Thus, the hypothesis proposes a positive feedback relation between the transport of auxin and cell polarization: auxin transport itself induces both new and continued polarization, while cell polarity determines oriented auxin transport. The passage of auxin through a vascular bundle keeps the tissue polarized along the axis of this bundle. Polarized cells are not labile to the influence of an inductive source of auxin coming from another direction. Therefore, when auxin is applied laterally near a longitudinal vascular bundle, the applied auxin flow and the attachment of its induced bundle to a longitudinal bundle may be delayed (Fig. 3.7B) or prevented (Fig. 3.7C) according to the amount of auxin flowing through the longitudinal bundle. The *canalization hypothesis* provides the tools for understanding how the process of progressively

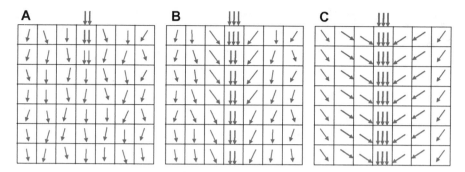

Fig. 3.6 Schematic diagrams illustrating the process of how increase in auxin flow promotes cell polarization and vascular tissue differentiation, as proposed by Sachs' *canalization hypothesis*. The arrows indicate the direction of auxin flow and their number in each cell the quantity of transported auxin, resulting from the distribution and quantity of the auxin efflux carrier in the plasma membrane. (**A**) Early stage of tissue development during which auxin flows by diffusion. (**B**) Auxin flow increases along a file of cells that start to transport auxin more efficiently. (**C**) The efficient transporting cell file become more polarized, transports more auxin and attracts auxin from neighboring cells. The continuous polar auxin transport through the canalized cell file will terminate in the formation of vascular conduit

improved pathways of auxin transport induces the differentiation of new vessel and sieve-tube patterns along the intact vascular system and also around wounds to connect the leaves with the roots. It explains why a new stream of auxin originating from a young leaf primordium is rejected by a recently formed vascular bundle (in a young internode) because the latter is loaded with auxin, and the new stream will become connected to an older bundle (in an older internode) because of its low-level flow of auxin (Sachs 1968, 1969, 1981). With this knowledge one can easily analyzed and predict the development of vascular networks in plants (Fig. 3.7).

Molecular evidence supports the *canalization hypothesis* (Sachs 1981) demonstrating that rearrangement of polar IAA flow changes tissue polarity through modification of the site of a PIN1 protein (an essential component involved in IAA efflux) on the plasma membrane (Sauer et al. 2006; Friml 2010; Kleine-Vehn et al. 2011; Ravichandran et al. 2020).

In addition, by comparative analysis of PIN1 localization and *MONOPTEROS (MP)* expression during vein pattern formation, Wenzel et al. (2007) provide evidence that *MP* expression is initially diffuse throughout the blade of *Arabidopsis* leaf primordia and progressively narrows into strands in a way very similar to Sachs (1981) proposed canalization. *MP* is activated by auxin, and therefore its distribution could possibly reflect and visualize auxin canalization into a vascular strand. Additionally, the pattern of the bioactive auxin flow (detected by *DR5::GUS* expression) from the auxin maximum at the leaf's margin to its induced differentiating vessel shown in Fig. 1.1, also visualizes the canalization of the hormone that start to move in a diffusion pattern and gradually becomes canalized to a narrow stream.

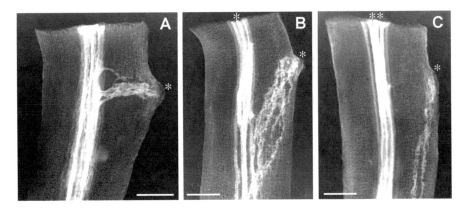

Fig. 3.7 Longitudinal views of thick sections of pea (*Pisum sativum*) cleared in lactic acid and stained with phloroglucinol. The vascular tissues in the micrographs have a white appearance and the other cleared tissues are darker. The three micrographs show the basal shoot internode after the shoot above it was decapitated, for demonstrating how a lateral source of auxin applied in lanolin (its application sites, on the internode's right side, is marked by the lateral *asterisk*) artificially induces new xylem strands: without auxin (**A**), by low (**B**), or high (**C**) auxin application on the decapitation cut (marked by one (**B**) or two (**C**) *asterisks* on the decapitation cut). The laterally applied auxin induced some tissue swelling at the application site. (**A**) The induced strand jointed directly, in the shortest way, to the central vascular cylinder, which was not supplied with auxin (and therefore attracted the lateral auxin stream to be efficiently transported through its polar well-transporting vascular cells). (**B**) The auxin-induced xylem strand connected the central vascular cylinder, which was supplied with auxin in lanolin (upper *asterisk*), but in a much lower point (due to a moderate rejection by the auxin-supplied central cylinder). (**C**) The induced lateral strand did not connect to the central vascular system (because it was well supplied with high auxin concentration, marked by the upper two *asterisks*), and therefore the artificially induced strand differentiated quite parallel to the pre-existing vascular system. All photomicrographs are at the same magnification, Bar = 1 mm. The figure was prepared from the original micrographs of Tsvi Sachs, which were published in: Sachs (1968) *Annals of Botany* 32: 781–790, with permission of Oxford University Press

3.1.3 Subcellular Mechanisms of Polar Auxin Transport: Cycling and Vesicle Trafficking of PIN1 and AUX1

Polar auxin movement is caused by asymmetric distribution and activity of flux carriers at the plasma membrane. The PIN1 auxin transport component usually functions on the plasma membrane located downward, toward the root tips, acting in auxin efflux out of the cell, whereas the AUXIN-RESISTANCE1 (AUX1) component is located on the opposite plasma membrane, toward the shoot tip, and is active in auxin influx into the cell. Studies of auxin transport inhibitors compared with the vesicle-trafficking inhibitor brefeldin A, which mimics the physiological effects of auxin transport inhibitors, revealed a continuous rapid actin-dependent cycling of PIN1 (at the lower membrane) (Fig. 3.8) and of AUX1 (at the upper membrane) between the plasma membrane and their endosomal components inside the cells. This vesicle cycling is dependent on GNOM (GN) guanine-nucleotide exchange factor for ADP-ribosylation-factor GTPases, which regulates vesicle formation and

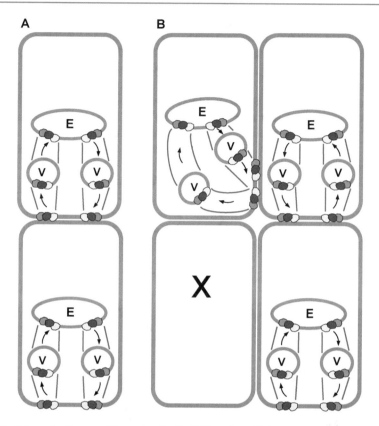

Fig. 3.8 Schematic diagrams illustrating the flexibility of modifying auxin pathway flow, due to the auxin actin-dependent PIN1 vesicle (V) cycling (PIN1 was stained *blue* in a protein complex *stained yellow and red*), which modifies PIN1 location along the plasma membrane. (**A**) PIN1 cycling movement along actin microfilaments between the lower plasma membrane and an endosomal (E) component in intact cells; and (**B**) around a damaged cell (marked by X) showing that the PIN1 is relocated to the lateral membrane (on the *right side*, in the cell above the damaged one), directing the auxin to the adjacent right cell, thus forming a new bypass of polar auxin flow around a damaged cell

controls the intracellular vesicle trafficking, thus regulating the localization of PIN1 and possibly other hormonal transporters (Geldner et al. 2001; Kleine-Vehn et al. 2006; Tal et al. 2016; Verna et al. 2019). Brefeldin A and auxin transport inhibitors stop polar auxin transport by interfering the continuous rapid cycling of vesicle trafficking between the plasma membrane and the endosomes, thus interrupting the continuous cytoplasmic streaming of auxin within the cell.

AUX1 uses a novel trafficking pathway that is distinct from PIN1 trafficking, providing an additional mechanism for the fine regulation of auxin transport. PIN1 streaming and AUX1 flowing dynamics display different sensitivities to auxin trafficking inhibitors. For example, PIN1 cycling is inhibited by the auxin efflux inhibitor 1-N-naphthylphthalamic acid (NPA) (see its effect in Fig. 7.12C, D), whereas

AUX1 is blocked by the auxin influx inhibitor 2,3,5-triiodobenzoic acid (TIBA) (Kleine-Vehn et al. 2006). These two inside cell vesicles cycling mechanisms enable to rapidly concentrate the auxin carriers on the lower (Figs. 3.8A and 3.9A), upper (at the SAM), or lateral (Figs. 3.8B and 3.9B) plasma membrane during the canalization process proposed by Sachs (1981) and confirmed by Sauer et al. (2006). Relocation of the auxin efflux carrier, PIN1, to new lateral sites on the plasma membrane promotes rearrangement of auxin polarity (Fig. 3.8B) resulting in new pathways of polar auxin flow around a wound, which is essential for the process of vascular regeneration (see Sect. 10.1). Likewise, the auxin influx carriers AUX1/LAX that transport auxin into the cell are important for vascular differentiation, as demonstrated by fewer and more spaced vascular bundles in the quadruple mutant plants of the auxin influx carriers *aux1lax1lax2lax3* (Fàbregas et al. 2015). Cytokinin that modulates endocytic trafficking of PIN1 in controlling plant organogenesis (Marhavý et al. 2011) may also be involved in directing PIN1 to the lateral plasma membrane during canalization.

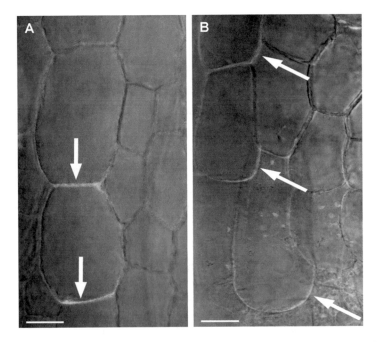

Fig. 3.9 (**A**) Downward PIN1 orientation (*arrows*) located on the lower membrane in cells adjacent to the midvein of intact young *Arabidopsis* rosette-leaf, directing polar auxin transport downward. (**B**) Lateral PIN1 orientation (*arrows*) observed 2 days after wounding of *Arabidopsis* rosette-leaf. PIN1 was detected as PIN1:GFP under GFP fluorescence, viewed with a CLSM. Bar = 10 μm. (J. Dayan and R. Aloni, unpublished)

3.1.4 IAA Transport Pathways

The polar movement of free auxin from leaves to roots occurs via defined transport pathways (Aloni 2004, 2010; Bennett et al. 2016), which are clarified below. Understanding these pathways is important for analyzing plant behavior and development. Most of the studies on auxin transport focused on the molecular mechanisms that control the transport of auxin (Gälweiler et al. 1998; Geldner et al. 2003; Teale et al. 2006; Kleine-Vehn et al. 2011). However, the complex routes of IAA transport (Uggla et al. 1996, 1998; Aloni 2010; Spicer et al. 2013; Bennett et al. 2016) were poorly investigated. Free auxin produced in young leaves moves polarly mainly in the vascular tissues (Sachs 1981; Roberts et al. 1988; Aloni 2001), specifically in the cambium (Uggla et al. 1996; Sundberg et al. 2000; Spicer et al. 2013; Fischer et al. 2019), and through xylem parenchyma (Gälweiler et al. 1998; Palme and Gälweiler 1999; Booker et al. 2003), starch sheath and the root's differentiating vascular tissues and pericycle (Friml and Palme 2002). In addition, there is evidence for rapid non-polar movement of IAA, upward and downward, through sieve tubes at the velocities of ca. 16–20 cm per hour (Eschrich 1968; Morris et al. 1973; Goldsmith et al. 1974). This non-polar auxin flow in the phloem conduits originates in mature leaves (Morris et al. 1973). Furthermore, there are indications that auxin flows in the epidermis (Barker-Bridgers et al. 1998; Friml and Palme 2002; Swarup et al. 2001; Reinhardt et al. 2003; Benková et al. 2003; Scarpella et al. 2006). These fragments of important information have been incorporated into a general concept model for explaining where IAA moves in the plant body (Aloni 2004). Recent experimental data (Figs. 3.10 and 3.11) support the **auxin pathway model** (Fig. 3.12).

All living cells in the plant body are capable of transporting IAA, but only those through which free auxin is canalized become specialized to transport the hormone rapidly, resulting in canalized files of cells (Sachs 1981). During plant development, initial auxin flows are canalized into three main longitudinal routes of IAA transport (Aloni 2004, 2010). These flow patterns can be visualized by different methods showing polar IAA flow along the epidermis-phellogen, in the procambium-cambium and the non-polar transport via the sieve-tubes (Figs. 3.10, 3.11 and 3.12).

These flow patterns start during embryogenesis. The first two polar pathways which originate in young leaves induce and control the vascular and protective tissues, while the third pathway, which is activated later during leaf maturation, controls the activity of the phloem conduits. The auxin transport model describes the following three main longitudinal routes:

1. **The internal route** – is a complex pathway that can be subdivided into the following components (Fig. 3.12): primary shoot (**i**), primary root (**iii**), and secondary body (**ii**) which is produced between the primary parts in woody dicotyledonous and gymnosperm species. Each of these components has its unique anatomy and physiology as follows:

    In the primary shoot (**i** in Fig. 3.12), IAA from young leaves moves polarly downward through the vascular bundles in three distinct streams: (a) via the

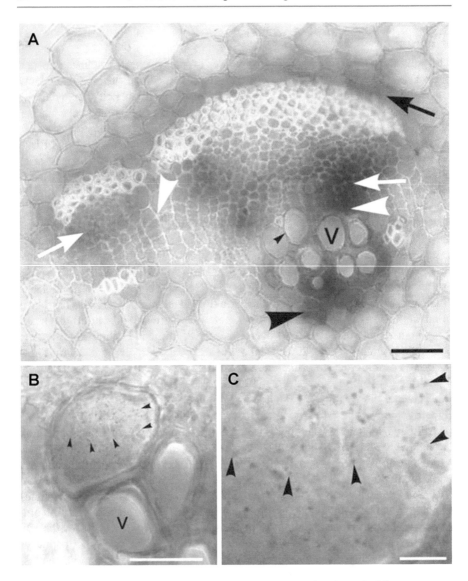

Fig. 3.10 Cross sections in stems of the *GH3* transformant of white clover, *Trifolium repens*, cv. Haifa, with the naturally occurring 749-bp auxin responsive promoter region of the soybean gene *GH3* (Hagen et al. 1991), contained at least three auxin-responsive elements (AuxRE) fused to the *GUS* gene, transformed according to Larkin et al. (1996) and grown as described by Schwalm et al. (2003). The histochemical blue staining for GUS activity in the sections shows auxin pathways in vascular bundles (**A**) and in a differentiating vessel (**B, C**). Auxin transports through two vascular bundles (small bundle, mainly phloem bundle on the left side, and large bundle on the right side) occurs in the cambium (*white arrowheads*), in the sieve tubes (*white arrows*), in a differentiating vessel (*small black arrowhead*), in the bundle sheath (*black arrow*) and in xylem parenchyma cells (*large black arrowhead*). Blue stained vesicles (*arrowheads*) are observed along actin filaments in a differentiating vessel (**B**) which is enlarged in (**C**). v = vessel. Bar = 10 μm in (**C**), 50 μm in **A**, **B**. (Photographed by R Aloni and CI Ullrich; published in Aloni 2013b)

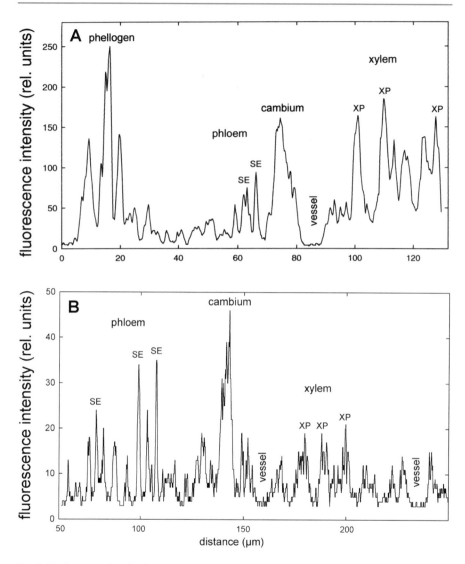

Fig. 3.11 Patterns of auxin distribution in the vascular and protective tissues of *Arabidopsis thaliana* viewed and recorded in CLSM fluorescence intensity profile of Alexa 568-labeled monoclonal-IAA antibodies along two (**A** and **B**) radial straight line in cross sections in two hypocotyls, excited with 568-nm light (emission 600–615 nm) using a 50-mW krypton/argon laser, revealing high total auxin peaks in the phellogen, phloem sieve tubes (SE), cambium, and xylem parenchyma (XP) cells. (Measured by M Langhans, CI Ullrich and R Aloni; **A** - was published in Aloni 2013b)

vascular meristem (*M* in Fig. 3.12), namely, the procambium or early stages of developing cambium, (b) in the surrounding (*B* in Fig. 3.12) bundle sheath at the phloem pole (*black arrow* in Fig. 3.10A), and (c) through parenchyma cells at the (*X* in Fig. 3.12) xylem pole (see in Spicer et al. 2013) (*large black arrowhead*

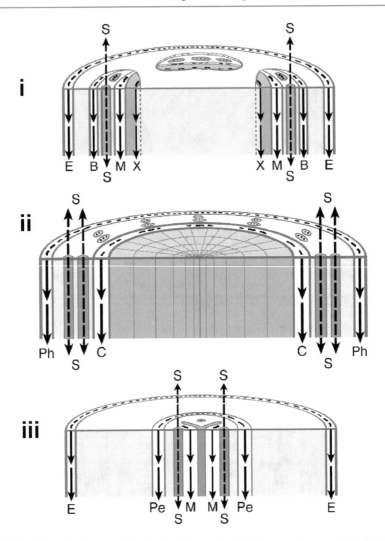

Fig. 3.12 Schematic diagrams showing the three major IAA transport pathways in the primary shoot (**i**), the secondary body (**ii**), and the primary root (**iii**). In the *internal route*, IAA moves polarly through the vascular meristem (M) (which is the procambium in the primary body), and continuously through the cambium (C) in the secondary body, the (X) xylem parenchyma cells, the (B) bundle sheath (in primary shoot); and (Pe) pericycle (in primary root). In the *non-polar route*, IAA moves up and down in the (S) sieve tubes. In the *peripheral route*, most of the IAA moves polarly downward through the (E) epidermis and the phellogen (Ph) (From Aloni 2010)

in Fig. 3.10A). This auxin flow through the xylem parenchyma cells inhibits lateral bud development in *Arabidopsis* (Booker et al. 2003) and possibly also in other species.

In the secondary body (**ii** in Fig. 3.12), the internal polar IAA streams descending from the primary shoot (Fig. 3.12-**i**) merge into one major pathway, which occurs in the cambium (*white arrowheads* in Fig. 3.10A; cambium peak in Fig. 3.11A, B; c in Fig. 3.12-**ii**) and differentiating vessel elements (*small black arrowhead* in Fig. 3.10A and the differentiating vessel with blueish cytoplasm in Fig. 3.10B, C). Additionally, auxin continues to flow also through parenchyma cells of the primary xylem (Spicer et al. 2013) and secondary xylem parenchyma (*XP* in Fig. 3.11A, B).

    In the primary root (**iii** in Fig. 3.12), IAA moves polarly in the vascular cylinder, through the vascular meristem (*M*) and the pericycle (*Pe*).

2. **The peripheral route** – courses through the protective and adjunct tissues. The polar auxin movement in this pathway originates in the subepidermal cells of young leaves and in auxin-producing trichomes and stomata (Aloni et al. 2003) and moving toward the root tips through the epidermis (see Fig. 1a, b, c in Aloni 2013a; phellogen in Fig. 3.11A; epidermis (*E*) in both Fig. 3.12-**i** and Fig. 3.12-**iii**). Through the phellogen (*Ph*) in Fig. 3.12-**ii** and phellogen in Fig. 3.11A, which produces the cork. When the peripheral auxin flow in the epidermal cells is locally interrupted by wounding, this site responds by promoting tangential cell divisions resulting in phellogen formation (in dicots) or by suberin deposition in the exposed cells for local protection (in monocots).

3. **The non-polar route** – which originates in mature leaves courses in the phloem conduits, where auxin moves rapidly (Morris et al. 1973; Goldsmith et al. 1974) up and down via the sieve tubes (*white arrows* in Fig. 3.10; *SE* in Fig. 3.11A, B; *S* in Fig. 3.12). This fast auxin flow is considered as a house-keeping signal that reduces callose levels in the sieve tubes (Aloni et al. 1991; Aloni 1995; Aloni and Peterson 1991). The non-polar IAA flow can also remove the dormancy callose and promotes the resumption of phloem activity in spring (Aloni and Peterson 1991), whereas a non-polar cytokinin flow in the sieve tubes (Kudo et al. 2010) increases callose levels on the sieve plates and can plug the sieve tubes for winter dormancy (Aloni et al. 1990; Aloni 1995 and references therein).

In addition to the above mentioned polar internal and peripheral IAA routes, there are minor connecting auxin pathways between the longitudinal bundles, which could induce phloem anastomoses in young internodes (Figs. 2.17 and 2.18) (Aloni and Peterson 1990; Aloni and Barnett 1996). Bennett et al. (2016) have elegantly demonstrated that in addition to the major auxin efflux carrier PIN1 that efficiently contributes to the high conductance polar auxin transport streams in the longitudinal bundles; in *Arabidopsis*, the following three auxin efflux carriers, PIN3, PIN4, and PIN7, are major contributors of low conductance, less polar auxin transport occurring between the large stem bundles, which was termed **connective auxin transport** (Bennett et al. 2016). This lateral auxin transport between the bundles occurs through small parenchyma cells, which could therefore slow auxin transport, because between the bundles the IAA has to flow through more cell membranes per distance resulting in low conductance, as already proposed by Aloni and Peterson (1990).

    A radial distribution pattern of endogenous IAA was detected across the cambium region in *Pinus sylvestris* showing a peak in the cambial zone where cell division takes place, steeply decreasing toward the mature xylem and phloem. The IAA content was measured, by gas chromatography-mass spectrometry technique, in serial 30-µm-thick longitudinal tangential sections obtained across the cambial region (Uggla et al. 1996). A technical sampling problem might have caused the absence of a clear IAA peak in the phloem. Because the circumference of a tree is rounded, it is possible that the tangential sections of the cambium included also the thin phloem tissue and, therefore, a clear peak of IAA in the phloem could not be detected in the pine trees, as might have been expected from studies of angiosperm species (Morris et al. 1973; Goldsmith et al. 1974). It is also possible that the quantity of IAA in the phloem conduits of *P. sylvestris* is extremely low and therefore could not be detected.

By studying free-auxin patters (by *GH3 and DR5::GUS* expression) in cross sections of *Trifolium* and *Arabidopsis*, Langhans, Ullrich, and Aloni (shown in Aloni 2013b) confirm Uggla et al. (1996) results of the presence of free auxin in the cambium (*white arrowhead* in Fig. 3.10A; *cambium* in Fig. 3.11A, B) and clearly detect also IAA in the sieve tubes (*white arrows* in Fig. 3.10; *SE* in Fig. 3.11A, B). Additionally, we detect IAA in the phellogen (Fig. 3.11A; and in Aloni 2013a, Fig. 2a). Furthermore, by analyzing the fluorescence pattern (Fig. 2c in Aloni 2013a) originating from cells analyzed by auxin antibodies (detecting total auxin of both bound and free auxin) along a radial line (Fig. 2c in Aloni 2013a) in a cross section, we found clear high-auxin concentration peaks in the phellogen, sieve tubes, cambium and secondary xylem parenchyma cells (Fig. 3.11), as was also found in the secondary xylem of poplar trees (Spicer et al. 2013). Thus, by using cross-section analysis of both *DR5::GUS* expression and auxin immunolocalization, we show the anatomy of each specific transporting cell and demonstrate the three IAA pathways in the cambium, sieve tubes, and phellogen (Fig. 3.11A), through which IAA moves continuously. However, it should be noted that during the short process of vessel differentiation (which is induced by IAA) the IAA moves polarly through differentiating vessel elements and can be visualized by *DR5::GUS* expression (Fig. 3.10B, C).

3.1.5 Polar Movement of IAA in a Wave-Like Pattern

Quantitative measurements of auxin concentrations along the cambium of the pine trees *Pinus sylvestris and P. contorta* revealed that the downward polar movement of auxin along the stem occurs in a wave-like pattern. To quantify the native IAA concentrations along the cambial zone, the auxin from the cambial region of the pine *stems was* collected into agar strips (during 10 min of contact with the stem cambial region) from axial series of 6 mm long sections and quantified (by Went *Avena* coleoptile curvature assay, high performance liquid chromatography (HPLC), gas chromatography-mass spectrometry (GC-MS), and the IAA was subsequently quantified by GC-MS-selected ion monitoring (SIM) using an internal standard of [(13)C]-(C(6))-IAA) (Wodzicki et al. 1987). The quantitative results from *P. sylvestris* and *P. contorta* showed a wave-like pattern of oscillating IAA concentrations along the vascular cambium, evidently demonstrating that auxin moved basipetally in a wave-like pattern along the cambium region of the pine *stems* (Wodzicki et al. 1984, 1987). This downward wavy pattern of polar IAA transport can provide morphogenetic information along the plant axis, which can inform the cambium initials and their differentiating phloem and xylem derivatives about their location along a decreasing polar auxin gradient from the auxin producing young leaves downward to the roots (Wodzicki et al. 1984, 1987; Zajaczkowski et al. 1984).

A study of auxin homeostasis and gradients in *Arabidopsis* roots, uncovered a major primary IAA catabolite formed in their tissues, namely, 2-oxindole-3-acetic acid (oxIAA). OxIAA had little biological activity and was formed rapidly and irreversibly in response to an increase in auxin level. The authors propose that oxIAA is an important element in the regulation of **auxin gradients** and, therefore, in the regulation of auxin homeostasis and response mechanisms. (Pěnčík et al. 2013) (see Chap. 13).

3.1.6 Sensitivity to Auxin

A novel putative auxin transport facilitator family, called PIN-LIKES (PILS), was uncovered (Barbez et al. 2012), showing that the PILS proteins are required for auxin-dependent regulation of plant growth by determining the cellular sensitivity to auxin. PILS proteins regulate intracellular auxin accumulation at the endoplasmic reticulum and thus the free IAA levels available for nuclear auxin signaling. The analysis of the PILS proteins suggests that intracellular auxin transport and, therefore, the auxin compartmentalization might be evolutionarily older than the directional cell-to-cell PIN-dependent auxin transport mechanism. The identification of a novel protein family for the regulation of intracellular auxin homeostasis highlights the evolutionary and developmental importance of intracellular auxin transport (Barbez et al. 2012).

During the evolution of temperate deciduous hardwood trees, the ring-porous trees have developed from diffuse-porous trees. The analysis of their developmental biology suggests that the cambium in ring-porous trees has become very sensitive to extremely low IAA stimulation (Aloni 1991, 2001), which will be clarified in Sect. 19.3 on the evolution of ring-porous trees.

3.1.7 Sugar from Photosynthesis Promotes Auxin Synthesis

Sugars from photosynthesis act both as an energy source and as signaling molecules promoting IAA synthesis (Lilley et al. 2012; Sairanen et al. 2012). This regulation of IAA synthesis by the availability of free sugar means that under favorable light conditions more IAA can be produced promoting more growth and increased vascular differentiation in response to improved irradiation in upgraded environmental conditions.

3.2   Gibberellins (GAs) Are the Mature Leaf Signals

Gibberellins are a large family of more than 125 tetracyclic (four-ringed) diterpenoid compounds (Fig. 3.1). Some of them are essential endogenous regulators which promote cell, shoot, and root elongation as well as many other developmental functions in plants such as seed germination, the transition to flowering, and flower development. Most of the GAs are inactive and may serve as precursors of the bioactive gibberellins, namely, GA_1, GA_3, GA_4, and GA_7. Gibberellic acid (GA_3) was first isolated in Japan by Teijiro Yabuta and Yusuke Sumuki in the 1930s as a natural product of the fungus *Gibberella fujikuroi*, which is a pathogen of rice (Taiz et al. 2018).

Gibberellic acid (GA_3) promotes cell elongation and cell width in primary phloem fibers (Sircar and Chakraverty 1960; Atal 1961; Stant 1961,1963). Gibberellins are the mature leaves signals, which stimulate cell divisions in the cambium and induce fiber differentiation along the stem (Hess and Sachs 1972; Aloni 1979, 2001; Dayan et al. 2012) and in roots (Singh et al. 2019). GA is the specific signal that induces phloem fibers (Aloni 1979) and xylem fiber differentiation (Aloni 1979; Saks et al. 1984; Dayan et al. 2012). GA translocation (Binenbaum et al. 2018; Wexler et al. 2018) from producing leaves (Dayan et al. 2012) through grafts between normal and dwarf seedlings of *Zea mays* (Kastumi et al. 1983) and via grafts of *Arabidopsis thaliana* (Ragni et al. 2011) demonstrates that GA is a mobile signal which is graft transmissible. Long-distance movement of GA seems to be mostly restricted to its nonbioactive precursors rather than to bioactive forms (Binenbaum et al. 2018).

Like auxin, the bioactive GAs are weak acids that can diffuse through cell membranes in the acidic apoplast (pH ~ 5.5) and are trapped in the alkaline environment of the cytosol (pH ~ 7.0) as negatively charged GAs. This ion-trap mechanism limits permeability and requires the activity of **GA transporters**. Proteins from the **nitrate transporter 1/peptide transporter family (NPF)** were recently identified and shown to transport GA *in planta* indicating that although GA transport is not polar, it is likely under an active membrane regulation control (Lacombe and Achard 2016; Tal et al. 2016; Park et al. 2017; Binenbaum et al. 2018). The NPF3 transporter is the first GA discovered influx carrier (Tal et al. 2016), which efficiently transports GA across cell membranes. The NPF3 is active at the plasma membrane and subject to rapid BFA-dependent recycling (like the vesicle-trafficking of PIN1 inhibited by brefeldin A, that was already described in Sect. 3.1.3). The NPF3 is expressed in root endodermis, which is likely a preferable GA transport pathway, and is repressed by GA. Abscisic acid (ABA), an antagonist of GA, is also transported by the NPF3 influx carrier (Tal et al. 2016). Also, the GTR1 protein is an influx carrier of gibberellin, as well as of the defense hormone jasmonate, specifically of jasmonoyl-l-isoleucine (JA-Ile) (Saito et al. 2015; Nguyen et al. 2017).

GA transport is not polar in the leaf blade and along the plant axis (Fig. 3.13); therefore, GA induces xylem fiber formation in both the acro- and basipetal directions (Dayan et al. 2012) (see Sect. 5.2.3 and Chap. 11, Fig. 11.3). A few of the GA precursors show long-distance movements. Thus, the precursor of the gibberellin hormone GA_1, namely, GA_{20}, produced in mature leaves of tobacco can flow

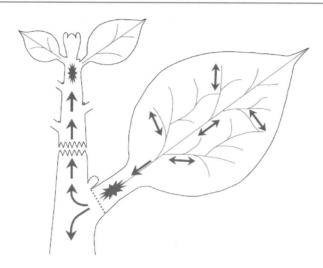

Fig. 3.13 Schematic diagram illustrating the source and translocation of the leaf-derived gibberellin signal. The GA originates in developing and maturing leaves (longer than 3 cm in tobacco (*Nicotiana tabacum*), but not leaves that are emerging at the shoot apex). GA movement is non-polar in the leaf blade and becomes polar only in the lower midvein toward the stem (*arrows* mark flow orientation). The unique anatomy at the base of the petiole (*blue dotted line*) of short cells, potentially retards the flow, which induces a local maximum (*star*), thereby acting as the leaf's elongation driving force. The signal flows in both directions along the stem; its upward movement from developing leaves reaches the young internodes (*star*) and induces stem elongation at the shoot apex. Throughout the flow along the stem, the signal results in bioactive GA signaling that controls cambial activity and fiber differentiation. (From Dayan et al. 2012)

non-polarly specifically via the phloem, from the mature leaves to sink organs. When the mature-leaf-induced GA_{20} precursor arrives to the cambium, it is converted, by local cambial activity of the GA 20-oxidase, to the bioactive gibberellin form (GA_1), which forms local GA maximum that activates the cambium. Therefore, the removal of mature leaves substantially depletes the endogenous GA concentrations in the stem, which impairs cambial activity, fiber differentiation, and shoots elongation (Dayan et al. 2012).

Additionally, an early GA precursor, the GA_{12} is a major mobile GA signal over long distances. It was detected in both the xylem and phloem exudates indicating that GA_{12} moves through the vascular tissues. The GA_{12} is functional in recipient tissues, supporting growth via the activation of the GA signaling cascade (Regnault et al. 2015). GA_{12} shows a significantly higher ability to freely penetrate into cells than any other of the tested GA forms. Therefore, it was speculated that the high membrane permeability of GA_{12} is a characteristic that allows it to serve as a long-distance GA mobile signal (Binenbaum et al. 2018).

The bioactive gibberellins (GA_1 and GA_4) were predominantly found in the expansion zone of differentiating xylem cells in *Populus* suggesting their main role in early stages of wood formation, including cell elongation (Israelsson et al. 2005). Stems of transgenic plants with elevated GA concentrations grow rapidly, produce longer fibers (Eriksson et al. 2000; Biemelt et al. 2004; Dayan et al. 2010) and enhanced wood production (Dayan et al. 2010). Light and GAs promoted lignin deposition and increased xylem fiber differentiation in tobacco stems (Falcioni et al. 2018); however, it is possible that the promoting irradiation effect resulted from an endogenous increase in auxin concentration, known to be stimulated by free sugars from photosynthesis (Lilley et al. 2012; Sairanen et al. 2012).

Gibberellins regulate the transition from juvenile to adult phases; thus, in conifers, exogenous GA application can induce the reproductive phase and cone production in young trees, which is practically used to obtain seeds from selected trees (Bockstette and Thomas 2019). Conversely, in woody angiosperms including many fruit trees, gibberellins promote vegetative growth by the inhibition of flowering (Goldschmidt and Samach 2004), and when bioactive GAs are applied to mature woody angiosperms, they may induce rejuvenation (Taiz and Zeiger 2006) (see Chap. 12).

3.3   Cytokinins (CKs) Are the Root Cap Signals

Cytokinins are adenine derivatives (Fig. 3.1) and the most common CK is zeatin. CKs were discovered as factors necessary for cell division in tissue culture (Skoog and Miller 1957). They are produced in the root tip, specifically in the root cap (Figs. 2.2 and 3.14A), and are major hormonal signals of the root, perceived by membrane-localized histidine-kinase receptors, and are transduced through a His-Asp phosphorelay signaling to activate a family of transcription factors in the nucleus (Aloni et al. 2004, 2005, 2006a; Miyawaki et al. 2004; Ursache et al. 2013). The CKs exist as free and bound molecules. The free forms induce cell divisions (cytokinesis) in the stem cells of meristematic tissues and promote fibers, sieve-tube, and vessel differentiation and regeneration along the plant (Aloni 1982; Saks et al. 1984; Aloni et al. 1990; Baum et al. 1991). Cytokinin signaling positively regulates cambial activity (Aloni et al. 1990; Baum et al. 1991; Nieminen et al. 2008; Matsumoto-Kitano et al. 2008; Ursache et al. 2013). CKs increase the sensitivity of the cambium to the auxin signal (Baum et al. 1991), which has promoted the development of ring-porous wood during the evolution of temperate deciduous hardwood trees (Aloni 1991, 2001) and will be discussed in Sect. 19.3.

The suggestion of Letham (1994) that the main site of CK synthesis is the root tip, specifically the root cap cells, was confirmed by Aloni et al. (2004, 2005) (see Chap. 2, Fig. 2.2B), showing that in seedlings as well as in plants grown under conditions of almost no transpiration, the highest concentration of free CK occurs in the root cap statocytes (Figs. 2.2B and 3.14A). From the root cap, the CK is transported upward through plasmodesmata, which provide symplastic continuity (Kim and Zambryski 2005) in the meristematic and elongation zones, and from the differentiation zone through the xylem (Aloni et al. 2005).

The upward CK transport in the shoot is regulated by the transpiration stream, where cytokinin moves mainly to developing shoot organs with high transpiration rates (Aloni et al. 2005). In ARR5::GUS transformants of *Arabidopsis* completely protected from any air movement grown under almost 100% humidity, the strongest *ARR5::GUS* expression was found in the root cap statocytes and spreading upward

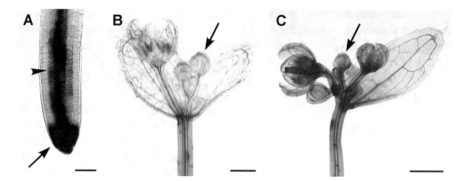

Fig. 3.14 Free-cytokinin distribution in a root (**A**), in young flowers protected from the wind (**B**), or exposed to gentle wind (**C**), as detected by free bioactive-CK-dependent *GUS* expression in the CK-responsive *ARR5::GUS* transformant of 5-week-old *Arabidopsis* grown under long-day conditions (described in Aloni et al. 2005). (**A**) Root of wind-protected plant with strong *ARR5::GUS* expression reflecting high bioactive CK concentration in the cap (*arrow*), considerable concentration in the vascular cylinder (*arrowhead*) and lower concentration in the cortex. (**B**) Wind-protected inflorescence (by a tight translucent plastic bag, under 95–100% humidity) showing young apical flower buds (*arrow*) without *GUS* expression. (**C**) Plant from the same experiment as in (**A**) and (**B**), but exposed to gentle wind (of 0.2–0.7 m s^{-1} under lower humidity of 60–70%) only for the last 3 h before harvesting for the GUS assay, resulting in very strong *ARR5::GUS* expression in the young apical flower buds (*arrow*). Scale bars = 40 μm (**A**), 500 μm (**B**, **C**). (From Aloni et al. 2006a)

in the vascular cylinder (Fig. 3.14A). This pattern in roots was congruent with that found by CK immunolocalization. In plants exposed to gentle wind at lower humidity for a short period (3 hours before harvesting), the *ARR5::GUS* expression was considerably increased in the shoots (Fig. 3.14C). These results demonstrate that the bulk of the CK is synthesized in the root cap, exported upward to the shoot through vessels in the xylem, and accumulates at sites of highest transpiration (Aloni et al. 2005).

Commonly, CKs are produced in the root cap and transported upward from root to shoot in the xylem, via vessels and tracheids (Aloni et al. 2005; Sakakibara et al. 2006). CK can also be biosynthesized in the shoot, at the nodes rather than in the roots, usually after shoot decapitation (Tanaka et al. 2006) (see Sect. 6.2).

Additionally, cells in the root can regulate the movement of CK upward (Ko et al. 2014; Zhang et al. 2014b). These studies show that a member of the ATP-binding cassette (ABC) family, the G-type ABC transporter, namely, ABCG14, is essential for the long-distance transport of *trans*-zeatin-type CKs from the roots to the shoots. The *ABCG14* is expressed in the pericycle and stellar cells of the root's differentiation zone. At the subcellular level, the ABCG14 is localized in the plasma membrane and likely functions as an efflux transporter (Ko et al. 2014; Zhang et al. 2014b; Durán-Medina et al. 2017).

Shoot-synthesized cytokinins can move downward via the phloem. This long-distance CK transport was suggested to regulate vascular pattern in the root meristem (Bishopp et al. 2011) and controls nodulation in legumes roots (Sasaki et al. 2014).

CKs have opposing roles in shoots and roots; in young shoot organs, the CKs positively regulate development and promote shoot growth (Rahayu et al. 2005; El-Showk et al. 2013), but in roots they are negative regulators of development and growth of lateral roots (Werner et al. 2003; Márquez et al. 2019) (see Chap. 9).

NO_3^-, the nitrate (but not NH_4^+) supply to nitrogen-depleted roots, causes a rapid up-regulation of isopentenyl transferase *(IPT)* genes resulting in an increase of cytokinins content in the root, which is transported via the xylem upward into the shoot (Geßler et al. 2004; Miyawaki et al. 2004; Sakakibara et al. 2006; Ruffel et al. 2011) (see Chap. 4). Under conditions of insufficient CK supply from the root, nitrate-responsive *IPT3* is expressed in the phloem (Takei et al. 2004; Miyawaki et al. 2004) and produces the CK in the shoot, which regulates cytokinesis in the shoot's cambium.

3.4   Ethylene (C_2H_4) the Gaseous Signal

The gas ethylene (Fig. 3.1) is a plant hormone which is synthesized locally in various plant tissues in response to stress. Wounding, flooding, wind, bending, high auxin levels, elevated cytokinin concentrations, and methyl jasmonate promote ethylene synthesis in plants. Elevated C_2H_4 concentrations inhibit stem elongation and may promote leaf and fruit abscission. Ethylene stimulates defense responses to injury or disease and reduces vessel width (Aloni et al. 1998; Hudgins and Franceschi 2004; Taiz et al. 2018).

Ethylene stimulates tracheary element (TE) differentiation in *Zinnia elegans* cell culture (Pesquet and Tuominen 2011). The C_2H_4 peaks at the time of TE maturation correlating with the activity of the ethylene biosynthetic 1-aminocyclopropane-1-carboxylic acid (ACC) oxidase, and the maturing *Zinnia* TEs accumulate ethylene (Pesquet and Tuominen 2011). Blocking ethylene signaling by using silver thiosulfate (STS) appears to block TE maturation (Bollhöner et al. 2012).

In wood, the ethylene synthesized in differentiating tracheary elements of the sapwood (the active water transporting secondary xylem) of trees (Eklund 1990; Ingemarsson et al. 1991) diffuses in the centrifugal direction and this radial ethylene flow through the cambium initials induces the **vascular ray** (by promoting cell divisions in the cambial fusiform initials) and the enlargement of existing rays (Lev-Yadun and Aloni 1995; Aloni et al. 2000; Aloni 2013a) (see Chap. 15). In poplar trees, ethylene and its ACC synthase promote cambial cell division and wood formation (Love et al. 2009) and are required for tension wood function (Seyfferth et al. 2019).

Ethylene serves as a plant sensor to analyze flooding. When the continuous regular centrifugal transport of ethylene outward to the plant environment is blocked by flooding water, the ethylene (which almost do not dissolve in water) accumulates in the cortex or the bark. This accumulation boosts local C_2H_4 concentrations, which may induce **aerenchyma** (a spongy tissue with large air spaces between the cells allowing gases circulation) (Li et al. 2006), which enables aeration of flooded stems and roots (Fig. 2.19). High ethylene concentrations also promote lateral and adventitious root formation (Aloni et al. 2006a; Aloni 2013b) by local interruptions of the polar auxin flow, causing local sites of high IAA concentrations above the interruption site, which induce lateral and adventitious root tips (Aloni 2013b) (see Chap. 9).

In conifer trees, the ethylene hormone promotes chemical defenses against insects and pathogens. Wounding and ethylene can mediate traumatic resin duct formation in conifer woods (Hudgins and Franceschi 2004; Hudgins et al. 2006) induced by jasmonates (Schmidt et al. 2011). The resin duct epithelial cells produce oleoresin terpenoids, which protect the tree from insects and their associated pathogens (Keeling and Bohlmann 2006; Ralph et al. 2007; Schmidt et al. 2011; Aloni 2013a; Foster et al. 2016) (see Chap. 17).

3.5   Abscisic Acid (ABA) Regulates Stress Responses

The abscisic acid is an isoprenoid hormone (Fig. 3.15), which is the long-distance stress signal produced in the meristematic cells of the root tip (Koiwai et al. 2004) in their meristematic endodermal cells (Duan et al. 2013; Bloch et al. 2019) when the soil is drying and in response to salt stress. ABA is transported upward through the xylem from roots to shoot to regulate the closure of stomata under stress and retarding vascular meristematic activity before dormancy (Taiz et al. 2018). However, ABA can also be produced in the phloem of the shoot and even in the stomata themselves (Koiwai et al. 2004). The flow rate of water through tracheids and vessels is crucially affected by stomata opening and closing; ABA closes the stomata under water stress conditions.

ABA, which is the universal stress hormone of higher plants, has a central role in plant developmental plasticity; it is involved in slowing down and stopping cell division in the cambium and wood differentiation in trees toward their winter dormancy by retarding shoot development and ending their cambium activity. In intact *Eucommia ulmoides* trees, ABA retards cambial activity and promotes dormancy (Hou et al. 2006). ABA initiates cambial dormancy by promoting the blockage of the cambial plasmodesmata (Tylewicz et al. 2018) (see Sect. 11.2). In addition, ABA plays a crucial role in promoting plant tolerance to cold (Galiba et al. 2009), which may be used to improve plant sustainable survival (Xue-Xuan et al. 2010).

Conversely, cytokinin from the root caps which is also transported upward via the xylem (Aloni et al. 2005) has a positive effect on stomata opening (Dodd 2003), on cambium activity (Nieminen et al. 2008; Matsumoto-Kitano et al. 2008; Bhalerao and Fischer 2017; Fischer et al. 2019) and shoot development (Aloni 2013a; Taiz et al. 2018).

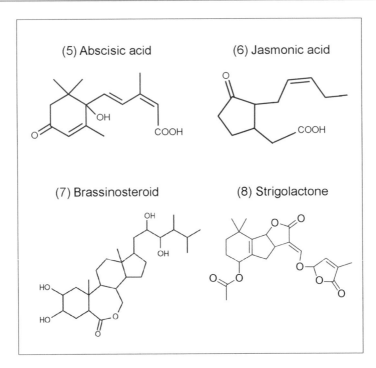

Fig. 3.15 Molecular structure of hormonal signals that regulate vascular differentiation in plants: abscisic acid, jasmonic acid, brassinosteroid, and strigolactone

The ABA long-distance transport from the root synthesizing cells to target tissues is regulated by plasma membrane activity. Several ABA transporters from the ATP-binding cassette (ABC) proteins and the nitrate transporter 1/peptide transporter (NPF) protein family function in the long-distance ABA transport (Boursiac et al. 2013; Lacombe and Achard 2016). Transporter activity studies have shown that some, like ABCG25 (Kuromori et al. 2010), acts as ABA efflux transporters, while others, like NPF3 (Tal et al. 2016), are influx carriers (Kuromori et al. 2011; Boursiac et al. 2013; Lacombe and Achard 2016).

ABA signaling is required for proper xylem development in *Arabidopsis* roots. On the other hand, upon external ABA application, or under limited water availability, extra protoxylem (PX) vessels were induced in *Arabidopsis* roots (Ramachandran et al. 2018). These results were confirmed in *Arabidopsis* by Bloch et al. (2019) also showing the ABA effect in tomato roots, indicating that the ABA-enhanced PX vessel differentiation by stress is an evolutionarily conserved mechanism. A similar effect was evident in stressed stems of corn (*Zea maize*), which probably indicates that the extra PX vessels in stressed corn stem (see Chap. 16, Fig. 16.2B, D) are induced by a mechanism involving ABA signaling.

Patterning of primary vessels in the root is regulated by auxin-cytokinin signaling and endodermally produced microRNA *miR165a/166b*-mediated suppression of genes encoding class III homeodomain-leucine zipper (HD-ZIPIII) transcription factors in the endodermis and stele periphery. The resulting differential distribution of target mRNA in the vascular cylinder determines xylem cell types in a dosage-dependent manner (Carlsbecker et al. 2010). In *Arabidopsis*, osmotic stress via ABA signaling in meristematic endodermal cells induces increased differentiation of protoxylem vessels (PX) in association with increased *VASCULAR-RELATED NAC-DOMAIN7 (VND7)* expression (Bloch et al. 2019). The transcription factor VND7 is known to induce PX vessels (Kubo et al. 2005) (see Sect. 5.2.4). Thus, ABA regulates increased PX vessels via increased *miR165a/166b*-regulated expression, resulting in reduced levels of the HD-ZIPIII RNAs, in association with increased VND7 levels, consequently promoting more protoxylem vessels (Bloch et al. 2019).

3.6   Jasmonates: Activate Defense Responses

The phytohormone jasmonic acid (JA) and its volatile methyl ester (MeJA) are fatty acid derived cyclopentanones (Fig. 3.15). A positive effect of jasmonate application on cambium activity indicated a stimulatory role of JA in secondary growth, suggesting that JA signaling can promote cambial cell divisions (Sehr et al. 2010) possibly accelerating vascular recovery after injury.

Jasmonates activate defense-related genes in dicotyledonous and monocotyledonous plants against insects and pathogens (Howe 2004; Gregg and Howe 2004; Sun et al. 2011; Okada et al. 2015; Schulze et al. 2019; Wang et al. 2019). Systemic wounding responses are induced by an electrical signal derived from a damaged leaf, followed by rapid signaling of JA and jasmonoyl-l-isoleucine (JA-Ile), which are translocated from wounded to undamaged leaves. GTR1/NPF2.10 is a transporter of JA-lle and possibly also of JA, which is involved in transferring the hormonal signal also to undamaged leaves (Saito et al. 2015; Ishimaru et al. 2017; Nguyen et al. 2017). The GTR1 is an influx carrier of JA-Ile as well as of gibberellin (Saito et al. 2015), while JAT1 (AtABCG16/JAT1) is an efflux carrier of JA (Nguyen et al. 2017). These transporters at the plasma membrane are actively involved in the early wounding response, preparing the neighboring intact leaves to a possible future attack and damage. In *Arabidopsis*, wounding induces systemic jasmonate elevations within less than 1.6 min. The JA and its active isoleucine conjugate (JA-Ile) are involved in the fast systemic wound response mainly through the vascular system, which enables long-distance signaling (Heyer et al. 2018). Recent grafting experiments and hormone profiling uncovered an additional long-distance mobile hormone precursor, cis-12-oxo-phytodienoic acid (OPDA) and its derivatives (but not the bioactive JA-Ile conjugate) that translocate through the phloem from wounded shoots into undamaged *Arabidopsis* roots. Upon root relocation, the mobile precursors cooperatively regulated JA responses through their conversion into JA-Ile and JA signaling activation (Schulze et al. 2019). Jasmonate responses decay gradually with increasing plant age, and signaling components vary between different tissues (Jin and Zhu 2017 and references therein).

Ultraviolet radiation exposure time and intensity increase tomato resistance against herbivorous arthropods through activation of the JA signaling (Escobar-Bravo et al. 2019). The MeJA moves in both the phloem and xylem pathways. The MeJA enters into the phloem and moves in the sieve tube sap. MeJA promotes its own transport; whole plant experiments suggest that enhanced transport of both MeJA and sugar may be due to MeJA enhancing the energy of the plasma membrane (Thorpe et al. 2007).

In the pine family (Pinaceae), application of MeJA can induce **traumatic resin duct** formation (Hudgins et al. 2003; Hudgins and Franceschi 2004; Huber et al. 2005). This MeJA-induced defense response is mediated by ethylene, which promotes the formation of traumatic resin-secreting epithelial cells (Hudgins and Franceschi 2004). In *Picea abies*, treatment with MeJA induces the expression of isoprenyl diphosphate synthase genes during the formation of traumatic resin ducts, promoting the terpenoid-based oleoresin accumulation in the ducts, which protect against herbivores and pathogens. The trunk response to the MeJA treatment was detected up to 60 cm above the site of application (Schmidt et al. 2011).

The ability of jasmonates to activate defense-related genes against pathogens can be practically applied in both horticulture and medicine. Postharvest application of MeJA to various cultivars of cut greenhouse grown rose flowers (*Rosa hybrid*) before sending the cut flowers to markets induces plant defense responses that significantly reduced lesion size and appearance of the known "grey mould" or "gray mold," caused by the infection of the necrotrophic fungus *Botrytis cinerea*, on the petals (Meir et al. 1998). The effective jasmonate treatment neither caused any phytotoxicity on the leaves and petals nor impaired flower quality and longevity, suggesting that MeJA application provides systemic protection against *Botrytis* rot by inducing natural resistance mechanisms in the treated cut roses without impairing flower quality (Meir et al. 1998). Similar defense responses were efficiently induced by application of MeJA to sweet cherry fruit, which significantly reduced the blue mold decay caused by *Penicillium expansum*. The fruit treated with MeJA had significantly lower disease incidence and smaller lesion diameter than the control fruit did (Wang et al. 2015).

Jasmonic acid was detected in early stages of crown gall tumor formation (Veselov et al. 2003; Aloni and Ullrich 2008; Ullrich et al. 2019). Jasmonates induced swelling of mitochondria isolated from human cancer cells, but not from healthy cells. This selectivity of jasmonates against human cancer cells may be based at mitochondrial differences between cancer and normal cells (Goldin et al. 2007), making jasmonates candidates for human cancer therapy (see Chap. 21).

3.7   Brassinosteroids (BRs)

Brassinosteroids are a group of naturally occurring plant polyhydroxysteroids (Fig. 3.15) with wide-range of biological activities that interact with other hormones including auxin, gibberellins, and ethylene (Zhang et al. 2009; Clouse 2011). Nanomolar levels of BRs stimulate tracheary element formation in isolated mesophyll cells of *Zinnia elegans* (Iwasaki and Shibaoka 1991) and regulate expression of genes associated with xylem formation (Fukuda 1997, 2004). BRs levels increased dramatically prior to the morphogenesis of tracheary elements in cultured *Zinnia* cells, showing that BRs are necessary for the initiation of the final stage of tracheary element differentiation (Yamamoto et al. 1997, 2001). BRs from the epidermis might influence the differentiation of vascular tissues (Savaldi-Goldstein et al. 2007) and possibly control tissue-type specificity of vascular cell proliferation (Carlsbecker and Helariutta 2005). BRs play a foundational role in the regulation of secondary growth and wood formation in *Populus*, through the regulation of cell differentiation and secondary cell wall biosynthesis. Elevated BR levels resulted in increases in secondary growth and tension wood formation, while inhibition of BR synthesis resulted in decreased growth and secondary vascular differentiation (Du et al. 2020). Recent studies have indicated that BRs can improve crops yield and quality especially under stress conditions (Khripach et al. 2000; Liu et al. 2017b).

3.8   Strigolactones (SLs) Are Root Hormones and Attracting Signals of Symbiotic Fungi and Root-Parasitic Weeds

Strigolactones, a group of terpenoid lactones (Fig. 3.15) derived from carotenoids (Al-Babili and Bouwmeester 2015), are plant hormones involved in the repression of shoot branching (Klee 2008; Gomez-Roldan et al. 2008; Umehara et al. 2008; Domagalska and Leyser 2011). The SLs are produced in the root and move upward via the xylem (Kohlen et al. 2011) to the stem, where they inhibit lateral bud development. Strigolactones positively regulate cambial activity and vascular differentiation (Agusti et al. 2011). The SL signaling in the vascular cambium itself is sufficient for cambium stimulation, and it interacts with the auxin signaling pathway (Agusti et al. 2011). SLs also regulate root architecture, root-hair length, and density (Kapulnik and Koltai 2016).

The strigolactones were known as root signals that stimulate the germination of root-parasitic weeds that pose a serious threat to agricultural crops (Cook et al. 1972). These SLs were recently found to attract arbuscular mycorrhizal fungi, which are plant-fungus symbionts that facilitating the uptake of soil nutrients (Akiyama et al. 2005). These fungi form the most widespread type of mycorrhiza, in which the fungus penetrates into the cortical cells of the root. Strigolactone exudation from roots involves a specific efflux carrier. In *Petunia hybrida*, the ATP-binding cassette (ABC) transporter PLEIOTROPIC DRUG RESISTANCE 1 (PDR1) has a key role in regulating the development of arbuscular mycorrhizae and axillary bud development. Its close homologue from *Nicotiana tabacum* is PDR6. The PDR1 transporter is located on the plasma membrane and functions as a cellular strigolactone exporter. The *P. hybrida* pdr1 mutants are defective in strigolactone exudation from their roots, resulting in reduced symbiotic interactions. While above ground, pdr1 mutants have an enhanced branching phenotype, indicating their impaired strigolactone allocation (Kretzschmar et al. 2012; Borghi et al. 2016).

From evolutionary point of view, Aloni (2013a) suggested that plant roots started to release strigolactones to their environment as a primary signal for attracting the endomycorrhizal symbiotic fungi, to improve their soil nutrient uptake of phosphorus and nitrogen. Much later during evolution, the root-parasitic plants have emerged, eventually adapting themselves to use this ecological root signal to stimulate their germination, thus ensuring the presence of a nearby host root, for their penetration and nutrient supply (Aloni 2013a; Borghi et al. 2016; Lumba et al. 2017) (see Sect. 20.1). Strigolactone concentration increases as a response to inorganic phosphate (PO_4^-) deficiency (López-Ráez et al. 2008; Kohlen et al. 2011), which promotes main stem elongation on the expenses of branch development, thus giving the plant an advantage in its competition for light against neighboring plants (see Chap. 4).

3.9   Florigen Is the Universal Trigger for Flowering and Associated Fiber Differentiation

The protein hormone florigen is a universal systemic inducer of flowering that also functions as a generic growth terminator across flowering plants. In *Arabidopsis* leaves, the florigen is encoded by the *FLOWERING LOCUS T* (*FT*) gene (Huang et al. 2005; Lifschitz et al. 2006; Lin et al. 2007) and acts as the long-distance flowering inducing signal from the leaves to shoot tips (see Sect. 8.1). Florigen synchronizes the shoot system toward flowering; it accelerates secondary cell wall biogenesis (SCWB), and hence vascular differentiation, independently of flowering. By accelerating SCWB mainly in differentiating xylem fibers, the florigen reprograms the distribution of resources, signals, and mechanical loads required for carrying the increasing weight of developing fruit (Shalit-Kaneh et al. 2019). Gibberellin (GA) can induce and promote flowering in many species of long-day plants (which flower during early summer as the days begin to get longer) that grow as a rosette under short-day conditions (Zeevaart 1983, 2006). GA is also the specific signal that induces fiber differentiation (Aloni 1979; Dayan et al. 2012). Therefore, it is likely that gibberellin mediates the role of florigen in promoting fiber differentiation in the upper internodes, below the developing fruits, which prepares the stem to carry fruit loads. By synchronizing mutually independent flowering and secondary cell wall biogenesis, the systemic florigen functions as a coordinator and great communicator of the reproductive phase (Shalit-Kaneh et al. 2019).

3.10   Hormonal Cross Talk and Interactions

It has long been known that hormones influence each other biosynthesis and activity (Taiz et al. 2018). Effects induced by one hormone may be mediated by another. For example, the influence of jasmonate on traumatic resin duct formation is mediated by ethylene (Hudgins and Franceschi 2004). Auxin interactions with other hormones are common. The auxin-cytokinin cross talk regulates various aspects of cell type specification, vascular differentiation, and shoot-root developmental relationships (Sachs and Thimann 1967; Chandler and Werr 2015); so are the combined interactions between auxin, cytokinin and ethylene (Aloni et al. 1998; El-Showk et al. 2013; Liu et al. 2017a), strigolactone and auxin (Koltai 2015), ethylene and cytokinin (Zdarska et al. 2015), and many more.

References and Recommended Readings[1]

Adamowski M, Friml J (2015) PIN-dependent auxin transport: action, regulation, and evolution. Plant Cell 27: 20–32.

*Agustí J, Blázquez MA (2020) Plant vascular development: mechanisms and environmental regulation. Cell Mol Life Sci 77: 3711–3728.

Agusti J, Herold S, Schwarz M, Sanchez P, Ljung K, Dun EA, Brewer PB, Beveridge CA, Sieberer T, Sehr EM, Greb T (2011) Strigolactone signaling is required for auxin-dependent stimulation of secondary growth in plants. Proc Natl Acad Sci USA 108: 20242–20247.

Akiyama, K., Matsuzaki K, Hayashi H (2005) Plant sesquiterpenes induce hyphal branching in arbuscular mycorrhizal fungi. Nature 435: 824–827.

*Al-Babili S, Bouwmeester HJ (2015) Strigolactones, a novel carotenoid-derived plant hormone. Annu Rev Plant Biol 66: 161–186.

Aloni R (1979) Role of auxin and gibberellin in differentiation of primary phloem fibers. Plant Physiol 63: 609–614.

Aloni R (1982) Role of cytokinin in differentiation of secondary xylem fibers. Plant Physiol 70: 1631–1633.

Aloni R (1987) Differentiation of vascular tissues. Annu Rev Plant Physiol 38: 179–204.

Aloni R (1991) Wood formation in deciduous hardwood trees. In: *Physiology of Trees*, AS Raghavendra (ed). John Wiley & Sons, New York, London, Sydney, pp. 175–197.

Aloni R (1995) The induction of vascular tissues by auxin and cytokinin. In: *Plant Hormones: Physiology, Biochemistry and Molecular biology*, PJ Davies (ed), Kluwer, Dordrecht, pp. 531–546.

Aloni R (2001) Foliar and axial aspects of vascular differentiation - hypotheses and evidence. J Plant Growth Regul 20: 22–34.

Aloni R (2004) The induction of vascular tissue by auxin. In: *Plant Hormones: Biosynthesis, Signal Transduction, Action!* PJ Davies (ed). Kluwer Academic Publishers, Dordrecht, Boston, London, pp. 471–492.

Aloni R (2007) Phytohormonal mechanisms that control wood quality formation in young and mature trees. In: *The Compromised Wood Workshop 2007*, K Entwistle, P Harris, J Walker (eds). The Wood Technology Research Centre, University of Canterbury, Christchurch, New Zealand, pp. 1–22.

*Aloni R (2010) The induction of vascular tissues by auxin. In: *Plant Hormones: Biosynthesis, Signal Transduction, Action!* PJ Davies (ed). Kluwer Academic Publishers, Dordrecht, pp. 485–506.

Aloni R (2013a) The role of hormones in controlling vascular differentiation. In: *Cellular Aspects of Wood Formation*, J Fromm (ed). Springer-Verlag, Berlin, pp. 99–139.

Aloni R (2013b) Role of hormones in controlling vascular differentiation and the mechanism of lateral root initiation. Planta 238: 819–830.

*Aloni R (2015) Ecophysiological implications of vascular differentiation and plant evolution. Trees 29: 1–16.

---

[1] Papers of particular interest for suggested reading have been highlighted (with *)

*Aloni, R, Aloni E, Langhans M, Ullrich CI (2006a) Role of cytokinin and auxin in shaping root architecture: regulating vascular differentiation, lateral root initiation, root apical dominance and root gravitropism. Ann Bot 97: 883–893.

Aloni R, Aloni E, Langhans M, Ullrich CI (2006b) Role of auxin in regulating *Arabidopsis* flower development. Planta 223: 315–328.

Aloni R, Barnett JR (1996) The development of phloem anastomoses between vascular bundles and their role in xylem regeneration after wounding in *Cucurbita* and *Dahlia*. Planta 198: 595–603.

Aloni R, Baum SF (1991) Naturally occurring regenerative differentiation of xylem around adventitious roots of *Luffa cylindrica* seedlings. Ann Bot 67: 379–382.

Aloni R, Baum SF, Peterson CA (1990) The role of cytokinin in sieve tube regeneration and callose production in wounded *Coleus* internodes. Plant Physiol 93: 982–989.

Aloni R, Feigenbaum P, Kalev N, Rozovsky S (2000) Hormonal control of vascular differentiation in plants: the physiological basis of cambium ontogeny and xylem evolution. In: *Cell and Molecular Biology of Wood Formation*, RA Savidge, JR Barnett, R Napier (eds). BIOS Scientific Publishers, Oxford, pp. 223–236.

Aloni R, Foster A, Mattsson J (2013) Transfusion tracheids in the conifer leaves of *Thuja plicata* (Cupressaceae) are derived from parenchyma and their differentiation is induced by auxin. Am J Bot 100: 1949–1956.

Aloni R, Langhans M, Aloni E, Dreieicher E, Ullrich CI (2005) Root-synthesized cytokinin in *Arabidopsis* is distributed in the shoot by the transpiration stream. J Exp Bot 56: 1535–1544.

Aloni R, Langhans M, Aloni E, Ullrich CI (2004) Role of cytokinin in the regulation of root gravitropism. Planta 220: 177–182.

Aloni R, Peterson CA (1990) The functional significance of phloem anastomoses in stems of *Dahlia pinnata* Cav. Planta 182: 583–590.

Aloni R, Peterson CA (1991) Seasonal changes in callose levels and fluorescein translocation in the phloem of *Vitis vinifera* L. IAWA Bull ns 12: 223–234.

Aloni R, Plotkin T (1985) Wound induced and naturally occurring regenerative differentiation of xylem in *Zea mays* L. Planta 163: 126–132.

Aloni R, Raviv A, Peterson CA (1991) The role of auxin in the removal of dormancy callose and resumption of phloem activity in *Vitis vinifera*. Can J Bot 69: 1825–1832.

*Aloni R, Schwalm K, Langhans M, Ullrich CI (2003) Gradual shifts in sites of free-auxin production during leaf-primordium development and their role in vascular differentiation and leaf morphogenesis in *Arabidopsis*. Planta 216: 841–853.

Aloni R, Ullrich CI (2008) Biology of crown gall tumors. In: *Agrobacterium, From Biology to Biotechnology*, Tzfira T, Citovsky VH (eds). Springer, New York, pp. 565–591.

Aloni R, Wolf A, Feigenbaum P, Avni A, Klee HJ (1998) The *Never ripe* mutant provides evidence that tumor-induced ethylene controls the morphogenesis of *Agrobacterium tumefaciens*-induced crown galls on tomato stems. Plant Physiol 117: 841–847.

Atal CK (1961) Effect of gibberellin on the fibers of hemp. Econ Bot 15: 133–139.

Bhalerao RP, Fischer U (2017) Environmental and hormonal control of cambial stem cell dynamics. J Exp Bot 68: 79–87.

Barbez E, Kubeš M, Rolčík J, Béziat C, Pěnčík A, Wang B, Rosquete MR, Zhu J, Dobrev PI, Lee Y, Zažímalovà E, Petrášek J, Geisler M, Friml J, Kleine-Vehn J (2012) A novel putative auxin carrier family regulates intracellular auxin homeostasis in plants. Nature 485: 119–122.

Barker-Bridgers M, Ribnicky DM, Cohen JD, Jones AM (1998) Red-light regulated growth. II. Changes in the abundance of indoleacetic acid in the maize mesocotyl. Planta 204: 207–211.

Baum SF, Aloni R, Peterson CA (1991) The role of cytokinin in vessel regeneration in wounded *Coleus* internodes. Ann Bot 67: 543–548.

*Baylis T, Cierlik I, Sundberg E, Mattsson J (2013) *SHORT INTERNODES/STYLISH* genes, regulators of auxin biosynthesis, are involved in leaf vein development in *Arabidopsis thaliana*. New Phytol 197: 737–750.

Benková E, Michniewicz M, Sauer M, Teichmann T, Seifertová D, Jürgens G, Friml J (2003) Local, efflux-dependent auxin gradients as a common module for plant organ formation. Cell 115: 591–602.

*Bennett T, Hines G, van Rongen M, Waldie T, Sawchuk MG, Scarpella E, Ljung K, Leyser O (2016) Connective auxin transport in the shoot facilitates communication between shoot apices. PLoS Biol 14: e1002446.

Berleth T, Mattsson J, Hardtke CS (2000) Vascular continuity and auxin signals. Trends Plant Sci 5: 387–393.

Biemelt S, Tschiersch H, Sonnewald U (2004) Impact of altered gibberellin metabolism on biomass accumulation, lignin biosynthesis, and photosynthesis in transgenic tobacco plants. Plant Physiol 135: 254–265.

*Binenbaum J, Weinstain R, Shani E (2018) Gibberellin localization and transport in plants. Trends Plant Sci 23: 410–421.

Bishopp A, Lehesranta S, Vatén A, Help H, El-Showk S, Scheres B, Helariutta K, Mähönen AP, Sakakibara H, Helariutta Y (2011) Phloem-transported cytokinin regulates polar auxin transport and maintains vascular pattern in the root meristem. Curr Biol 21: 927–932.

*Bloch D, Puli MR, Mosquna A, Yalovsky S (2019) Abiotic stress modulates root patterning via ABA-regulated microRNA expression in the endodermis initials. Development 146: dev177097.

Bockstette SW, Thomas BR (2019) Impact of genotype and parent origin on the efficacy and optimal timing of $GA_{4/7}$ stem injections in a lodgepole pine seed orchard. New Forest 51: 421–434.

*Bollhöner B, Prestele J, Tuominen H (2012) Xylem cell death: emerging understanding of regulation and function. J Exp Bot 63: 1081–1094.

Booker J, Chatfield S, Leyser O (2003) Auxin acts in xylem-associated or medullary cells to mediate apical dominance. Plant Cell 15: 495–507.

Borghi L, Liu GW, Emonet A, Kretzschmar T, Martinoia E (2016) The importance of strigolactone transport regulation for symbiotic signaling and shoot branching. Planta 243: 1351–1360.

Boursiac Y, Léran S, Corratgé-Faillie C, Gojon A, Krouk G, Lacombe B (2013) ABA transport and transporters. Trends Plant Sci 18: 325–333.

*Caño-Delgado A, Lee JY, Demura T (2010) Regulatory mechanisms for specification and patterning of plant vascular tissues. Annu Rev Cell Dev Biol 26: 605–637.

Carlsbecker A, Helariutta Y (2005) Phloem and xylem specifications: pieces of the puzzle emerge. Curr Opin Plant Biol 8: 512–517.

*Carlsbecker A, Lee JY, Roberts CJ, Dettmer J, Lehesranta S, Zhou J, Lindgren O, Moreno-Risueno MA, Vatén A, Thitamadee S, Campilho A, Sebastian J, Bowman JL, Helariutta Y, Benfey PN (2010) Cell signalling by microRNA165/6 directs gene dose-dependent root cell fate. Nature 465: 316–321.

Chandler JW, Werr W (2015) Cytokinin-auxin crosstalk in cell type specification. Trends Plant Sci 20: 291–300.

Clouse SD (2011) Brassinosteroids. In: The Arabidopsis Book, Vol 9, Amer Soc Plant Biol Rockville, MD: Amer Soc Plant Biol, doi: 10.1199/tab.0073, http://www.aspb.org/publications/arabidopsis/

Cook CE, Whichard LP, Wall ME (1972) Germination stimulants. II. The structure of strigol - a potent seed germination stimulant for witchweed (Striga lutea Lour.). J Am Chem Soc 94: 6198–6199.

Darwin C (1880) The Power of Movement in Plants. John Murray Publishers, London.

Dayan J, Schwarzkopf M, Avni A, Aloni R (2010) Enhancing plant growth and fiber production by silencing GA 2-oxidase. Plant Biotechnol J 8: 425–435.

*Dayan J, Voronin N, Gong F, Sun T-p, Hedden P, Fromm H, Aloni R (2012) Leaf-induced gibberellin signaling is essential for internode elongation, cambial activity, and fiber differentiation in tobacco stems. Plant Cell 24: 66–79.

De Coninck B, De Smet I (2016) Plant peptides - taking them to the next level. J Exp Bot 67: 4791–4795.

*De Rybel B, Mähönen AP, Helariutta Y, Weijers D (2016) Plant vascular development: from early specification to differentiation. Nat Rev. Mol Cell Biol 17: 30–40.

Dodd IC (2003) Hormonal interactions and stomatal responses. J Plant Growth Regul 22: 32–46.

*Domagalska MA, Leyser O (2011) Signal integration in the control of shoot branching. Nat Rev Mol Cell Biol 12: 211–221.

Du J, Gerttula S, Li Z, Zhao ST, Liu YL, Liu Y, Lu MZ, Groover AT (2020) Brassinosteroid regulation of wood formation in poplar. New Phytol 225: 1516–1530.

Duan L, Dietrich D, Ng CH, Chan PMY, Bhalerao R, Bennett MJ, Dinneny JR (2013) Endodermal ABA signaling promotes lateral root quiescence during salt stress in Arabidopsis seedlings. Plant Cell 25: 324–341.

Durán-Medina Y, Díaz-Ramírez D, Marsch-Martínez N (2017) Cytokinins on the move. Front Plant Sci 8: 146.

Eklund L (1990) Endogenous levels of oxygen, carbon dioxide and ethylene in stems of Norway spruce trees during one growing season. Trees 4: 150–154.

*El-Showk S, Ruonala R, Helariutta Y (2013) Crossing paths: cytokinin signalling and crosstalk. Development 140: 1373–1383.

Eriksson ME, Israelsson M, Olsson O, Moritz T (2000) Increased gibberellin biosynthesis in transgenic trees promotes growth, biomass production and xylem fiber length. Nat Biotechnol 18: 784–788.

Eschrich W (1968) Translokation radioaktiv markierter Indolyl-3-essigsäure in Siebröhren von *Vicia faba*. Planta 78: 144–157.

Escobar-Bravo R, Cheng G, Kyong Kim H, Grosser K, van Dam NM, Leiss KA, Klinkhamer PGL (2019) Ultraviolet radiation exposure time and intensity modulate tomato resistance against herbivory through activation of the jasmonic acid signaling. J Exp Bot 70: 315–327.

Fàbregas N, Formosa-Jordan P, Confraria A, Siligato R, Alonso JM, Swarup R, Bennett MJ, Mähönen AP, Caño-Delgado AI, Ibañes M (2015) Auxin influx carriers control vascular patterning and xylem differentiation in *Arabidopsis thaliana*. PLoS Genet 11: e1005183.

Falcioni R, Moriwaki T, de Oliveira DM, Andreotti GC, de Souza LA, Dos Santos WD, Bonato CM, Antunes WC (2018) Increased gibberellins and light levels promotes cell wall thickness and enhance lignin deposition in xylem fibers. Front Plant Sci 9: 1391.

*Fischer U, Kucukoglu M, Helariutta Y, Bhalerao RP (2019) The dynamics of cambial stem cell activity. Annu Rev Plant Biol 70: 293–319.

Foster AJ, Aloni R, Fidanza M, Gries R, Gries G, Mattsson J (2016) Foliar phase changes are coupled with changes in storage and biochemistry of monoterpenoids in western redcedar (*Thuja plicata*). Trees 30: 1361–1375.

*Friml J (2010) Subcellular trafficking of PIN auxin efflux carriers in auxin transport. Eur J Cell Biol 89: 231–235.

Friml J, Palme K (2002) Polar auxin transport - old questions and new concepts? Plant Mol Biol 49: 273–284.

Friml J, Wisniewska J, Benková E, Mendgen K, Palme K (2002) Lateral relocation of auxin efflux regulator PIN3 mediates tropism in Arabidopsis. Nature 415: 806–809.

Fukuda H (1997) Tracheary element differentiation. Plant Cell 9: 1147–1156.

Fukuda H (2004) Signals that control vascular cell differentiation. Nat Rev Mol Cell Biol 5: 379–391.

*Fukuda H, Ohashi-Ito K (2019) Vascular tissue development in plants. Curr Top Dev Biol 131: 141–160.

Furuta KM, Hellmann E, Helariutta Y (2014) Molecular control of cell specification and cell differentiation during procambial development. Annu Rev Plant Biol 65: 604–638.

Galiba G, Vagujfalvi A, Li CX, Soltesz A, Dubcovsky J 2009. Regulatory genes involved in the determination of frost tolerance in temperate cereals. Plant Sci 176: 12–19.

Gälweiler L, Guan C, Müller A, Wisman E, Mendgen K, Yephremov A, Palme K (1998) Regulation of polar auxin transport by AtPIN1 in *Arabidopsis* vascular tissue. Science 282: 2226–2230.

Geldner N, Anders N, Wolters H, Keicher J, Kornberger W, Muller P, Delbarre A, Ueda T, Nakano A, Jürgens G (2003) The *Arabidopsis* GNOM ARF-GEF mediates endosomal recycling, auxin transport, and auxin-dependent plant growth. Cell 112: 219–230.

Geldner N, Friml J, Stierhof YD, Jürgens G, Palme K (2001) Auxin transport inhibitors block PIN1 cycling and vesicle trafficking. Nature 413: 425–428.

Geßler A, Kopriva S, Rennenberg H (2004) Regulation of nitrate uptake at the whole-tree level: interaction between nitrogen compounds, cytokinins and carbon metabolism. Tree Physiol 24: 1313–1321.

Goldin N, Heyfets A, Reischer D, Flescher E (2007) Mitochondria-mediated ATP depletion by anti-cancer agents of the jasmonate family. J Bioenerg Biomembr 39: 51–57.

Goldschmidt EE, Samach A (2004) Aspects of flowering in fruit trees. In: *Proc 9th IS on Plant Bioregulators*, SM Kang et al. (eds). Acta Hort 653, pp 23–27.

Goldsmith MHM, Cataldo DA, Karn J, Brenneman T, Trip P (1974) The nonpolar transport of auxin in the phloem of intact *Coleus* plants. Planta 116: 301–317.

Gomez-Roldan V, Fermas S, Brewer PB, Puech-Pagès V, Dun EA, Pillot JP, Letisse F, Matusova R, Danoun S, Portais JC, Bouwmeester H, Bécard G, Beveridge CA, Rameau C, Rochange SF (2008) Strigolactone inhibition of shoot branching. Nature 455: 189–194.

Gregg A, Howe GA (2004) Jasmonates as signals in the wound response. J Plant Growth Regul 23: 223–237.

Hagen G, Martin G, Li Y, Guilfoyle TJ (1991) Auxin-induced expression of the soybean GH3 promoter in transgenic tobacco plants. Plant Mol Biol 17: 567–579.

Ham BK, Lucas WJ (2017) Phloem-mobile RNAs as systemic signaling agents. Annu Rev Plant Biol 68: 173–195.

Hellmann E, Ko D, Ruonala R, Helariutta Y (2018) Plant vascular tissues-connecting tissue comes in all shapes. Plants (Basel) 7: E109.

Hess T, Sachs T (1972) The influence of a mature leaf on xylem differentiation. New Phytol 71: 903–914.

Heyer M, Reichelt M, Mithöfer A (2018) A holistic approach to analyze systemic jasmonate accumulation in individual leaves of *Arabidopsis* rosettes upon wounding. Front Plant Sci 9: 1569.

Holland JJ, Roberts D, Liscum E (2009) Understanding phototropism: from Darwin to today. J Exp Bot 60: 1969–1978.

Hou H-W, Zhou Y-T, Mwange K-N, Li W-F, He X-Q, Cui K-M (2006) ABP1 expression regulated by IAA and ABA is associated with the cambium periodicity in *Eucommia ulmoides* Oliv. J Exp Bot 57: 3857–3867.

Howe GA (2004) Jasmonates. In: *Plant Hormones: Biosynthesis, Signal Transduction, Action!* PJ Davies (ed). Kluwer Academic Publishers, Dordrecht, pp 610–634.

Huang T, Böhlenius H, Eriksson S, Parcy F, Nilsson O (2005) The mRNA of the *Arabidopsis* gene *FT* moves from leaf to shoot apex and induces flowering. Science 309: 1694–1696.

Huber DP, Philippe RN, Madilao LL, Sturrock RN, Bohlmann J (2005) Changes in anatomy and terpene chemistry in roots of Douglas-fir seedlings following treatment with methyl jasmonate. Tree Physiol 25: 1075–1083.

Hudgins JW, Christiansen E, Franceschi VR (2003) Methyl jasmonate induces changes mimicking anatomical defenses in diverse members of the Pinaceae. Tree Physiol 23: 361–371.

Hudgins JW, Franceschi VR (2004) Methyl jasmonate-induced ethylene production is responsible for conifer phloem defense responses and reprogramming of stem cambial zone for traumatic resin duct formation. Plant Physiol 135: 2134–2149.

Hudgins JW, Ralph SG, Franceschi VR, Bohlmann J (2006) Ethylene in induced conifer defense: cDNA cloning, protein expression, and cellular and subcellular localization of 1-aminocyclopropane-1-carboxylate oxidase in resin duct and phenolic parenchyma cells. Planta 224: 865–877.

Ingemarsson BSM, Lundqvist E, Eliasson L (1991) Seasonal variation in ethylene concentration in wood of *Pinus sylvestris* L. Tree Physiol 8: 273–279.

Ishimaru Y, Oikawa T, Suzuki T, Takeishi S, Matsuura H, Takahashi K, Hamamoto S, Uozumi N, Shimizu T, Seo M, Ohta H, Ueda M (2017) GTR1 is a jasmonic acid and jasmonoyl-l-isoleucine transporter in *Arabidopsis thaliana*. Biosci Biotechnol Biochem 81: 249–255.

Israelsson M, Sundberg B, Moritz T (2005) Tissue-specific localization of gibberellins and expression of gibberellin-biosynthetic and signaling genes in wood-forming tissues in aspen. Plant J 44: 494–504.

Iwasaki, T, Shibaoka, H (1991). Brassinosteroids act as regulators of tracheary-element differentiation in isolated *Zinnia* mesophyll cells. Plant Cell Physiol 32: 1007–1014.

Jacobs WP (1952) The role of auxin in differentiation of xylem around a wound. Am J Bot 39: 301–309.

Jacobs WP (1984) Function of hormones in tissue level organization. In: *Hormonal regulation of development. II. The functions of hormones from the level of the cell to the whole plant*. Scott TK (ed), Encyclopedia of Plant Physiology (New Series) vol 10. Springer, Berlin, pp. 149–171.

Jacobs WP (1979) *Plant Hormones and Plant Development*. Cambridge University Press, Cambridge.

Jin H, Zhu Z (2017) Temporal and spatial view of jasmonate signaling. Trends Plant Sci 22: 451–454.

Kapulnik Y, Koltai H (2016) Fine-tuning by strigolactones of root response to low phosphate. J Integr Plant Biol 58: 203–212.

Kastumi M, Foard DE, Phinney RO (1983) Evidence for the translocation of gibberellin A3 and gibberellin-like substances in grafts between normal, dwarf1 and dwarft5 seedlings of *Zea mays* L. Plant Cell Physiol 24: 379–388.

Keeling CI, Bohlmann J (2006) Genes, enzymes and chemicals of terpenoid diversity in the constitutive and induced defence of conifers against insects and pathogens. New Phytol 170: 657–75.

Khripach V, Zhabinskii V, de Groot A (2000) Twenty years of brassinosteroids: steroidal plant hormones warrant better crops for the XXI century. Ann Bot 86: 441–447.

Kim I, Zambryski PC (2005) Cell-to-cell communication via plasmodesmata during *Arabidopsis* embryogenesis. Curr Opin Plant Biol 8: 593–599.

Klee H (2008) Evidence points to the existence of a hitherto uncharacterized type of hormone that controls different aspects of plant growth and interaction. The hunt for that hormone is heating up. Nature 455: 176–177.

Kleine-Vehn J, Dhonukshe P, Swarup R, Bennett M, Friml J (2006) Subcellular trafficking of the *Arabidopsis* auxin influx carrier AUX1 uses a novel pathway distinct from PIN1. Plant Cell 18: 3171–3181.

Kleine-Vehn J, Wabnik K, Martinière A, Łangowski Ł, Willig K, Naramoto S, Leitner J, Tanaka H, Jakobs S, Robert S, Luschnig C, Govaerts W, Hell SW, Runions J, Friml J (2011) Recycling, clustering, and endocytosis jointly maintain PIN auxin carrier polarity at the plasma membrane. Mol Syst Biol 7: 540.

Ko D, Kang J, Kiba T, Park J, Kojima M, Do J, Kim KY, Kwon M, Endler A, Song WY, Martinoia E, Sakakibara H, Lee Y (2014) Arabidopsis ABCG14 is essential for the root-to-shoot translocation of cytokinin. Proc Natl Acad Sci USA 111: 7150–7155.

Kohlen W, Charnikhova T, Liu Q, Bours R, Domagalska MA, Beguerie S, Verstappen F, Leyser O, Bouwmeester H, Ruyter-Spira C (2011) Strigolactones are transported through the xylem and play a key role in shoot architectural response to phosphate deficiency in nonarbuscular mycorrhizal host *Arabidopsis*. Plant Physiol 155: 974–987.

Koiwai H, Nakaminami K, Seo M, Mitsuhashi W, Toyomasu T, Koshiba T (2004) Tissue-specific localization of an abscisic acid biosynthetic enzyme, AAO3, in *Arabidopsis*. Plant Physiol 134: 1697–1707.

Koltai H (2015) Cellular events of strigolactone signalling and their crosstalk with auxin in roots. J Exp Bot 66: 4855–4861.

Kretzschmar T, Kohlen W, Sasse J, Borghi L, Schlegel M, Bachelier JB, Reinhardt D, Bours R, Bouwmeester HJ, Martinoia E (2012) A petunia ABC protein controls strigolactone-dependent symbiotic signalling and branching. Nature 483: 341–344.

*Kubo M, Udagawa M, Nishikubo N, Horiguchi G, Yamaguchi M, Ito J, Mimura T, Fukuda H, Demura T (2005) Transcription switches for protoxylem and metaxylem vessel formation. Genes Dev 19: 1855–1860.

Kudo T, Kiba T, Sakakibara H (2010) Metabolism and long-distance translocation of cytokinins. J Integr Plant Biol 52: 53–60.

Kuromori T, Miyaji T, Yabuuchi H, Shimizu H, Sugimoto E, Kamiya A, Moriyama Y, Shinozaki K (2010) ABC transporter AtABCG25 is involved in abscisic acid transport and responses. Proc Natl Acad Sci USA 107: 2361–2366.

Kuromori T, Sugimoto E, Shinozaki K (2011) Arabidopsis mutants of AtABCG22, an ABC transporter gene, increase water transpiration and drought susceptibility. Plant J 67: 885–894.

Lacombe B, Achard P (2016) Long-distance transport of phytohormones through the plant vascular system. Curr Opin Plant Biol 34: 1–8.

Larkin PJ, Gibson JM, Mathesius U, Weinmann JJ, Gärtner E, Hall E, Tanner GJ, Rolfe BG, Djordjevic MA (1996) Transgenic white clover. Studies with the auxin-responsive promoter, GH3, in root gravitropism and lateral root development. Transgenic Res 5: 325–335.

Letham DS (1994) Cytokinins as phytohormones – sites of biosynthesis, translocation and function of translocated cytokinin. In: *Cytokinins: chemistry, activity and function,* DWS Mok, MC Mok (eds). CRC Press, Boca Raton, FL, pp. 57–80.

Lev-Yadun S, Aloni R (1995) Differentiation of the ray system in woody plants. Bot Rev 61: 45–84.

*Leyser O (2018) Auxin signaling. Plant Physiol 176: 465–479.

Li S, Pezeshki SR, Shields FD (2006) Partial flooding enhances aeration in adventitious roots of black willow (*Salix nigra*) cuttings. J Plant Physiol 163: 619–628.

*Lifschitz E, Eviatar T, Rozman A, Shalit A, Goldshmidt A, Amsellem Z, Alvarez JP, Eshed Y (2006) The tomato FT ortholog triggers systemic signals that regulate growth and flowering and substitute for diverse environmental stimuli. Proc Natl Acad Sci USA 103: 6398–6403.

Lilley JL, Gee CW, Sairanen I, Ljung K, Nemhauser JL (2012) An endogenous carbon-sensing pathway triggers increased auxin flux and hypocotyl elongation. Plant Physiol 160: 2261–2270.

Lin MK, Belanger H, Lee YJ, Varkonyi-Gasic E, Taoka K, Miura E, Xoconostle-Cázares B, Gendler K, Jorgensen RA, Phinney B, Lough TJ, Lucas WJ (2007) FLOWERING LOCUS T protein may act as the long-distance florigenic signal in the cucurbits. Plant Cell 19: 1488–1506.

Liu J, Moore S, Chen C, Lindsey K (2017a) Crosstalk complexities between auxin, cytokinin, and ethylene in *Arabidopsis* root development: from experiments to systems modeling, and back again. Mol Plant 10: 1480–1496.

Liu J, Zhang D, Sun X, Ding T, Lei B, Zhang C (2017b) Structure-activity relationship of brassinosteroids and their agricultural practical usages. Steroids 124: 1–17.

*Ljung K (2013) Auxin metabolism and homeostasis during plant development. Development 140: 943–950.

Ljung K, Hull AK, Kowalczyk M, Marchant A, Celenza J, Cohen JD, Sandberg G (2002) Biosynthesis, conjugation, catabolism and homeostasis of indol-3-acetic acid in *Arabidopsis*. Plant Mol Biol 50: 309–332.

López-Ráez JA, Charnikhova T, Gómez-Roldán V, Matusova R, Kohlen W, De Vos R, Verstappen F, Puech-Pages V, Bécard G, Mulder P, Bouwmeester H (2008) Tomato strigolactones are derived from carotenoids and their biosynthesis is promoted by phosphate starvation. New Phytol 178: 863–874.

Love J, Björklunda S, Vahalab J, Hertzbergc M, Kangasjärvib J, Sundberg B (2009) Ethylene is an endogenous stimulator of cell division in the cambial meristem of *Populus*. Proc Natl Acad Sci USA 106: 5984–5989.

*Lucas WJ, Groover A, Lichtenberger R, Furuta K, Yadav SR, Helariutta Y, He XQ, Fukuda H, Kang J, Brady SM, Patrick JW, Sperry J, Yoshida A, López-Millán AF,Grusak MA, Kachroo P (2013) The plant vascular system: evolution, development and functions. J Integr Plant Biol 55: 294–388.

Ludwig-Müller J (2011) Auxin conjugates: their role for plant development and in the evolution of land plants. J Exp Bot 62: 1757–1773.

Lumba S, Holbrook-Smith D, McCourt P (2017) The perception of strigolactones in vascular plants. Nat Chem Biol 13: 599–606.

Marchant A, Bhalerao R, Casimiro I, Eklöf J, Casero PJ, Bennett M, Sandberg G (2002) AUX1 promotes lateral root formation by facilitating indole-3-acetic acid distribution between sink and source tissues in the Arabidopsis seedling. Plant Cell 14: 589–597.

Marhavý P, Bielach A, Abas L, Abuzeineh A, Duclercq J, Tanaka H, Pařezová M, Petrášek J, Friml J, Kleine-Vehn J, Benková E (2011) Cytokinin modulates endocytic trafficking of PIN1 auxin efflux carrier to control plant organogenesis. Dev Cell 21: 796–804.

Márquez G, Alarcón MV, Salguero J (2019) Cytokinin inhibits lateral root development at the earliest stages of lateral root primordium initiation in maize primary root. J Plant Growth Regul 38: 83–92.

Matsumoto-Kitano M, Kusumoto T, Tarkowski P, Kinoshita-Tsujimura K, Václavíková K, Miyawaki K, and Kakimoto T (2008) Cytokinins are central regulators of cambial activity. Proc Natl Acad Sci USA 105: 20027–20031.

Mattsson J, Sung ZR, Berleth T (1999) Responses of plant vascular systems to auxin transport inhibition. Development 126: 2979–2991.

Meir S, Droby S, Davidson H, Alsevia S, Cohen L, Horev C, Philosoph-Hadas S (1998) Suppression of *Botrytis* rot in cut rose flowers by postharvest application of methyl jasmonate. Postharvest Biol Technol 13: 235–243.

Miyawaki K, Matsumoto-Kitano M, Kakimoto T (2004) Expression of cytokinin biosynthetic iso-pentenyltransferase genes in *Arabidopsis*: tissue specificity and regulation by auxin, cytokinin, and nitrate. Plant J 37: 128–138.

Morris DA, Kadir GO, Barry AJ (1973) Auxin transport in intact pea seedlings (*Pisum sativum* L.): the inhibition of transport by 2,3,5-triiodobenzoic acid. Planta 110: 173–182.

Nguyen CT, Martinoia E, Farmer EE (2017) Emerging jasmonate transporters. Mol Plant 10: 659–661.

*Nieminen K, Immanen J, Laxell M, Kauppinen L, Tarkowski P, Dolezal K, Tähtiharju S, Elo A, Decourteix M, Ljung K, Bhalerao R, Keinonen K, Albert VA, Helariutta Y (2008) Cytokinin signaling regulates cambial development in poplar. Proc Natl Acad Sci USA 105: 20032–20037.

*Okada K, Abe H, Arimura G (2015) Jasmonates induce both defense responses and communication in monocotyledonous and dicotyledonous plants. Plant Cell Physiol 56: 16–27.

Palme P, Gälweiler L (1999) PIN-pointing the molecular basis of auxin transport. Curr Poin Plant Biol 2: 375–381.

Park J, Lee Y, Martinoia E, Geisler M (2017) Plant hormone transporters: what we know and what we would like to know. BMC Biol 15: 93.

*Pesquet E, Tuominen H (2011) Ethylene stimulates tracheary element differentiation in *Zinnia elegans* cell cultures. New Phytol 190: 138–149.

Ragni L, Nieminen K, Pacheco-Villalobos D, Sibout R, Schwechheimer C, Hardtke CS (2011) Mobile gibberellin directly stimulates *Arabidopsis* hypocotyl xylem expansion. Plant Cell 23: 1322–1336.

Ralph SG, Hudgins JW, Jancsik S, Franceschi VR, Bohlmann J (2007) Aminocyclopropane carboxylic acid synthase is a regulated step in ethylene-dependent induced conifer defense. Full-length cDNA cloning of a multigene family, differential constitutive, and wound- and insect-induced expression, and cellular and subcellular localization in spruce and Douglas fir. Plant Physiol 143: 410–424.

Ramachandran P, Wang G, Augstein F, de Vries J, Carlsbecker A (2018) Continuous root xylem formation and vascular acclimation to water deficit involves endodermal ABA signalling via miR165. Development 145: dev159202.

*Ravichandran SJ, Linh NM, Scarpella E (2020) The canalization hypothesis – challenges and alternatives. New Phytol, doi: https://doi.org/10.1111/nph.16605.

Regnault T, Davière JM, Wild M, Sakvarelidze-Achard L, Heintz D, Carrera Bergua E, Lopez Diaz I, Gong F, Hedden P, Achard P (2015) The gibberellin precursor GA12 acts as a long-distance growth signal in *Arabidopsis*. Nat Plants 1: 15073.

Reinhardt D, Pesce ER, Stieger P, Mandel T, Baltensperger K, Bennett M, Traas J, Friml J, Kuhlemeier C (2003) Regulation of phyllotaxis by polar auxin transport. Nature 426: 255–260.

*Pěnčík A, Simonovik B, Petersson SV, Henyková E, Simon S, Greenham K, Zhang Y, Kowalczyk M, Estelle M, Zazímalová E, Novák O, Sandberg G, Ljung K (2013) Regulation of auxin homeostasis and gradients in *Arabidopsis* roots through the formation of the indole-3-acetic acid catabolite 2-oxindole-3-acetic acid. Plant Cell 25: 3858–3870.

Rahayu YS, Walch-Liu P, Neumann G, Römheld V, von Wirén N, Bangerth F (2005) Root-derived cytokinins as long-distance signals for NO_3^- −induced stimulation of leaf growth. J Exp Bot 56: 1143–1152.

Robert HS, Crhak Khaitova L, Mroue S, Benková E (2015) The importance of localized auxin production for morphogenesis of reproductive organs and embryos in *Arabidopsis*. J Exp Bot 66: 5029–5042.

Roberts LW, Gahan BP, Aloni R (1988) *Vascular Differentiation and Plant Growth Regulators.* Springer-Verlag, Berlin.

Ruffel S, Krouk G, Ristova D, Shasha D, Birnbaum KD, Coruzzi GM (2011) Nitrogen economics of root foraging: Transitive closure of the nitrate–cytokinin relay and distinct systemic signaling for N supply vs. demand. Proc Natl Acad Sci USA 108: 18524–18529.

Sabatini S, Beis D, Wolkenfelt H, Murfett J, Guilfoyle T, Malamy J, Benfey P, Leyser O, Bechtold N, Weisbeek P, Scheres B (1999) An auxin-dependent distal organizer of pattern and polarity in the *Arabidopsis* root. Cell 99: 463–472.

*Sachs T (1968) On the determination of the pattern of vascular tissues in pea. Ann Bot 32: 781–790.

Sachs T (1969) Polarity and the induction of organized vascular tissues. Ann Bot 33: 263–275.

*Sachs T (1981) The control of patterned differentiation of vascular tissues. Adv Bot Res 9: 151–262.

Sachs T (1991) The canalization of vascular differentiation. In: *Pattern Formation in Plant Tissues*, T Sachs. Cambridge University Press, Cambridge, pp. 52–69.

Sachs T (2000) Integrating cellular and organismal aspects of vascular differentiation. Plant Cell Physiol 41: 649–656.

Sachs T, Thimann KV (1967) The role of auxins and cytokinins in the release of buds from dominance. Am J Bot 54: 136–144.

Sairanen I, Novák O, Pěnčík A, Ikeda Y, Jones B, Sandberg G, Ljung K (2012) Soluble carbohydrates regulate auxin biosynthesis via PIF proteins in *Arabidopsis*. Plant Cell 24: 4907–4916.

Saito H, Oikawa T, Hamamoto S, Ishimaru Y, Kanamori-Sato M, Sasaki-Sekimoto Y, Utsumi T, Chen J, Kanno Y, Masuda S, Kamiya Y, Seo M, Uozumi N, Ueda M, Ohta H (2015) The jasmonate-responsive GTR1 transporter is required for gibberellin-mediated stamen development in *Arabidopsis*. Nat Commun 6: 6095.

Sakakibara H, Takei K, Hirose N (2006) Interactions between nitrogen and cytokinin in the regulation of metabolism and development. Trends in Plant Sci 11: 440–448.

Saks Y, Feigenbaum P, Aloni R (1984) Regulatory effect of cytokinin on secondary xylem fiber formation in an in vivo system. Plant Physiol 76: 638–642.

Sasaki T, Suzaki T, Soyano T, Kojima M, Sakakibara H, Kawaguchi M (2014) Shoot-derived cytokinins systemically regulate root nodulation. Nat Commun 5: 4983.

Sauer M, Balla J, Luschnig C, Wisniewska J, Reinöhl V, Friml J, Benková E (2006) Canalization of auxin flow by Aux/IAA-ARF-dependent feedback regulation of PIN polarity. Genes Dev 20: 2902–2911.

*Sauer M, Robert S, Kleine-Vehn J (2013) Auxin: simply complicated. J Exp Bot 64: 2565–2577.

Savaldi-Goldstein S, Peto C, Chory J (2007) The epidermis both drives and restricts plant shoot growth. Nature 446: 199–202.

Scarpella E (2017) The logic of plant vascular patterning. Polarity, continuity and plasticity in the formation of the veins and of their networks. Curr Opin Genet Dev 45: 34–43.

*Scarpella E, Helariutta Y (2010) Vascular pattern formation in plants. Curr Top Dev Biol 91: 221–265.

*Scarpella E, Marcos D, Friml J, Berleth T (2006) Control of leaf vascular patterning by polar auxin transport. Genes Dev 20: 1015–1027.

Schmidt A, Nagel R, Krekling T, Christiansen E, Gershenzon J, Krokene P (2011) Induction of isoprenyl diphosphate synthases, plant hormones and defense signalling genes correlates with traumatic resin duct formation in Norway spruce (*Picea abies*). Plant Mol Biol 77: 577–590.

Schulze A, Zimmer M, Mielke S, Stellmach H, Melnyk CW, Hause B, Gasperini D (2019) Wound-induced shoot-to-root relocation of JA- Ile precursors coordinates *Arabidopsis* growth. Mol Plant 12: 1383–1394.

Schwalm K, Aloni R, Langhans M, Heller W, Stich S, Ullrich CI (2003) Flavonoid-related regulation of auxin accumulation in *Agrobacterium tumefaciens*-induced plant tumors. Planta 218: 163–178.

Sehr EM, Agusti J, Lehner R, Farmer EE, Schwarz M, Greb T (2010). Analysis of secondary growth in the Arabidopsis shoot reveals a positive role of jasmonate signalling in cambium formation. Plant J 63: 811–822.

Seyfferth C, Wessels BA, Gorzsás A, Love JW, Rüggeberg M, Delhomme N, Vain T, Antos K, Tuominen H, Sundberg B, Felten J (2019) Ethylene signaling Is required for fully functional tension wood in hybrid Aspen. Front Plant Sci 10: 1101 eCollection.

Shalit-Kaneh A, Eviatar-Ribak T, Horev G, Suss N, Aloni R, Eshed Y, Lifschitz E (2019) The flowering hormone florigen accelerates secondary cell wall biogenesis to harmonize vascular maturation with reproductive development. Proc Natl Acad Sci USA 116: 16127–16136.

Singh V, Sergeeva L, Ligterink W, Aloni R, Zemach H, Doron-Faigenboim A, Yang J, Zhang P, Shabtai S and Firon N (2019) Gibberellin Promotes Sweetpotato Root Vascular Lignification and Reduces Storage-Root Formation. Front Plant Sci 10: 1320.

Sircar SM, Chakraverty R (1960) The effect of gibberellic acid on jute (*corchorus capsularis* Linn). Sci Cult 26: 141–143.

Skoog F, Miller (1957). Chemical regulation of growth and organ formation in plant tissue culture in vitro. Symposium of the Society for Experimental Biology 11: 118–131.

Spicer R, Tisdale-Orr T, Talavera C (2013) Auxin-responsive DR5 promoter coupled with transport assays suggest separate but linked routes of auxin transport during woody stem development in *Populus*. PLoS One 8: e72499.

Stant MY (1961) The effect of gibberellic acid on fibre-cell length. Ann Bot 25: 453–462.

Stant MY (1963) The effect of gibberellic acid on cell width and the cell wall of some phloem fibres. Ann Bot 27: 185–196.

Sun JQ, Jiang HL, Li CY (2011) Systemin/Jasmonate-mediated systemic defense signaling in tomato. Mol Plant 4: 607–615.

Sundberg B, Uggla C, Tuominen H (2000) Cambial growth and auxin gradients. In: *Cell and Molecular Biology of Wood Formation*, RA Savidge, JR Barnett, R Napier (eds). BIOS Scientific Publishers, Oxford, pp. 169–188.

Swarup R, Friml J, Marchant A, Ljung K, Sandberg G, Palme K, Bennett M (2001) Localization of the auxin permease AUX1 suggests two functionally distinct hormone transport pathways operate in the *Arabidopsis* root apex. Genes Dev 15: 2648–2653.

Swarup R, Péret B (2012) AUX/LAX family of auxin influx carriers-an overview. Front Plant Sci 3: 225.

Taiz L, Zeiger E (2006) *Plant Physiology*, 4th edn. Sinauer, Sunderland MA.

Taiz L, Zeiger E, Møller IM, Murphy A (2018) *Fundamentals of Plant Physiology*. Sinauer, Sunderland, MA.

Takei K, Ueda N, Aoki K, Kuromori T, Hirayama T, Shinozaki K, Yamaya T, Sakakibara H (2004) *AtIPT3* is a key determinant of nitrate-dependent cytokinin biosynthesis in *Arabidopsis*. Plant Cell Physiol 45: 1053–1062.

*Tal I, Zhang Y, Jørgensen ME, Pisanty O, Barbosa IC, Zourelidou M, Regnault T, Crocoll C, Olsen CE, Weinstain R, Schwechheimer C, Halkier BA, Nour-Eldin HH, Estelle M, Shani E (2016) The *Arabidopsis* NPF3 protein is a GA transporter. Nat Commun 7: 11486.

Tanaka M, Takei K, Kojima M, Sakakibara H, Mori H (2006) Auxin controls local cytokinin biosynthesis in the nodal stem in apical dominance. Plant J 45: 1028–1036.

Teale WD, Paponov IA, Palme K (2006) Auxin in action: signalling, transport and the control of plant growth and development. Nature Rev Mol Cell Biol 7: 847–859.

Thimann KV (1935) Growth-substances in plants. Annu Rev Biochem 4: 545–568.

Thorpe MR, Ferrieri AP, Herth MM, Ferrieri RA (2007) 11C-imaging: methyl jasmonate moves in both phloem and xylem, promotes transport of jasmonate, and of photoassimilate even after proton transport is decoupled. Planta 226: 541–551.

*Tylewicz S, Petterle A, Marttila S, Miskolczi P, Azeez A, Singh RK, Immanen J, Mähler N, Hvidsten TR, Eklund DM, Bowman JL, Helariutta Y, Bhalerao RP (2018) Photoperiodic control of seasonal growth is mediated by ABA acting on cell-cell communication. Science 360: 212–215.

Uggla C, Mellerowicz EJ, Sundberg B (1998) Indole-3-acetic acid controls cambial growth in Scots pine by positional signaling. Plant Physiol 117: 113–121.

Uggla C, Moritz T, Sandberg G, Sundberg B (1996) Auxin as a positional signal in pattern formation in plants. Proc Nat Acad Sci USA 93: 9282–9286.

*Ullrich CI, Aloni R, Kleespie RG, Saeed MEM, Ullrich W, Efferth T (2019) Comparison between tumors in plants and human beings: mechanisms of tumor development and therapy with secondary plant metabolites. Phytomedicine 64: 153081.

Ulmasov T, Murfett J, Hagen G, Guilfoyle TJ (1997) Aux/IAA proteins repress expression of reporter genes containing natural and highly active synthetic auxin response elements. Plant Cell 9: 1963–1971.

Umehara M, Hanada A, Yoshida S, Akiyama K, Arite T, Takeda-Kamiya N, Magome H, Kamiya Y, Shirasu K, Yoneyama K, Kyozuka J, Yamaguchi S (2008) Inhibition of shoot branching by new terpenoid plant hormones. Nature 455: 195–200.

Ursache R, Nieminen K, Helariutta Y (2013) Genetic and hormonal regulation of cambial development. Physiol Plant 147: 36–45.

*Verna C, Ravichandran SJ, Sawchuk MG, Linh NM, Scarpella E (2019) Coordination of tissue cell polarity by auxin transport and signaling. Elife 8: e51061.

Veselov D, Langhans M, Hartung W, Aloni R, Feussner I, Götz C, Veselova S, Sclomski S, Dickler C, Bächmann K, Ullrich CI (2003) Development of *Agrobacterium tumefaciens* C58-induced plant tumors and impact on host shoots are controlled by a cascade of jasmonic acid, auxin, cytokinin, ethylene and abscisic acid. Planta 216: 512–522.

Wang L, Jin P, Wang J, Jiang L, Shan T, Zheng Y (2015) Methyl jasmonate primed defense responses against *Penicillium expansum* in sweet cherry fruit. Plant Mol Biol Reporter 33: 1464–1471.

Wang J, Wu D, Wang Y, Xie D (2019) Jasmonate action in plant defense against insects. J Exp Bot 70: 3391–3400.

Went FW (1935) Auxin, the plant growth-hormone. Bot Rev. 1: 162–182.

Went FW, Thimann KV (1937) *Phytohormones*. Macmillan, New York.

*Wenzel CL, Schuetz M, Yu Q, Mattsson J (2007) Dynamics of *MONOPTEROS* and PIN-FORMED1 expression during leaf vein pattern formation in *Arabidopsis thaliana*. Plant J 49: 387–398.

Werner T, Motyka V, Laucou V, Smets R, Van Onckelen HV, Schmülling T (2003) Cytokinin-deficient transgenic *Arabidopsis* plants show multiple developmental alterations indicating opposite functions of cytokinins in the regulation of shoot and root meristem activity. Plant Cell 15: 2532–2550.

Wexler S, Schayek H, Rajendar K, Tal I, Shani E, Meroz Y, Dobrovetsky R, Weinstain R (2018) Characterizing gibberellin flow *in planta* using photocaged gibberellins. Chem Sci 10: 1500–1505.

*Wodzicki TJ, Abe H, Wodzicki AB, Pharis RP, Cohen JD (1987) Investigations on the nature of the auxin-wave in the cambial region of pine stems: validation of IAA as the auxin component by the *Avena* coleoptile curvature assay and by gas chromatography-mass spectrometry-selected ion monitoring. Plant Physiol 84: 135–143.

Wodzicki TJ, Knegt E, Wodzicki AB, Bruinsma J (1984) Is indolyl-3-acetic acid involved in the wave-like pattern of auxin efflux from *Pinus sylvestris* stem segments? Physiol Plant 44: 122–126.

Xue-Xuan X, Hong-Bo S, Yuan-Yuan M, Gang X, Jun-Na S, Dong-Gang G, Cheng-Jiang R (2010) Biotechnological implications from abscisic acid (ABA) roles in cold stress and leaf senescence as an important signal for improving plant sustainable survival under abiotic-stressed conditions. Crit Rev. Biotechnol 30: 222–230.

Yamamoto R, Demura T, Fukuda H (1997) Brassinosteroids induce entry into the final stage of tracheary element differentiation in cultured *Zinnia* cells. Plant Cell Physiol 38: 980–983.

Yamamoto R, Fujioka S, Demura T, Takatsuto S, Yoshida S, Fukuda H (2001) Brassinosteroid levels increase drastically prior to morphogenesis of tracheary elements. Plant Physiol 125: 556–563.

Zajaczkowski S, Wodzicki TJ, Romberger JA (1984) Auxin waves and plant morphogenesis. In: *Hormonal regulation of development. II. The functions of hormones from the level of the cell to the whole plant.* Scott TK (ed), Encyclopedia of Plant Physiology (New Series) vol 10. Springer, Berlin, pp. 244–262.

Zazímalová E, Murphy AS, Yang H, Hoyerová K, Hosek P (2010) Auxin transporters - why so many? Cold Spring Harb Perspect Biol 2: a001552.

Zeevaart, JAD (1983). Gibberellins and flowering. In: *The Biochemistry and Physiology of Gibberellins*, A Crozier (ed), Vol 2. Praeger Scientific, New York pp 333–374.

Zdarska M, Dobisová T, Gelová Z, Pernisová M, Dabravolski S, Hejátko J (2015) Illuminating light, cytokinin, and ethylene signalling crosstalk in plant development. J Exp Bot 66: 4913–4931.

Zeevaart JAD (2006) Florigen coming of age after 70 years. Plant Cell 18: 1783–1789.

Zhang J, Nieminen K, Serra JA, Helariutta Y (2014a) The formation of wood and its control. Curr Opin Plant Biol 17: 56–63.

Zhang K, Novak O, Wei Z, Gou M, Zhang X, Yu Y, Yang H, Cai Y, Strnad M, Liu CJ (2014b) *Arabidopsis* ABCG14 protein controls the acropetal translocation of root-synthesized cytokinins. Nat Commun 5: 3274.

Zhang S, Wei Y, Lu Y, Wang X (2009) Mechanisms of brassinosteroids interacting with multiple hormones. Plant Signal Behav 12: 1117–1120.

Zhao Y (2018) Essential roles of local auxin biosynthesis in plant development and in adaptation to environmental changes. Ann Rev Plant Biol 69: 417–435.

Importance of NO$_3^-$ and PO$_4^-$ for Development, Differentiation, and Competition

4

PO$_4^-$ and NO$_3^-$ deserve special attention because their availability in the soil influences strigolactone (SL) and cytokinin (CK) synthesis and consequently affects plant development, shoot architecture, and vascularization. Plants can regulate which meristems are active according to their environmental conditions. Inhibition of lateral bud development and shoot branching occurs in response to the root hormone SL (Shinohara et al. 2013). The strigolactone is produced in the root and moves upward via the xylem (Kohlen et al. 2011; Waters et al. 2017) to the stem, where it inhibits lateral bud development. SL can inhibit shoot branching by triggering rapid depletion of the auxin efflux protein PIN1 from the plasma membrane of the stem's xylem parenchyma cells (Shinohara et al. 2013). SL positively regulates cambial activity and vascular differentiation (Agusti et al. 2011). The SL signaling in the vascular cambium itself is sufficient for cambium stimulation, and it interacts with the auxin signaling pathway to promote xylem differentiation (Agusti et al. 2011). By inhibiting lateral branching and promoting vascular differentiation in the leading stem, the SL improves nutrient supply to the apical bud and the primary stem that grow faster, which is a benefit in the competition with other plants for light exposure.

Strigolactone concentration increases as a response to inorganic phosphate (PO$_4^-$) deficiency (Kohlen et al. 2011; Kapulnik and Koltai 2016). Accordingly, in response to phosphate stress, the SL promotes rapid main stem elongation by inhibiting branch development, thus giving the plant an advantage in its competition for light against neighboring plants. Environmental conditions of mean annual precipitation of 2100 mm can likely wash out minerals and cause inorganic phosphate deficiency in the soil, which might be an important factor for the intense competition for light of *Eucalyptus* trees observed in the moist forests of south-eastern Australia, causing their rapid height growth (Prior and Bowman 2014) (see Chap. 16).

Conversely, the cytokinin produced in the root cap (Aloni et al. 2004, 2005, 2006) is a general promoting signal for buds and leaf development; its concentration increases following an increase in nitrate (NO$_3^-$) supply (Takei et al. 2004; Ruffel

© Springer Nature Switzerland AG 2021
R. Aloni, *Vascular Differentiation and Plant Hormones*,
https://doi.org/10.1007/978-3-030-53202-4_4

et al. 2011), whereas the SL promotes only main stem elongation by inhibiting lateral bud development. Thus, root-specific signals shape shoot developmental architecture and vascular differentiation in response to phosphate and nitrate concentrations in the soil. This basic information should be considered in planning experiments.

Below ground, SLs promote the elongation of the primary root and adventitious roots but repress lateral root formation and promote root-hair elongation and density in establishing plant-fungal mycorrhizal symbiosis (Borghi et al. 2016; Kapulnik and Koltai 2016; Sun et al. 2016). CKs produced by the leading root cap inhibit lateral root initiation and development (Aloni et al. 2006; Laplaze et al. 2007; Chang et al. 2013; Márquez et al. 2019), which led to the suggested mechanism of cytokinin-dependent root apical dominance (see Sect. 6.3). When studying either NO$_3^-$ and PO$_4^-$ or SLs and CKs (e.g., see Bellini et al. 2014), attention should be given to the possible influences of these mineral nutrition elements on SLs and CKs synthesis that regulate shoot and root development, which have been ignored in many previous studies.

References and Recommended Reading[1]

Agusti J, Herold S, Schwarz M, Sanchez P, Ljung K, Dun EA, Brewer PB, Beveridge CA, Sieberer T, Sehr EM, Greb T (2011) Strigolactone signaling is required for auxin-dependent stimulation of secondary growth in plants. Proc Natl Acad Sci USA 108: 20242–20247.

*Aloni, R, Aloni E, Langhans M, Ullrich CI (2006) Role of cytokinin and auxin in shaping root architecture: regulating vascular differentiation, lateral root initiation, root apical dominance and root gravitropism. Ann Bot 97: 883–893.

Aloni R, Langhans M, Aloni E, Dreieicher E, Ullrich CI (2005) Root-synthesized cytokinin in *Arabidopsis* is distributed in the shoot by the transpiration stream. J Exp Bot 56: 1535–1544.

Aloni R, Langhans M, Aloni E, Ullrich CI (2004) Role of cytokinin in the regulation of root gravitropism. Planta 220: 177–182.

*Bellini C, Pacurar DI, Perrone I (2014) Adventitious roots and lateral roots: similarities and differences. Annu Rev Plant Biol 65: 639–666.

Borghi L, Liu GW, Emonet A, Kretzschmar T, Martinoia E (2016) The importance of strigolactone transport regulation for symbiotic signaling and shoot branching. Planta 243: 1351–1360.

Chang L, Ramireddy E, Schmülling T (2013) Lateral root formation and growth of *Arabidopsis* is redundantly regulated by cytokinin metabolism and signalling genes. J Exp Bot 64: 5021–5032.

Kapulnik Y, Koltai H (2016) Fine-tuning by strigolactones of root response to low phosphate. J Integr Plant Biol 58: 203–212.

*Kohlen W, Charnikhova T, Liu Q, Bours R, Domagalska MA, Beguerie S, Verstappen F, Leyser O, Bouwmeester H, Ruyter-Spira C (2011) Strigolactones are transported through the xylem and play a key role in shoot architectural response to phosphate deficiency in nonarbuscular mycorrhizal host *Arabidopsis*. Plant Physiol 155: 974–987.

Laplaze L, Benkova E, Casimiro I, Maes L, Vanneste S, Swarup R, Weijers D, Calvo V, Parizot B, Herrera-Rodriguez MB, Offringa R, Graham N, Doumas P, Friml J, Bogusz D, Beeckman T, Bennett M (2007) Cytokinins act directly on lateral root founder cells to inhibit root initiation. Plant Cell 19: 3889–3900.

---

[1] Papers of particular interest for suggested reading have been highlighted (with *).

Márquez G, Alarcón MV, Salguero J (2019) Cytokinin inhibits lateral root development at the earliest stages of lateral root primordium initiation in maize primary root. J Plant Growth Regul 38: 83–92.

Prior LD, Bowman DMJS (2014) Across a macro-ecological gradient forest competition is strongest at the most productive sites. Front Plant Sci 5: 260.

Ruffel S, Krouk G, Ristova D, Shasha D, Birnbaum KD, Coruzzi GM (2011) Nitrogen economics of root foraging: Transitive closure of the nitrate–cytokinin relay and distinct systemic signaling for N supply vs. demand. Proc Natl Acad Sci USA 108: 18524–18529.

*Shinohara N, Taylor C, Leyser O (2013) Strigolactone can promote or inhibit shoot branching by triggering rapid depletion of the auxin efflux protein PIN1 from the plasma membrane. PLoS Biol 11: e1001474.

Sun H, Tao J, Gu P, Xu G, Zhang Y (2016) The role of strigolactones in root development. Plant Signal Behav 11: e1110662.

*Takei K, Ueda N, Aoki K, Kuromori T, Hirayama T, Shinozaki K, Yamaya T, Sakakibara H (2004) *AtIPT3* is a key determinant of nitrate-dependent cytokinin biosynthesis in *Arabidopsis*. Plant Cell Physiol 45: 1053–1062.

Waters MT, Gutjahr C, Bennett T, Nelson DC (2017) Strigolactone signaling and evolution. Annu Rev Plant Biol 68: 291–322.

Phloem and Xylem Differentiation

5

5.1 Low- and High-Concentration Polar Auxin Streams Induce Phloem and Xylem

The plant vascular system is composed of two conducting systems, the phloem for photosynthetic assimilates, and the xylem for water and minerals, which can produce complex patterns, being composed of several types of cells, which are induced by hormonal signals (Roberts et al. 1988; Evert 2006; Aloni 2013a). The hormonal signals that regulate the differentiation of the specialized vascular elements (sieve, and vessel elements, tracheids, and fibers) in these systems, in both organized and cultured tissues, will be clarified below.

Primitive phloem tissue has appeared in the large brown algae (Phaeophyta) grown in the ocean for transporting assimilates along these very long organisms (Esau 1969; Schmitz 1990; Raven 2003). Adaptation to land, to water uptake and body support, started in the transition of the nonvascular plants to terrestrial habitats. Genome analysis of the moss *Physcomitrella patens* that encode NAC proteins (Xu et al. 2014) revealed the development of the auxin and abscisic acid signaling pathways for coordinating multicellular growth and dehydration response (Rensing et al. 2008). Likewise, genome analysis of the basal land plant, the liverwort *Marchantia polymorpha*, showed that in comparison to the relative Charophycean (green algae), the liverwort is characterized by novel biochemical pathways, including the auxin signaling pathway (Bowman et al. 2017).

The earliest primary molecular signal that stimulated the differentiation of primitive transporting tissues in early plants (Raven 2003) is likely the plant hormone **auxin** (Xu et al. 2014). Its polar movement from cell to cell along plants created committed files of cells that started programed processes of partial (losing the nucleus in the nutrient conducting cells) or complete (losing the entire cytoplasm forming hollow water conducting elements) organized **programmed cell death** (PCD) that created narrow files of conducting cells specialized in assimilates translocation or water transport, respectively. Evolutionary and genetic evidence indicates that vascular development evolved originally as a PCD that allowed enhanced

© Springer Nature Switzerland AG 2021
R. Aloni, *Vascular Differentiation and Plant Hormones*,
https://doi.org/10.1007/978-3-030-53202-4_5

movement of water and nutrients in the early plants, while secondary wall formation in the water-conducting cells evolved afterward, providing mechanical support for effective long-distance transport of water (Bollhöner et al. 2012; Kabbage et al. 2017; Meents et al. 2018). The PCD process during vascular differentiation induced functioning files of transporting conduits that enabled the transition of plants from life in water to colonize land habitats. The vascular systems adapted plants to dry conditions and extreme surroundings, by developing complex vascular tissues that improved plant adaptation and evolution (Aloni 1991, 2015), which will be clarified in the book.

Experimental evidence shows that low-auxin concentration induces sieve elements (with no xylem) in tissue cultures of a few plant species, while there is need for higher auxin concentrations to induce xylem (Aloni 1980). These experimental results indicate that during evolution it is likely that a single signal, namely, a low-concentration stream of auxin from the upper growing parts of the brown algae to their basal portions was the stimulus that induced the early primitive phloem in these algae. During evolution, it is likely that an increase in the concentration of the polar auxin streams induced the xylem during the adaptation to land.

It was widely believed that **auxin and sugar** are vital for controlling the differentiation of xylem and phloem (Esau 1965; Cutter 1978). However, the evidence in the literature from growing plants (in vivo) and tissue culture (in vitro) studies was rather confusing. In *Coleus blumei* stems, Jacobs (1952) showed that the auxin, indol-3-acetic acid (IAA), from young leaves, was the limiting factor for xylem regeneration around a wound. Subsequently, LaMotte and Jacobs (1963) found that IAA was also the limiting factor for phloem regeneration around a wound and that sucrose did not affect phloem regeneration in *Coleus* internodes. Thompson and Jacobs (1966) confirm these earlier findings indicating that IAA was the common controlling hormonal signal for both phloem and xylem regeneration, confirming a positive correlation between auxin concentration and the quantity of regenerated vascular tissues.

Conversely, in tissue culture, Wetmore and Rier (1963) reported that the differentiation of phloem and xylem induced by buds grafted into the callus of *Syringa* was fully replaceable by sugar plus auxin. They were unable to observe the correlation between auxin concentration and the quantity of vascular tissues, which had been described for *Coleus* stems. With auxin concentration kept constant, 1.5–2.5% sucrose sufficed to induce strong xylem differentiation with little or no accompanying phloem differentiation, whereas differentiation of phloem with little or no xylem differentiation was obtained with 3–4% sucrose. Intermediate sucrose concentrations (2.5–3.5%) favored the production of both xylem and phloem, usually with cambium between them. Furthermore, Wetmore et al. (1964) reported that in fern protalhalli, low sugar concentrations (1.5–3%) plus auxin produced xylem, while higher sugar concentrations (4.5–5%) induced phloem. However, the results of Rier and Beslow (1967) with callus of *Parthenocissus tricuspidata* appear more contradictory than supportive of the earlier in vitro findings, showing that the number of xylem elements was directly proportional to the sucrose concentration in the medium up to 8%; unfortunately, the effect of sugar concentration on phloem differentiation was not reported in this study. In *Phaseolus*, the most advanced stage of both phloem and xylem differentiation was obtained with 2% sucrose (Jeffs and Northcote 1966); judging from the results of Wetmore and associates, this sugar concentration is rather low and one would expect it to favor xylem formation only. Cronshaw and Anderson (1971) were unable to induce the differentiation of sieve elements with 4% sucrose without also inducing the formation of xylem cells in stem-pith cultures of *Nicotiana*. It is possible that differences in the results of different authors stem from the use of different systems or plant species; nevertheless there are doubts as to whether sugar has a specific role as a controlling factor favoring phloem differentiation, as suggested by the Wetmore and Rier's hypothesis that sugar and auxin determine the differentiation of phloem and xylem.

A major problem when studying the differentiation of phloem in tissue culture is the difficulty to detect the sieve elements. In his review on phloem differentiation, Jacobs (1970, p. 265) emphasized this fact by noting that: "We should remember that there are special difficulties in trying to study sievetube differentiation in callus culture: not only are the cells apt to be very small, but they do not form such predictable locations as in regenerating internodes, nor do they usually differentiate as a strand or sievetube elements. The difficulty of searching through a whole callus for a few cells that might show sieve plates or slime plugs undoubtedly explains why the work published so far has not induced actual counts of sieve elements, nor shown a photograph of recognizable sieve element induced by chemicals." To overcome this technical difficulty and to accomplish a quantitative study of phloem differentiation, I have made use of the lacmoid clearing technique that was developed by Aloni and Sachs (1973), which permits a quick search of both the phloem sieve elements and the xylem tracheary elements throughout the entire callus (Aloni 1980) without a need for thin sectioning (Fig. 5.1).

The study of vascular tissues in callus of a few plant species, including *Syringa*, with the lacmoid clearing technique, revealed that in not one of the calluses was differentiation of tracheary elements observed in the absence of sieve elements. Interestingly, **low-auxin** (IAA) concentration resulted in the differentiation of only sieve elements (Figs. 5.2B and 5.3) with no tracheary elements (Aloni 1980). **High-IAA** concentrations resulted in the differentiation of both phloem and xylem (Fig. 5.3). However, even in tissue cultures grown at a high-IAA concentration (Aloni 1980), only phloem developed at the surface further away from the high-auxin containing medium (Fig. 5.2D). IAA concentration controlled the number of sieve and tracheary elements in the callus; increase in auxin concentration boosted the number of both cell types (Fig. 5.3). Changes in sucrose concentration (while the IAA concentration was kept constant) did not have any specific effect either on sieve element differentiation or on the ratio between phloem and xylem. Sucrose did, however, affected the quantity of the polysaccharide callose deposited on the sieve plates, which marks these plates and makes it easy to detect the callose-rich sieve plates (Aloni 1980). Therefore, it is likely that when Wetmore and Rier (1963) reported xylem with no phloem on low sugar concentrations, this was because of a failure to detect the callose-poor sieve elements in their thin sections.

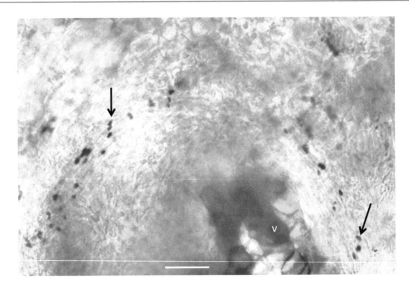

Fig. 5.1 Micrograph of a thick tissue of soybean (*Glycine max*) cleared and stained with lacmoid (according to Aloni and Sachs 1973), which was grown on a solid agar medium in culture, on 1 mg/l IAA and 6% sucrose, for 21 days. The figure shows a typical pattern of vascular differentiation in a nodule. The vessel elements (v) are in the center of the nodule (lower center) surrounded by circular sieve tubes. *Arrows* mark groups of callose-rich sieve plates (of three sieve tubes) stained blue by the lacmoid. Bar = 100 μm. (From Aloni 1980)

In shoots of many plant species, during early stages of a vascular bundle formation, phloem differentiates before the xylem (Esau 1965) and many young longitudinal bundles remain phloem only bundles (see Chap. 2, Fig. 2.6) (Roberts et al. 1988). The differentiation of phloem before xylem in young stem internodes occurs at early stages of leaf development, at the time when only low-concentration auxin streams are descending from the young leaf. The formation of networks of phloem anastomoses inside (in a bicollateral vascular bundle, between its external and internal phloem) and between bundles (Aloni and Sachs 1973; Aloni and Peterson 1990; Aloni and Barnett 1996), which might form dense patterns (Aloni and Peterson 1990; Aloni and Barnett 1996) (see Chap. 2, Figs. 2.9B, 2.17, 2.18B), is induced by low-grade auxin streams inside and between the bundles, whereas xylem differentiation requires high-auxin concentration streams (Aloni 1980, 1987, 1995). When high-auxin concentration was applied directly, or by wounding (which locally increases auxin concentration), vessels started to differentiate inside the phloem anastomoses (Aloni 1995; Aloni and Barnett 1996). It is likely that the low-IAA concentration streams that could induce phloem anastomoses among the longitudinal vascular bundles would be transported in *Arabidopsis* by the three auxin efflux carriers: PIN3, PIN4, and PIN7, which are major contributors of low conductance of the less polar auxin transport between longitudinal bundles (Bennett et al. 2016); these three auxin efflux carriers might be active also in other plant species. The lateral auxin transport occurs through small parenchyma cells, which likely slow

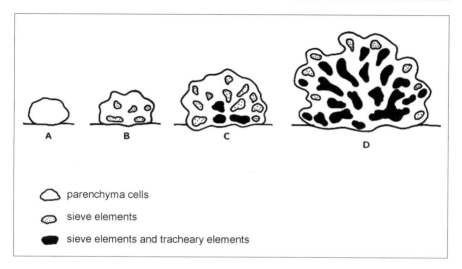

Fig. 5.2 Schematic diagrams of longitudinal median sections in tissue cultures grown on a solid medium, illustrating stages in the developmental patterns of phloem and xylem in the growing tissues. (**A**) Callus tissue consisting of homogenous parenchyma cells with no vascular elements grown on low-auxin medium (0.03 or 0.05 mg/l IAA). (**B**) Nodules of only sieve elements with no xylem developed either at early stage of vascular differentiation after transferring homogenous parenchyma to a medium containing a low-auxin concentration (0.1 mg/l IAA), or in callus tissue kept on this medium. (**C**) Intermediate stage in which tracheary elements start to appear in some of the phloem nodules located in the center of the callus. (**D**) Typical pattern of callus tissue grown on high-auxin concentration (0.5 or 1.0 mg/l IAA). In the periphery of the callus, there are either new developed nodules of phloem with no xylem, or nodules of phloem with little xylem. The center of the callus comprises nodules or short strands of well-developed phloem and xylem. (From Aloni 1980)

auxin transport because the IAA has to flow through more membranes (per distance) resulting in low conductance (Aloni and Peterson 1990). This is the reason why the short sieve elements building the phloem anastomose in *Dahlia pinnata* start to function in assimilate translocation (analyzed by fluorescein movement) mainly after wounding a longitudinal bundle, thus enabling immediate assimilate transport around the injury via already existing phloem anastomoses (Aloni and Peterson 1990).

In stems, the general developmental pattern in which the phloem precedes the xylem follows the evolutionary development of these tissues in early plants. This demonstrates that the polar auxin flow is the same controlling signal for both vascular tissues and only changes in its concentration induces and regulates the two different conducting vascular tissues: the phloem (with no xylem) is induced by low-concentration auxin streams, while the xylem is induced by high-concentration auxin streams (Aloni 2001, 2010). Therefore, it is likely that the low- versus high-concentration polar auxin streams selectively control the activation of the evolutionary conserved NAC-domain genes that encode transcription proteins, which are **master switches**, like the transcription factor ALTERED PHLOEM

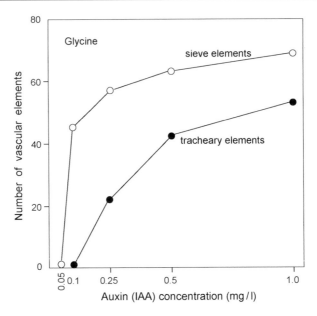

Fig. 5.3 Effect of IAA concentration on the differentiation of sieve and tracheary elements in tissue culture of *Glycine max* after 21 days. Sucrose concentration was 3%. Each point represents an average of three replicates from five callus tissues. (From Aloni 1980)

DEVELOPMENT (APL) that activates the phloem signaling cascade resulting in sieve-element differentiation (Bonke et al. 2003) versus the VASCULAR-RELATED NAC-DOMAIN (VND) and its homologs for inducing tracheary-element development (Xu et al. 2014; Kondo et al. 2016; Ohtani et al. 2017).

It should be mentioned that cytokinin (CK) is also an important controlling factor in sieve tube differentiation and regeneration around wounded collateral bundles in an in vivo system in which the endogenous cytokinin level has been minimized (Aloni et al. 1990a). CK is also a limiting and controlling factors of vessel differentiation and regeneration as will be discussed below (Baum et al. 1991). Both kinetin and zeatin were applied in aqueous solutions to the bases of excised, mature internodes of *Coleus blumei* that had an active vascular cambium. Each internode also received indoleacetic acid (IAA) in lanolin at its apical end. Both kinetin and zeatin induced a significant increase in sieve-tube regeneration around the wound. The two CKs also promoted callose production, which was most evident on the sieve plates of the regenerated sieve tube elements and on the walls of the parenchyma cells around the wound (Aloni et al. 1990a).

There is a major difference between **foliar and axial organs** regarding phloem and xylem differentiation (Aloni 2001). In leaves, the proximity between the major sites of high-auxin concentrations in the differentiating hydathodes (Aloni 2001, 2010; Aloni et al. 2003) and at the leaf-primordium margin (Mattsson et al. 1999; Sieburth 1999) promotes xylem differentiation with no phloem, due to exposing the cells to auxin maxima (highest local auxin concentrations) at these differentiating sites, which may explain why in leaves the xylem can differentiate in the absence of phloem at the hydathodes (Fahn 1990; Evert and Eichhorn 2013) and freely ending veinlets (Lersten 1990; Horner et al. 1994). Therefore, in the leaf periphery, the differentiation of xylem in the absence of phloem is a common feature and occurs in freely ending veinlets (Lersten 1989, 1990; Lersten and Curtis 1993; Horner et al. 1994) and hydathodes (Fahn 1990). Thus, in *Oxalis stricta* there are virtually no sieve tubes in any terminal vein, while *Polygonum convolvulus*, at the other extreme, has sieve tubes extending to the tips of most terminal veins (Horner et al. 1994). In most of the studied species, freely ending veinlets may display disparate relations between phloem and xylem in the same plant, e.g., leaf's vein endings lacking sieve tubes, or having them up to the tip, and sieve tubes that end at some intermediate point (Lersten 1990; Lersten and Curtis 1993; Horner et al. 1994). While in the stem, located away from the leaf's auxin maxima, the decreased auxin concentrations promote the differentiation of phloem before the xylem (Aloni 2001).

5.2 Hormonal Control of Xylem Cell Differentiation

In the xylem, the conducting cells are the tracheary elements (TE). They function in long-distance water transport (Tyree and Zimmermann 2002; Lucas et al. 2013) as nonliving vascular cells after autolysis of their cytoplasm. The autolysis of TE is determined as a specific process of vascular program cell death (PCD) through which the differentiating cells undergo a well-coordinated sequence of events resulting in either tracheids or vessels that become functional post-mortem as hollow cells through which water is transported (Iakimova and Woltering 2017).

TEs and fibers are characterized by thick secondary cell wall thickenings, which strengthen the cell walls with the phenolic polymer lignin (Oda and Fukuda 2012) enabling the TE to retain their shape when dead, despite the pressure of the surrounding cells. The cells that synthesize the secondary walls around their protoplast undergo a dramatic commitment to cellulose, hemicellulose, and lignin production, which are deposited under the guidance of cortical microtubules. Complexes of cellulose synthase produce cellulose microfibrils consisting of 18–24 glucan chains. These microfibrils are extruded into a cell wall matrix rich in hemicelluloses, typically xylan and mannan. The final step of secondary wall biosynthesis is lignification, with monolignols secreted by the lignifying cell and, in some cases, by neighboring parenchyma cells as well (Meents et al. 2018). Secondary cell walls are usually composed of cellulose (40–50%), hemicellulose (25–30%) and the phenolic polymer lignin (20–30%). Metabolic and environmental stresses can reduce the thickness of the lignified secondary walls, resulting in narrow and deformed vessels (see Figs. 2.20B and 2.32A, B).

The production of **lignin polymers** starts from the amino acid phenylalanine, from which the plant cells produce the monolignols: the paracoumaryl alcohol, the coniferyl alcohol and the sinapyl alcohol, which, respectively, form the H- (hydroxyphenyl), G- (guaiacyl), and S- (syringyl) subunits that build the lignin polymer. The ratio of these units varies with cell types and plant species. The lignin in the secondary walls of tracheid in gymnosperm wood is typically composed of G-units with a minor contribution of H-units, while in angiosperm wood, the secondary walls of fibers are built of both the G- and S-units, and vessel elements contain mainly G-units (Ek et al. 2009; Barros et al. 2015).

Lignin biosynthesis in the secondary walls of fibers and vessels is regulated by auxin and gibberellin (Aloni et al. 1990b; Tokunaga et al. 2006). In primary phloem fibers of *Coleus blumel* stems, auxin alone, or a combination of high IAA/low GA_3 (w/w in lanolin), induced short primary phloem fibers with thick secondary walls (Fig. 5.4B, C), that contained lignin rich in S-units (high ratio of syringyl/guaiacyl). On the other hand, a combination of high GA_3/low IAA, which promoted the differentiation of long primary phloem fibers with thin walls (Fig. 5.4D) decreased the relative content of the S-units (resulting in low syringyl/guaiacyl ratio) (Aloni et al. 1990b). In *Zinnia elegans* xylogenic liquid culture, exogenous application of GA_3 substantially increased the lignin content in the TE (Tokunaga et al. 2006).

Evidence from *Zinnia elegans* TE cell cultures revealed that TE lignification occurs *post-mortem* (i.e., after TE programmed cell death) (Pesquet et al. 2013). Suggesting that living, parenchyma xylem cells contribute to TE lignification in a non-cell-autonomous manner, thus enabling the *post-mortem* lignification of TEs. Results from *Arabidopsis* plants confirm the findings of non-cell-autonomous *post-mortem* lignification (Pesquet et al. 2013) of vessel elements *in Planta*, showing that nonlignifying xylem parenchyma cells can contribute lignin precursors (monolignols) to the lignification of their already dead neighboring vessel elements, whereas among the vascular bundles, the interfascicular fibers undergo cell autonomous lignification (i.e., are not dependent on monolignols from neighboring cells) (Smith et al. 2013). These interesting experimental results show that the lignification processes in the xylem tissue ranges from full autonomy in fibers (Smith et al. 2013) to a non-cell-autonomous process in TEs, thus enabling the ***post-mortem* lignification** of vessel elements that are dependent on their neighboring parenchyma cells to build their lignin polymers (Pesquet et al. 2013; Smith et al. 2013; Serk et al. 2015; Barros et al. 2015; Meents et al. 2018).

The two fundamental types of xylem conduits are tracheid (typical to gymnosperms) and vessel (in the developed angiosperms) that is built of vessel elements. Among the vessels are the fibers, which are the supporting cells of the plant body (Evert 2006). Tracheids appeared in ancient land plants in about 470 million years ago, while vessel elements were recorded about 140 million years ago, or even 250 million years ago (Li et al. 1996), and became dominant in angiosperms (Gerrienne et al. 2011; Evert and Eichhorn 2013; Harrison and Morris 2017).

5.2.1 Tracheid Differentiation

A tracheid is a non-perforated, usually long cell with bordered pits. Short tracheids can be found in conifer leaves (see Sect. 2.6). The tracheids have dual roles, namely, they are both the water conducting and physical supporting cells typically building the "softwood" of gymnosperms and the secondary xylem of monocotyledonous

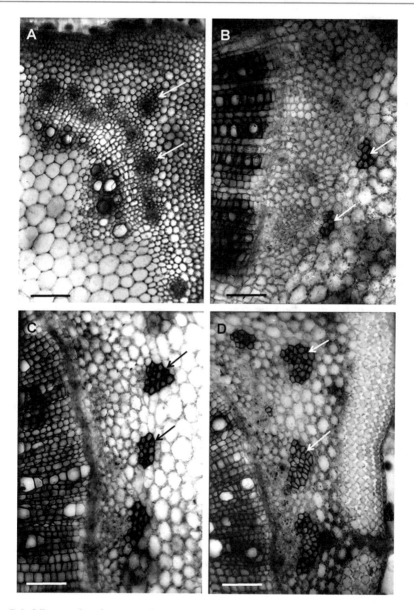

Fig. 5.4 Micrographs of cross sections taken from the middle of young internodes of *Coleus blumei* stems from which the leaves were excised, showing the pattern of primary phloem fiber differentiation induced by hormonal treatments during 20 days. (**A**) No fiber differentiation occurred when only 0.1% GA$_3$ (w/w in lanolin) (with no auxin) was applied. The *arrows* show the primary phloem fiber initials that remain unchanged during the experiment, demonstrating the need for an auxin background in order to induce fiber differentiation. (**B**) Small bundles of primary phloem fibers were induced by 0.1% IAA (w/w). These fibers were very short. (**C**) Fibers induced by the combination of 1.0% IAA + 0.05% GA$_3$ (w/w), characterized by thick secondary walls. (**D**) Fibers induced by 0.05% IAA + 1.0% GA$_3$, showing thin-walled fibers in the largest primary phloem bundles. These fibers were the longest in the experiment. *Arrows* mark primary phloem fibers in **B**, **C**, **D**. Bar = 100 μm. (From Aloni 1979)

(Evert 2006). However, the tracheids in conifers regularly extend slightly more than the cambial fusiform initials they originated from, as they are surrounded by other lignified rigid secondary tracheids that limit their elongations. Whereas tracheids of monocotyledons can undergo considerable intrusive elongation during their development, which makes them very long, e.g., up to about 55-fold longer in the dragon tree (*Dracaena draco*) than the mother cells from which they are derived (Jura-Morawiec 2017). The tracheids in monocots elongate inside bundles surrounded by parenchyma cells that allow their intrusive elongation.

Auxin movement through the cambium of pine trees induces the differentiation of tracheids from the cambium stem cells (Uggla et al. 1996). In young pine seedlings, tracheids can also redifferentiate from young parenchyma cells following the application of both auxin and gibberellin (Kalev and Aloni 1998). When only IAA is applied, it induces short tracheid, while GA, in the presence of IAA, promotes tracheid elongation by stimulating intrusive growth of both the upper and lower ends of the differentiating tracheids. This experimental result demonstrates that both **auxin and gibberellin** are the stimulating and controlling signals of tracheid differentiation (see Chap. 19, Figs. 19.3, 19.4 and 19.6).

There are contradictory hypotheses concerning the control of tracheid size. *The hypothesis of tracheid diameter regulation* (Larson 1969) proposes that gradients of tracheid diameter throughout the stem are positively regulated by parallel gradients of auxin. This hypothesis, which was based on photoperiodic studies of growth ring formation in *Pinus*, ascribes the formation of wide tracheids (earlywood) to high levels of auxin associated with shoot extension and leaf development and the differentiation of narrow tracheids (latewood) to low levels of auxin associated with cessation of shoot growth (Larson 1964, 1969). On the other hand, the *auxin gradient hypothesis* (originally named the *six-point hypothesis*, Aloni and Zimmermann 1983) proposes inverse relationships between auxin levels and vascular element dimensions. This hypothesis ascribes the general increase in tracheid size from leaves to roots (Zimmermann 1983) to a gradient of decreasing auxin concentration from leaves to roots. Accordingly, the distance from the source of auxin induction (the young leaves) controls the duration of cell expansion, and, consequently, the latter determines the final size of the differentiating cells (Aloni and Zimmermann 1983) (see details in Chap. 13).

To clarify the validity of the *auxin gradient hypothesis* (Aloni and Zimmermann 1983) for tracheid differentiation, the size patterns, and duration of cell expansion of primary and secondary tracheids was studied quantitatively in the hypocotyl of *Pinus pinea* seedlings (Saks and Aloni 1985). The results show that the number of primary and secondary tracheids decreased and their radial diameter increased from cotyledons to roots. Rapid differentiation immediately beneath the cotyledons resulted in narrow tracheids, while slow development at the root end yielded wide tracheids. The metaxylem elements in the root direction were the widest tracheids as a result of the longer time of their cell expansion. Differentiation of secondary tracheids started immediately below the cotyledons, where the highest auxin concentration is expected, and continued basipetally toward the root (Saks and Aloni 1985). These results are supported by studying the duration of secondary tracheid differentiation along the stem of a *Picea abies* tree, demonstrating that the extent of the expansion phase is positively correlated with the lumen area of the tracheids and that the lumen area of the conduits from the top of the tree to its base was linearly dependent on the time during which the differentiating tracheids remained in the expansion phase (Anfodillo et al. 2012); at the top of the tree's trunk (9 m above ground) the tracheid expansion time was 7 days, at 6 m above ground the cells expended for 14 days, and at 3 m for 19 days (Anfodillo et al. 2012). Therefore, the tracheids at the base of the tree, which have the longest period of cell expansion, became the widest conduits (Anfodillo et al. 2012). Similarly, in *Picea mariana*, the wide earlywood tracheids undergo a longer expansion phase of widening that lasted for 12 days on average, while the short expansion phase of 7 days results in narrow latewood tracheids (Buttò et al. 2019). The above results are in accordance with the *auxin gradient hypothesis* (Aloni and Zimmermann 1983) showing that the rate of cell expansion (controlled by auxin signaling) determines the width of the conduits; short differentiation phase results in narrow tracheids, whereas longer time of cell expansion forms wide conduits (see Chap. 13).

Tracheids are considered safe vascular elements in regards to embolism, in comparison to the efficient water transporting vessels (Tyree and Zimmermann 2002), which explains why conifers that their conduits are the safe tracheids can survive at the highest and coldest mountain slopes, while most of the angiosperm trees containing vessels cannot. For a discussion on "safety zones" in organ junctions and the hydraulic segmentation of branches and roots from the main stem, see Chap. 14.

5.2.2 Vessel Differentiation

Vessel elements have evolved from tracheids independently in several diverse groups of plants, making them an excellent example of parallel evolution. A vessel is a long continuous tube usually made up of many vessel elements connected end-to-end by perforation plates (openings) and limited in length by imperforated walls at both extremities. A vessel is the physiologically operating conduit and not vessel elements (Tyree and Zimmermann 2002; Cai and Tyree 2014) (already discussed in Chap. 2).

Auxin, namely, indole-3-acetic acid (IAA) is the primary signal and the limiting factor that induces vessel differentiation in plants. The polar flow of IAA originating in young leaves moves along the plant toward the roots and induces vessels along the auxin pathways (Jacobs 1952; Sachs 1981; Aloni 2010; Scarpella and Helariutta 2010). Auxin is involved in all aspects of vessel differentiation (Aloni 1987, 2001), which have already been discussed in Chap. 3 and will be further discussed in the entire book. High-auxin concentration applied to hypocotyls of young pine seedlings induces perforations in differentiating tracheids (Aloni et al. 2000; Aloni 2013a) (see Sect. 19.1, Fig. 19.2). Accordingly, these tracheids with experimentally induced perforations support the general view about the evolutionary origin of vessels from tracheids (Aloni 2013a, b).

Cytokinins (CKs) are limiting and controlling factors of vessel differentiation and regeneration as will be clarified below. Most of the knowledge on how CKs influence vessel element differentiation originated from studies done in tissue cultures (e.g., Fosket and Torrey 1969; Dalessandro 1973; Fukuda and Komamine 1980; Minocha 1984; Phillips 1987; Roberts et al. 1988; Fukuda 1992; Demura 2014). Relatively little information has been gained about the role of CK from studies done with intact plants. Torrey et al. (1971) suggested that the difficulty in proving the involvement of CK in in vivo systems is probably a result of considerable levels of root-supplied CK in tissues and organs of higher plants. This suggestion has been supported by experiments in which the endogenous CK level had been minimized (Saks et al. 1984; Aloni et al. 1990a; Baum et al. 1991). The difficulty in studying CK participation in differentiation within the plant is further increased by the fact that excised plant organs or tissues are capable of synthesizing CK under experimental conditions. It is also relatively difficult to follow the effects of CK on the differentiation of primary and secondary sieve tubes and vessels inside vascular bundles. This technical problem can be solved by studying the regenerative differentiation of phloem and xylem between the vascular bundles (Aloni et al. 1990a; Baum et al. 1991). In addition, sometimes it is impossible to evaluate quantitatively the direct effect of CK on vascular differentiation in the whole plant because CK also affects the development of buds and leaves, which are the main sources of auxin. To avoid this indirect CK influence on vascular differentiation, the buds and leaves should be removed and replaced by a source of auxin (Aloni et al. 1990a; Baum et al. 1991).

The role of CK in regenerative differentiation of vessels around a wound was studied in excised internodes of *Coleus* receiving IAA in lanolin at their apical ends and CK in aqueous solution at their base (Baum et al. 1991). This in vivo system enables an experimental study of the role of the growth regulators in organized tissues which maintain the natural gradients and polarities inherent in the intact plant. In order to minimize endogenous CK production due to adventitious root initiation, 1 mm slices were cut daily from the basal end of the excised internodes (Baum et al. 1991).

CK absence and CK application to the aqueous solution at the base of the excised internodes revealed that CK is a limiting and controlling factor in the regeneration of vessels around a wound (Baum et al. 1991). Zeatin was most effective and displayed its maximum promoting effect under low auxin level (0.1% IAA in lanolin), while kinetin showed its maximum effect under high IAA level (1.0%) (Baum et al. 1991). Studies of transgenic plants with altered levels of cytokinin confirm the involvement of cytokinin as a controlling factor in the differentiation of vessels and fibers (Medford et al. 1989; Li et al. 1992).

CK induced polar regeneration in *Coleus* internodes. The maturation of regenerated vessels occurred in the acropetal (upward) direction. In the internodes treated with CK, the regenerated vessel elements below the wound were fully differentiated, while secondary wall formation on the vessel elements at the upper end above the wound was incomplete (Baum et al. 1991). In addition, more regenerated vessel elements differentiated below the wound than above it. In extreme cases, the regenerated vessel elements above the wound were absent, while below the wound there was usually complete regeneration (Baum et al. 1991).

A major question concerning the role of CK in inducing the upward polar pattern of vascular differentiation is whether root apices can stimulate the development of an upward pattern of vessel differentiation similar to those induced by applied CK to excised internodes (Baum et al. 1991). Experimental evidence from *Cucurbita pepo* seedlings shows that adventitious roots stimulate differentiation of regenerated vessels in the upward direction (Aloni 1993), similar to the pattern induced by applied CK to the excised *Coleus* internodes (Baum et al. 1991). When adventitious roots are induced along the hypocotyl of young *Cucurbita pepo* seedlings, following main root removal, they develop either near a vascular strand or between strands (Aloni 1993). The first tracheary elements of the very young adventitious root develop at its base. From this site, primary vessels develop into the growing root, while in the opposite direction, regenerated vessels develop from the root toward the neighboring hypocotyl strands, usually with an upward tendency towards the leaves (Aloni 1993). In later developmental stages, the regenerated vessels connect the vessel system of the adventitious roots to the neighboring vascular strands.

5.2.3   Fiber Differentiation

Fibers are usually long, narrow cells with thick secondary walls that are heavily lignified, providing mechanical support to the plant body. The study of fiber differentiation has enormous economic importance in terms of increasing fiber yield in industrial plants and trees, as fibers are the raw material for paper and related wood industries. It has been shown that fiber differentiation is dependent on signals originating in both the leaves (Sachs 1972; Hess and Sachs 1972; Aloni 1976, 1978, 1979, 1987; Dayan et al. 2012) and the roots (Saks et al. 1984).

The differentiation of primary phloem fibers was studied in *Coleus blumei* stems on a quantitative basis in order to develop an experimental system for studying fibers differentiation (Aloni 1976). The natural pattern of fiber differentiation and maturation in intact, untreated plants was found to occur in the upward direction (from a mature internode to a young one). The youngest internodes to differentiate mature primary phloem fibers were those with cambial activity (Aloni 1976). Although primary phloem fibers develop along the stem in a general acropetal direction, fiber differentiation starts in a discontinuous pattern, occurring first at a node and only later in the internode below it (Aloni 1978), likely due to higher hormonal concentrations at the nodes which stimulate faster fiber differentiation in nodes. Differentiation of fibers around a wound was polar, primary phloem fibers differentiated above the wound but not below it (Fig. 5.5). The wounding enabled to detect the source of signals that induced the fibers. There was a need for the presence of leaves for inducing primary phloem fibers. Stems from which all the leaves were excised did not produced fibers (Fig. 5.6B). Young leaves did not induce differentiation of primary phloem fibers in young internodes (Aloni 1978). Wounding concentrated the fiber-inducing signals from maturing leaves resulting in primary

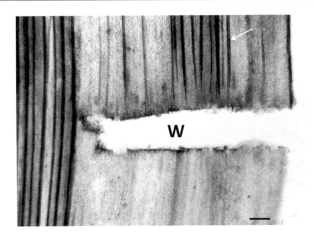

Fig. 5.5 Typical polar pattern of primary phloem fiber differentiation detected 10 days after wounding in the middle of a mature internode of *Coleus blumei*. For observation, the phloem was separated from the xylem at the cambium and stained by the lacmoid clearing technique (Aloni and Sachs 1973). The photograph shows strands of fibers (*arrow*) that differentiated above and beside the wound (W) and absence of fibers immediately below the wound (separating the tissues above and below with parafilm), demonstrating that the fiber-inducing signals arrived from the leaves above the wound (there is no fiber differentiation when the leaves above the wound are excised; see Fig. 5.6E). Bar = 100 μm. (From Aloni 1976)

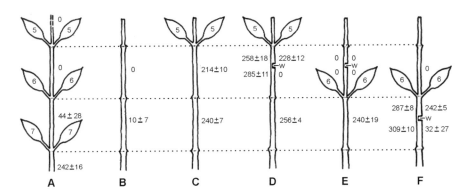

Fig. 5.6 Schematic diagrams illustrating the differentiation of primary phloem fibers in internodes 5 and 6 of *Coleus blumei* stems. The numbers beside the internodes are averages (from 5 stems per treatments, plus and minus SE) of the primary phloem fibers in the adjacent half cross section, whether taken from the middle of the internode, or 3 mm above and below the wound made in the middle of the internode. The wound (−w) is shown in Fig. 5.5. (**A**) was analyzed at 0 time from untreated stems. All the other treatments (**B–F**) were harvested after 14 days; their youngest internodes, lateral buds and the absent leaves illustrated in the experimental design were excised at 0 time. (From Aloni 1976)

phloem differentiation in young internodes, which at this developmental stage did not produce primary phloem fibers (Aloni 1978). The signals from the leaves could move downward along a few internodes and induce fibers in mature internodes. The leaves had a polar inducing effect, producing primary phloem fibers only below them (Fig. 5.6C–F) but not above them (Fig. 5.6E, F) (Aloni 1978, 1979).

Based on early studies on the role of **gibberellin** in promoting fiber differentiation (Sircar and Chakraverty 1960; Atal 1961; Stant 1961,1963) and the results from *Coleus* stems showing polar patterns of fiber differentiation (Aloni 1976, 1978), a working hypothesis was suggested that auxin and gibberellin control fiber differentiation. The hypothesis was confirmed experimentally in *Coleus* stems showing that the role of the leaves in primary phloem fiber differentiation can be fully replaced, both qualitatively and quantitatively, by exogenous application of auxin (IAA) and gibberellic acid (GA₃) (Aloni 1979). These signals also control fiber lignification (Aloni et al. 1990b). Indoleacetic acid (IAA) alone sufficed to cause the differentiation of a few short primary phloem fibers. GA₃ by itself did not exert any effect on fiber differentiation. Combinations of IAA with GA₃ completely replaced the role of the leaves in primary phloem fiber differentiation qualitatively and quantitatively (Figs. 5.4 and 5.7D, E) (Aloni 1979).

When various combinations of both hormones were applied, high concentrations of IAA stimulated rapid differentiation of fibers with thick secondary walls, while high levels of GA₃ resulted in long fibers with thin walls (Fig. 5.4). The size of the primary phloem fibers correlated with the dimensions of the differentiating internode, thereby providing evidence that both growth regulators also regulate stem extension. High IAA/low GA₃ concentrations have an inhibitory effect on fiber and internode elongation, resulting in short fibers and short internodes; whereas low

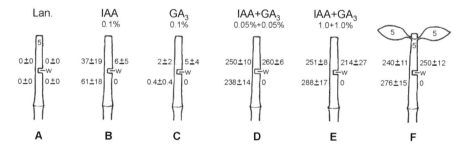

Fig. 5.7 Schematic diagrams illustrating the role of auxin (IAA) and gibberellic acid (GA₃) in differentiation of primary phloem fibers around a wound (−W) in the mature internode 5 (counted downward from the shoot apical bud) of *Coleus blumei*, after 10 days. The values beside the internodes are the mean ± SE of the number of primary phloem fibers in the adjacent half cross section taken 3 mm above and 3 mm below the wound in the middle of the internode. Sample size was five in all cases. All the plant leaves were removed, except for pair 5 in treatment **F**. The lanolin pastes were applied to node 5 (the node above internode 5) in the various hormonal concentrations (w/w) shown at the top of each diagram (in treatment **A**, the Lan. is plain lanolin; and in **F**, the leaves were the source of signals). Treatment **D** shows that the low hormonal combination was more effective that auxin (**B**), or gibberellin (**C**) alone. The high combination (**E**) did not have an additional promoting effect. Gibberellin by itself was not effective (**C**). (From Aloni 1979)

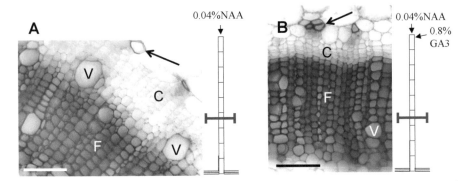

Fig. 5.8 Micrographs of cross-sections taken from the middle of mature internodes of *Nicotiana tabacum* stems from which all the leaves were excised; the location of the cross-sections are marked by blue bars, in the adjacent schematic diagrams illustrating the experimental design, showing the long-distance effect of hormonal application to a young internode on vascular differentiation in a mature internode. (**A**) Showing the effect of low auxin concentration (0.04% NAA in lanolin w/w) resulting in slow differentiation rate that enable longer time of vessel (V) and xylem fiber (F) expansion, resulting in wide xylem elements. (**B**) Mixture of gibberellic acid (0.8% GA$_3$ in lanolin w/w) with the auxin (0.04% NAA in lanolin w/w) promoted cell divisions in the cambium, faster cell differentiation resulting in narrow vascular elements, and increased lignification of fibers and vessels. The *arrows* mark primary phloem fibers. Cambium (C), fibers (F), vessels (V). Bars = 50 μm. (J Dayan and R Aloni, unpublished)

IAA/high GA$_3$ concentrations promote maximal fiber and stem elongation (Aloni 1979).

Both auxin and gibberellin have a long-distance controlling role (Figs. 5.8 and 5.9). In tobacco (*Nicotiana tabacum*) stems with excised leaves, application of low auxin concentration on a young internode induced wide vessels in a mature internode, away from the application site (Fig. 5.8A), while adding gibberellin with the auxin promoted fiber lignification and reduced width of both the vessels and fibers (Fig. 5.8B) similar to their combined effect in latewood. When auxin was applied to a young internode of tobacco stem from which all the leaves were excised, only vessels differentiated (Fig. 5.9A). Interestingly, when gibberellin was added and applied to a mature internode, the gibberellin induced in the young internode lignified fibers in both the xylem and the phloem (Fig. 5.9B); evidently proving that the **GA signal is nonpolar** and can move upward from mature leaves to young internodes, when they are supplied with auxin (J Dayan and R Aloni, unpublished results) (see also Chap. 11, Fig. 11.3C showing that gibberellin can induce fiber differentiation in the upward direction). This important evidence shows that the strong basipetal (downward) polarity of fiber differentiation observed in the *Coleus* stem experiments (Aloni 1976, 1978, 1979) was due to the polarity of auxin movement, occurring only toward the roots. Because GA by itself (without auxin) does not induce fibers (Aloni 1979), fibers were produced only below the leaves and not above them.

The evidence that gibberellin can be applied to the base of the plant for promoting fiber differentiation in the entire stem enables to effectively apply gibberellin to

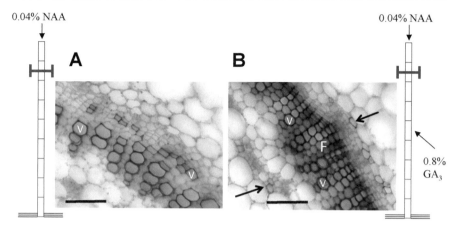

Fig. 5.9 Micrographs of cross sections taken from the middle of young internodes of *Nicotiana tabacum* stems from which the leaves were excised; the location of the cross sections is marked by blue bars, in the adjacent schematic diagrams illustrating the experimental design. (**A**) Showing the pattern of vessel differentiation induced in the young internode by low auxin concentration (0.04% NAA in lanolin) applied on the upper end; the results demonstrate that auxin alone induces only vessels, and the late formed ones are narrow vessels (v), with no any fiber differentiation. (**B**) The addition of gibberellin (0.8% GA₃ in lanolin) to a mature internode demonstrates the nonpolar effect of GA on a young internode, namely, that GA can move upward and induce primary fibers in both the inside and outside phloem (*arrows*) as well as secondary xylem fibers (F) among the vessels (v). Bars = 50 μm. (J Dayan and R Aloni, unpublished)

the roots. Application of gibberellin to sunflower (*Helianthus annuus*) roots during irrigation resulted in substantially increased stem growth and secondary fiber production (Fig. 5.10). The induced fibers have modified lignin content, judged by color change in the stained fiber, as was found with *Coleus* plants (Aloni et al. 1990b). Similar treatment can be applied to fast-growing industrial fiber crops to modify their fiber production and lignin structure.

The role of the root signaling in fiber differentiation was first studied in cultured hypocotyl segments of *Helianthus annuus* (Fig. 5.11). In this experimental system, **cytokinin** was a limiting and controlling factor in early stages of xylem fiber differentiation, at the time when many cell divisions occurred in the explants (Aloni 1982), evidently showing that CK controls the early stages of fiber differentiation. Later stages of fiber differentiation occurred in the absence of CK (Table 5.1). Aloni (1982) proposed that the mechanism which stimulates and controls the early stages of fiber differentiation is based on the interaction of three major hormonal signals, namely, auxin and gibberellin from the leaves with cytokinin from the root apices. Furthermore, the basis for the correlation between the development of the plant body and the differentiation of its supportive tissues is based on their common dependence on the same shoot-root feedback control signals (Aloni 1982) (see Sect. 6.1).

The role of CK in fiber differentiation in organized tissues was demonstrated in intact young plants of *Helianthus* (Saks et al. 1984). Reducing the endogenous CK

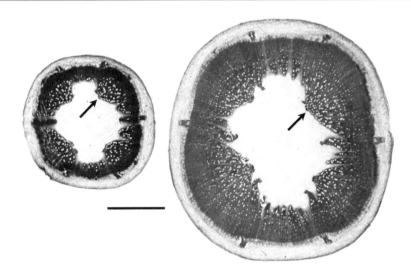

Fig. 5.10 Photographs of cross sections taken from the middle of mature internodes of *Helianthus annuus* stems at the same distance from roots. The intact control is shown on the left side and the GA-treated stem on the right. Showing the effect of gibberellic acid (100 ppm GA$_3$) applied for 6 weeks to the root system by irrigation. The GA treatment substantially increased stem size and secondary xylem fiber formation. The different color of the xylem fibers (*arrows*) stained with lacmoid indicates modification in lignin composition. Bar = 10 mm

supply, either by side root removal, or by lowering the transpiration rate, decreased the formation of secondary xylem fibers in the hypocotyl of the *Helianthus* plants. On the other hand, application of kinetin to intact plant via the roots promoted the differentiation of secondary xylem fibers in the hypocotyl, both in the bundles and between them (Fig. 5.12) (Saks et al. 1984). Early short-time exposure (for 3 days) of young *Helianthus* plants to low concentration of kinetin had a temporary promoting effect (for about 2 weeks) on the differentiation of secondary xylem fibers. However, this promoting effect was detected only a few days after the exposure to kinetin. This delayed response to the short-time exposure emphasizes that CK affects mainly the early stages of fiber differentiation. This promoting effect of CK on fiber differentiation was not a result of promoting shoot development (Saks et al. 1984). The nature of CK regulatory activity on fiber differentiation is probably not single tracked. The influence of CK on early stages of fiber differentiation can be either promotion or inhibition, depending on its physiological levels (Saks et al. 1984). Taking into account that fiber differentiation is affected by auxin and gibberellin (Aloni 1979), the ability of CK to influence shoot development, which is the source of these inducing signals (Sachs 1972; Aloni 1978) parallel to its direct effect on fiber differentiation (Aloni 1982), makes CK a tool of major importance in the regulation of fiber differentiation.

Over-expression of GA 20-oxidase, a gene encoding the enzyme responsible for the rate limiting step involved in gibberellin synthesis, enhances fiber yield (Eriksson et al. 2000). The transgenic tobacco plants and poplar trees showed higher levels of

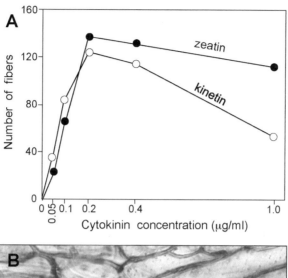

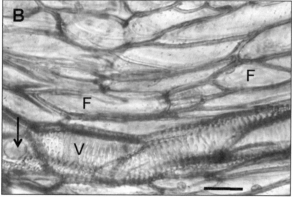

Fig. 5.11 Effect of zeatin and kinetin on the differentiation of secondary xylem fibers in cultured hypocotyl segments of *Helianthus annuus* after 30 days in the presence of 0.5 μg/ml IAA and 0.1 μg/ml GA₃. (**A**) The maximal promoting effect of both cytokinins was at 0.2 μg/ml, each point represents an average of five replicates from 10 explants. (**B**) Photograph of the zeatin-induced secondary xylem fibers in *H. annuus* cultured segment, stained with lacmoid. The fibers (F) are characterized by small simple pits compare with the large perforation (*arrow*) and reticulated secondary wall thickenings of the vessel elements (V). Bar = 25 μm. (From Aloni 1982)

GAs in their shoots and an increase in fiber length (Eriksson et al. 2000; Eriksson and Moritz 2002). However, this genetic manipulation also increased the GA inactivation, due to GA 2-oxidase catalysis. Dayan et al. (2010) have therefore used another approach to elevate GA concentrations by silencing the GA 2-oxidase (i.e., preventing deactivation of the bioactive gibberellin), which elevates the bioactive GA concentrations in tobacco and promoted rapid shoot elongation, increased fiber production and enlarged fiber size. Endogenous bioactive gibberellin concentrations could be boosted up by inducing both the over-expression of GA 20-oxidase and silencing the GA 2-oxidase genes (Dayan et al. 2010), which resulted in

Table 5.1 Effect of cytokinin on secondary xylem fiber differentiation in cultured hypocotyl seg-ments of *Helianthus annuus* after 30 days, demonstrating the need for cytokinin during the first 2 weeks for fiber formation (A, C), while absence of CK in the early stage prevent fiber differentia-tion (B, D). Values are mean ± SE. Sample size was 10 for each treatment. There was no significant difference in the number of secondary xylem fiber under treatments A and C. This was also true for the final fresh weight under any of the four treatments. Kinetin concentration was 0.2 μg/ml. All culture media contained 0.5 μg/ml IAA and 0.1 μg/ml GA$_3$

Treatment	First 15 days	Last 15 days	Number of fibers	Final fresh weight (mg)
A	+ kinetin	+ kinetin	123.4 ± 13.7	994 ± 291
B	− kinetin	+ kinetin	0 ± 0	613 ± 179
C	+ kinetin	− kinetin	117.5 ± 15.3	736 ± 256
D	− kinetin	− kinetin	0 ± 0	595 ± 164

From Aloni (1982)

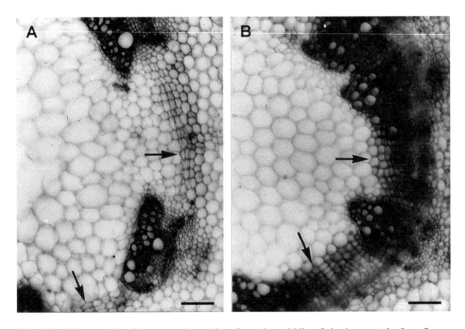

Fig. 5.12 Photographs of cross sections taken from the middle of the hypocotyl of sunflower (*Helianthus annuus*), showing the pattern of secondary xylem fibers induced by kinetin applied to the roots of intact plants during 21 days. The two photographs were taken from the same experi-ment and under the same magnification. (**A**) The intact control shows the typical bundles with interfascicular cambium (*arrows*) between them at the end of the 21-day period of growth in Hoagland medium (Hoagland and Arnon 1950) without kinetin. (**B**) Intact plants treated with 0.25 μg/ml kinetin applied in the medium to their roots show secondary xylem fibers induced between the bundles (*arrows*). Note that there are also more fibers within the bundles. Bars = 100 μm. (From Saks et al. 1984)

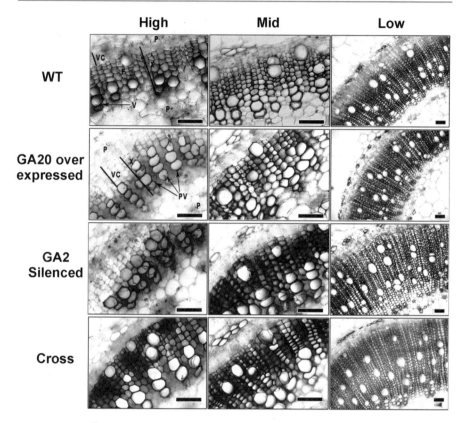

Fig. 5.13 Stem anatomy of wild-type control and tobacco (*N. tabacum*) plants over-expressing GA 20-oxidase, silencing GA 2-oxidase, and their crosses. Three columns separate between different regions of the dissected stems (corresponding to young (High), middle (Mid), and basal (Low) internodes). It is evident that the silenced plants have higher cambial activity with thicker xylem zone. In the young cross sections, a delay in xylem differentiation is depicted in the transgenic lines, prolonging the developmental stage, and allowing cell elongation. High xylem fiber production is observed in the basal internodes of silenced lines and their cross-fertilization with GA 20-oxidase over-expressing plants. The middle internodes do not exhibit any significant difference in cambial activity and xylem development. X xylem; P phloem; VC vascular cambium; V vessels; PV primary vessels; F fibers. Bars = 80 μm. (From Dayan et al. 2010)

synergistic effects (Fig. 5.13). These molecular manipulations could also modify lignin metabolism and change lignin structure and content.

Recent experiments on *Arabidopsis* show that the early GA precursor GA_{12}, although biologically inactive by itself, is a major mobile GA signal over long distances and that it moves via the vascular tissues. The GA_{12} is functional in recipient tissues, supporting growth via the activation of the GA signaling cascade (Regnault et al. 2015). Similarly, both the gibberellin hormone (GA_1) and its precursor (GA_{20}) produced in mature leaves of tobacco can flow non-polarly via the phloem, from the mature leaves to sink organs. When the mature-leaf-induced GA_{20} precursor arrives to the cambium, it is converted, by local cambial activity of the GA 20-oxidase, to the bioactive gibberellin form (GA_1), which activates the cambium and induces fiber differentiation. Therefore, the removal of mature leaves, which are major source of gibberellin, substantially depletes the endogenous GA concentrations in the cambium which impairs cytokinesis and fiber differentiation (Dayan et al. 2012).

Stems of transgenic plants with elevated GA concentrations grow rapidly, produce longer fibers (Eriksson et al. 2000; Biemelt et al. 2004; Dayan et al. 2010), and enhanced wood production (Dayan et al. 2010). This knowledge and technology should be applied in agriculture and forestry.

5.2.4   Insights from Transdifferentiation of Isolated Single Parenchyma Cells into Tracheary Elements in Liquid Cultures

Kohlenbach and Schmidt (1975) finding that mechanically isolated single mesophyll parenchyma cells of *Zinnia elegans* (isolated from the first pair of leaves of young seedlings) can directly transform (transdifferentiate) into tracheary elements (TEs), motivated Fukuda and Komamine (1980) to establish a more effective experimental system with a higher differentiation frequency (about 30% of the isolated parenchyma cells kept in a rotation liquid culture can directly transdifferentiate into TEs, in a synchronous developmental process) (5.14A). This powerful in vitro experimental system enables to investigate TE differentiation at the single cell level (Fukuda 1996, 2016; Demura 2014). The system has been widely used for uncovering detailed roles of the hormonal signaling in vessel element formation, i.e., auxin and cytokinin (Fukuda and Komamine 1980), brassinosteroids (Iwasaki and Shibaoka 1991), gibberellins (Tokunaga et al. 2006), and ethylene (Pesquet and Tuominen 2011).

Because the TEs are kept as single isolated cells, they do not develop perforations. However, when two or more differentiating TEs are attached to each other, they may form perforations between them. Therefore, these TE are a type of vessel elements that do not form perforations because they are experimentally retained as single cells in a rotation liquid culture.

The TEs are characterized by formation of a visible secondary cell wall (annular, spiral, and/or reticulate patterns) and autolysis, which makes them a suitable system for studying programmed cell death (PCD) (Fukuda 1996, 2004; Iakimova and Woltering 2017). These two major processes, of formation of secondary wall thickening and cytoplasm autolysis by PCD, permitted the differentiation of the water-conducting system in plants and enabled their transition from water to colonize the land, as mentioned above (Xu et al. 2014; Ohtani et al. 2017).

Molecular biological approaches using liquid tissue cultures of *Zinnia* and *Arabidopsis* have identified genes which are key regulators of TE differentiation. E.g., a microarray analysis of TE differentiation in *Arabidopsis* liquid single cells culture and promoter analysis revealed the possible involvement of some plant-specific **NAC-domain transcription factors** in vessel formation. These are VASCULAR-RELATED NAC-DOMAIN6 (VND6) and VND7 that can induce transdifferentiation of parenchymatic cells into metaxylem- and protoxylem-like vessel elements, respectively, in *Arabidopsis* and in poplar. A dominant repression of VND6 and VND7 specifically inhibits metaxylem and protoxylem vessel formation in roots. The results suggest that the VND6 and VND7 genes are master transcription switches for plant metaxylem and protoxylem vessel formation (Kubo et al. 2005; Yamaguchi et al. 2010; Ohtani et al. 2017; Fukuda and Ohashi-Ito 2019). The conserved genetic basis of the NAC transcription factors regulation and cellular function is evident in the moss *Physcomitrella patens* that encode NAC proteins and in *Arabidopsis thaliana* (Xu et al. 2014) shows evolutionary conserved adaptation of plants to land. Similarly, in the phloem, the ALTERED PHLOEM DEVELOPMENT (APL) is the master transcription regulator of vascular cell differentiation in the phloem (Bonke et al. 2003; Kondo et al. 2016; Ohtani et al. 2017; Agustí and Blázquez 2020).

Polar auxin transport is essential for the formation of a continuous long vessel along the plant body. PIN1 cycling and vesicle trafficking (Geldner et al. 2001; Kleine-Vehn et al. 2006; Taiz et al. 2018) is a requirement for the active polar auxin efflux (see Sect. 3.1.3, Fig. 3.8). Interestingly, the polar auxin transport inhibitor 1-N-naphthylphthalamic acid (NPA), which blocks PIN1 cycling and auxin efflux from cells, prevented TE formation in *Zinnia* single cell liquid culture. This result possibly indicates the necessity for active polar auxin efflux from cells during vessel differentiation within the plant body and surprisingly also when the cells are kept as isolated single cells. This NPA-induced suppression of TE formation was overcome by an increase in auxin concentration (Fig. 5.14) (Yoshida et al. 2005).

Summary

- Genome analysis revealed that the auxin signaling pathway has already developed in the early nonvascular plants.

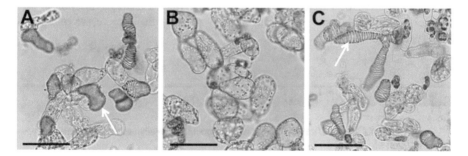

Fig. 5.14 Effect of the auxin transport inhibitor 1-N-naphthylphthalamic acid (NPA) on tracheary element (TE) differentiation in *Zinnia elegans* liquid culture. (**A**) Showing isolated tracheary elements (*arrows*) of *Zinnia* cells that were cultured on a xylogenic inducing medium for 96 h without NPA. (**B**) TE differentiation was inhibited by 20 µM NPA. (**C**) The NPA-induced suppression of TE formation was overcome by high-auxin concentration (10.7 µM NAA). Scale bars = 100 µm. (Reproduced from Yoshida et al. (2005) *Plant and Cell Physiology* 46: 2019–2028, with permission of Oxford University Press)

- Auxin is the primary hormonal signal that moves downward from cell to cell to induce organized programmed cell death (PCD) forming narrow files of conducting cells specialized either in assimilates translocation or water transport.
- From evolutionary perspective, phloem with no xylem appeared first in brown algae. In vascular plants, low-auxin concentration induces sieve elements (with no xylem) in tissue cultures and along the stem. During collateral bundle differentiation, phloem differentiates before the xylem.
- High-auxin concentrations induce xylem elements in the phloem strands; while auxin maxima in the leaf's periphery induce xylem with no phloem in the hydathods.
- Low- versus high-auxin concentrations likely promote the activation of master switches, the transcription factor ALTERED PHLOEM DEVELOPMENT (APL) that activates the phloem signaling cascade resulting in sieve-element differentiation versus the VASCULAR-RELATED NAC-DOMAIN (VND) that induces tracheary-element development.
- Cytokinin produced in root tips is an important limiting and controlling factor in sieve-tube and vessel differentiation and regeneration around wounds.
- Gibberellin is the specific signal that induces fiber differentiation, only in the presence of auxin. Cytokinin is needed for fiber formation during early stage of fiber differentiation. GA promotes fiber differentiation in both the upward and downward directions.
- Over-expression of GA 20-oxidase or/and silencing the GA 2-oxidase increase GA concentration, which promote longer fibers, increase in fiber production and xylem differentiation.
- Parenchyma cells of *Zinnia elegans* in liquid cultures can directly transform (transdifferentiate) into tracheary elements (TEs), creating a powerful experimental system for analyzing various issues regarding TE differentiation.
- Promoter analysis revealed the involvement of some plant-specific NAC-domain transcription factors in TE differentiation. These are the VASCULAR-RELATED NAC-DOMAIN6 (VND6) and VND7 that can induce transdifferentiation of parenchymatic cells into metaxylem- and protoxylem-like vessel elements, respectively.

References and Recommended Readings[1]

*Agustí J, Blázquez MA (2020) Plant vascular development: mechanisms and environmental regulation. Cell Mol Life Sci 77: 3711–3728.

Aloni R (1976) Polarity of induction and pattern of primary phloem fiber differentiation in *Coleus*. Am J Bot 63: 877–889.

Aloni R (1978) Source of induction and site of primary phloem fibre differentiation in *Coleus blumei*. Ann Bot 42: 1261–1269.

[1] Papers of particular interest for suggested reading have been highlighted (with *)

*Aloni R (1979) Role of auxin and gibberellin in differentiation of primary phloem fibers. Plant Physiol 63: 609–614.

*Aloni R (1980) Role of auxin and sucrose in the differentiation of sieve and tracheary elements in plant tissue cultures. Planta 150: 255–263.

Aloni R (1982) Role of cytokinin in differentiation of secondary xylem fibers. Plant Physiol 70: 1631–1633.

Aloni R (1987) Differentiation of vascular tissues. Annu Rev Plant Physiol 38: 179–204.

Aloni R (1991) Wood formation in deciduous hardwood trees. In: *Physiology of Trees*. AS Raghavendra (ed). Wiley & Sons, New York, pp. 175–197.

Aloni R (1993) The role of cytokinin in organised differentiation of vascular tissues. Aust J Plant Physiol 20: 601–608.

*Aloni R (1995) The induction of vascular tissues by auxin and cytokinin. In: *Plant Hormones: Physiology, Biochemistry and Molecular biology*, PJ Davies (ed). Kluwer, Dordrecht, pp 531–546.

Aloni R (2001) Foliar and axial aspects of vascular differentiation - hypotheses and evidence. J Plant Growth Regul 20: 22–34.

Aloni R (2010) The induction of vascular tissues by auxin. In: *Plant Hormones: Biosynthesis, Signal Transduction, Action!* PJ Davies (ed). Kluwer Academic Publishers, Dordrecht, pp 485–506.

Aloni R (2013a) The role of hormones in controlling vascular differentiation. In: *Cellular Aspects of Wood Formation*, J Fromm (ed). Springer-Verlag, Berlin, pp 99–139.

Aloni R (2013b) Role of hormones in controlling vascular differentiation and the mechanism of lateral root initiation. Planta 238: 819–830.

*Aloni R (2015) Ecophysiological implications of vascular differentiation and plant evolution. Trees 29: 1–16.

Aloni R, Alexander JD, Tyree MT (1997) Natural and experimentally altered hydraulic architecture of branch junctions in *Acer saccharum* Marsh. and *Quercus velutina* Lam. trees. Trees 11: 255–264.

Aloni R, Barnett JR (1996) The development of phloem anastomoses between vascular bundles and their role in xylem regeneration after wounding in *Cucurbita* and *Dahlia*. Planta 198: 595–603.

*Aloni R, Baum SF, Peterson CA (1990a) The role of cytokinin in sieve tube regeneration and callose production in wounded *Coleus* internodes. Plant Physiol 93: 982–989.

Aloni R, Feigenbaum P, Kalev N, Rozovsky S (2000) Hormonal control of vascular differentiation in plants: the physiological basis of cambium ontogeny and xylem evolution. In: *Cell and Molecular Biology of Wood Formation*. RA Savidge, JR Barnett, R Napier (eds). BIOS Scientific Publishers, Oxford, pp 223–236.

Aloni R, Griffith M (1991) Functional xylem anatomy in root-shoot junctions of six cereal species. Planta 184: 123–129.

Aloni R, Peterson CA (1990) The functional significance of phloem anastomoses in stems of *Dahlia pinnata* Cav. Planta 182: 583–590.

Aloni R, Sachs T (1973) The three-dimensional structure of primary phloem systems. Planta 113: 343–353.

Aloni R, Schwalm K, Langhans M, Ullrich CI (2003) Gradual shifts in sites of free-auxin production during leaf-primordium development and their role in vascular differentiation and leaf morphogenesis in *Arabidopsis*. Planta 216: 841–853.

Aloni R, Tollier T, Monties B (1990b) The role of auxin and gibberellin in controlling lignin formation in primary phloem fibers and in xylem of *Coleus blumei* stems. Plant Physiol 94: 1743–1747.

Aloni R, Zimmermann MH (1983) The control of vessel size and density along the plant axis - a new hypothesis. Differentiation 24: 203–208.

*Anfodillo T, Deslauriers A, Menardi R, Tedoldi L, Petit G, Rossi S (2012) Widening of xylem conduits in a conifer tree depends on the longer time of cell expansion downwards along the stem. J Exp Bot 63: 837–845.

Atal CK (1961) Effect of gibberellin on the fibers of hemp. Econ Bot 15:133–139.

Barros J, Serk H, Granlund I, Pesquet E (2015) The cell biology of lignification in higher plants. Ann Bot 115: 1053–1074.

Baum SF, Aloni R, Peterson CA (1991) The role of cytokinin in vessel regeneration in wounded *Coleus* internodes. Ann Bot 67: 543–548.

Bennett T, Hines G, van Rongen M, Waldie T, Sawchuk MG, Scarpella E, Ljung K, Leyser O (2016) Connective auxin transport in the shoot facilitates communication between shoot apices. PLoS Biol 14: e1002446.

Biemelt S, Tschiersch H, Sonnewald U (2004) Impact of altered gibberellin metabolism on biomass accumulation, lignin biosynthesis, and photosynthesis in transgenic tobacco plants. Plant Physiol 135: 254–265.

*Bollhöner B, Prestele J, Tuominen H (2012) Xylem cell death: emerging understanding of regulation and function. J Exp Bot 63: 1081–1094.

*Bonke M, Thitamadee S, Mähönen AP, Hauser MT, Helariutta Y (2003) APL regulates vascular tissue identity in *Arabidopsis*. Nature 426: 181–186.

Bowman JL, Kohchi T, Yamato KT, Jenkins J, Shu S, Ishizaki K, Yamaoka S, Nishihama R, Nakamura Y, Berger F, Adam C, Aki SS, Althoff F, Araki T, Arteaga-Vazquez MA, Balasubrmanian S, Barry K, Bauer D, Boehm CR, Briginshaw L, Caballero-Perez J, Catarino B, Chen F, Chiyoda S, Chovatia M, Davies KM, Delmans M, Demura T, Dierschke T, Dolan L, Dorantes-Acosta AE, Eklund DM, Florent SN, Flores-Sandoval E, Fujiyama A, Fukuzawa H, Galik B, Grimanelli D, Grimwood J, Grossniklaus U, Hamada T, Haseloff J, Hetherington AJ, Higo A, Hirakawa Y, Hundley HN, Ikeda Y, Inoue K, Inoue SI, Ishida S, Jia Q, Kakita M, Kanazawa T, Kawai Y, Kawashima T, Kennedy M, Kinose K, Kinoshita T, Kohara Y, Koide E, Komatsu K, Kopischke S, Kubo M, Kyozuka J, Lagercrantz U, Lin SS, Lindquist E, Lipzen AM, Lu CW, De Luna E, Martienssen RA, Minamino N, Mizutani M, Mizutani M, Mochizuki N, Monte I, Mosher R, Nagasaki H, Nakagami H, Naramoto S, Nishitani K, Ohtani M, Okamoto T, Okumura M, Phillips J, Pollak B, Reinders A, Rövekamp M, Sano R, Sawa S, Schmid MW, Shirakawa M, Solano R, Spunde A, Suetsugu N, Sugano S, Sugiyama A, Sun R, Suzuki Y, Takenaka M, Takezawa D, Tomogane H, Tsuzuki M, Ueda T, Umeda M, Ward JM, Watanabe Y, Yazaki K, Yokoyama R, Yoshitake Y, Yotsui I, Zachgo S, Schmutz J (2017) Insights into land plant evolution garnered from the *Marchantia polymorpha* genome. Cell 171: 287–304.

Buttò V, Rossi S, Deslauriers A, Morin H (2019) Is size an issue of time? Relationship between the duration of xylem development and cell traits. Ann Bot 123: 1257–1265.

Cai J, Tyree MT (2014) Measuring vessel length in vascular plants: can we divine the truth? History, theory, methods, and contrasting models. Trees 28: 643–655.

Cronshaw J, Anderson R (1971) Phloem differentiation in tobacco pith culture. J Ultrastruct Res 34: 244–259.

Cutter EG (1978) *Plant Anatomy: Experiment and Interpretation. P.I. Cells and Tissues.* 2nd edn, Edward Arnold, London.

Dalessandro G (1973) Interaction of auxin, cytokinin and gibberellin on cell division and xylem differentiation in cultured explants of Jerusalem artichoke. Plant Cell Physiol 14: 1167–1176.

*Dayan J, Schwarzkopf M, Avni A, Aloni R (2010) Enhancing plant growth and fiber production by silencing GA 2-oxidase. Plant Biotechnol J 8: 425–435.

*Dayan J, Voronin N, Gong F, Sun T-p, Hedden P, Fromm H, Aloni R (2012) Leaf-induced gibberellin signaling is essential for internode elongation, cambial activity, and fiber differentiation in tobacco stems. Plant Cell 24: 66–79.

*Demura T (2014) Tracheary element differentiation. Plant Biotechnol Rep 8: 17–21.

Ek M, Gellerstedt G, Henriksson G (2009) Lignin, In: *Pulp and paper Chemistry and Technology, Volume 1: Wood Chemistry and Biotechnology*, G Henriksson ed. Walter de Gruyter GmbH & Co. kg, Berlin, pp 121–124.

Eriksson ME, Israelsson M, Olsson O, Moritz T (2000) Increased gibberellin biosynthesis in transgenic trees promotes growth, biomass production and xylem fiber length. Nat Biotechnol 18: 784–788.

Eriksson ME, Moritz T (2002) Daylength and special expression of gibberellin 20-oxidase isolated from hybrid aspen (*Populus termulata* x *P. termuloides* Michx.). Planta 214: 920–930.

Esau K (1965) *Vascular Differentiation in Plants*. Holt, Rinhart & Winston, New York.

Esau K (1969) *The Phloem*. In: *Encyclopedia of Plant Anatomy*, Gebrüder Borntraeger, Berlin.

Evert RF (2006) *Esau's Plant Anatomy, Meristems, Cells, and Tissues of the Plant Body – their Structure, Function and Development*. 3rd edn, Wiley & Sons, Hoboken, NJ.

Evert RF, Eichhorn SE (2013) *Raven Biology of Plants*, 8th edn. Freeman, New York.

Fahn A (1990) *Plant Anatomy*, 4th edn. Pergamon Press, Oxford.

Fosket DE, Torrey JG (1969). Hormonal control of cell proliferation and xylem differentiation in cultured tissues of *Glycine max* var. Biloxi. Plant Physiol 44: 781–880.

Fukuda H (1992) Tracheary element formation as a model system of cell differentiation. Inter Rev Cytol 136: 289–332.

*Fukuda H (1996) Xylogenesis: initiation, progression, and cell death. Annu Rev Plant Physiol Plant Mol Biol 47: 299–325.

Fukuda H (2004) Signals that control vascular cell differentiation. Nat Rev Mol Cell Biol 5: 379–391.

*Fukuda H (2016) Signaling, transcriptional regulation, and asynchronous pattern formation governing plant xylem development. Proc Jpn Acad Ser B Phys Biol Sci 92: 98–107.

Fukuda H Komamine A (1980) Establishment of an experimental system for the study of tracheary element differentiation from single cells isolated from the mesophyll of *Zinnia elegans*. Plant Physiol. 65: 57–60.

Fukuda H, Ohashi-Ito K (2019) Vascular tissue development in plants. Curr Top Dev Biol 131: 141–160.

Geldner N, Friml J, Stierhof YD, Jürgens G, Palme K (2001) Auxin transport inhibitors block PIN1 cycling and vesicle trafficking. Nature 413: 425–428.

Gerrienne P, Gensel PG, Strullu-Derrien C, Lardeux H, Steemans P, Prestianni C (2011) A simple type of wood in two early Devonian plants. Science 333: (6044) 837.

Harrison CJ, Morris JL (2017) The origin and early evolution of vascular plant shoots and leaves. Phil Trans R Soc B 373: 20160496.

Hess T, Sachs T (1972) The influence of a mature leaf on xylem differentiation. New Phytol 71: 903–914.

Hoagland DR, Arnon DI (1950) The water-culture method for growing plants withouth soil. Calif Agric Exp Stn Circ No 347: 1–32.

Horner HT, Lersten NR, Wirth CL (1994) Quantitative survey of sieve tube distribution in foliar terminal veins of ten dicot species. Am J Bot 81: 1267–1274.

Iakimova ET, Woltering EJ (2017) Xylogenesis in zinnia (Zinnia elegans) cell cultures: unravelling the regulatory steps in a complex developmental programmed cell death event. Planta 245: 681–705.

Iwasaki T, Shibaoka H (1991). Brassinosteroids act as regulators of tracheary-element differentiation in isolated *Zinnia* mesophyll cells. Plant Cell Physiol 32: 1007–1014.

Jacobs WP (1952) The role of auxin in differentiation of xylem around a wound. Am J Bot 39: 301–309.

*Jacobs WP (1970) Regeneration and differentiation of sieve tube elements. Int Rev Cytol 28: 239–273.

Jeffs RA, Northcote DH (1966) Experimental induction of vascular tissue in an undifferentiated plant callus. Biochem J 101: 146–152.

Jura-Morawiec J (2017) Atypical origin, structure and arrangement of secondary tracheary elements in the stem of the monocotyledonous dragon tree, *Dracaena draco*. Planta 245: 93–99.

*Kabbage M, Kessens R, Bartholomay LC, Williams B (2017) The Life and Death of a Plant Cell. Annu Rev Plant Biol 68: 375–404.

Kalev N, Aloni R (1998) Role of auxin and gibberellin in regenerative differentiation of tracheids in *Pinus pinea* seedlings. New Phytol 138: 461–468.

Kleine-Vehn J, Dhonukshe P, Swarup R, Bennett M, Friml J (2006) Subcellular trafficking of the *Arabidopsis* auxin influx carrier AUX1 uses a novel pathway distinct from PIN1. Plant Cell 18: 3171–3181.

Kohlenbach HW and Schmidt B (1975) Cytodifferenzierung in Form einer direkten Umwandlung isolierter Mesophyllzellen zu Tracheiden. Z Pflanzenphysiol 75: 369–374.

Kondo Y, Nurani AM, Saito C, Ichihashi Y, Saito M, Yamazaki K, Mitsuda N, Ohme-Takagi M, Fukuda H (2016) Vascular cell induction culture system using *Arabidopsis* Leaves (VISUAL) reveals the sequential differentiation of sieve element-like cells. Plant Cell 28: 1250–1262.

*Kubo M, Udagawa M, Nishikubo N, Horiguchi G, Yamaguchi M, Ito J, Mimura T, Fukuda H, Demura T (2005) Transcription switches for protoxylem and metaxylem vessel formation. Gene Dev 19: 1855–1860.

LaMotte CE, Jacobs WP (1963) A role of auxin in phloem regeneration in *Coleus* internodes. Dev Biol 8: 80–98.

Larson PR (1964) Some indirect effects of environment on wood formation, In: *Formation of Wood in Forest Trees*, MH Zimmermann (ed). Academic Press, New York, pp 345–65.

Larson PR (1969) *Wood Formation and the Concept of Wood Quality*. Yale University, School of Forestry, Bulletin 74, New Haven.

Lersten NR (1989) Paraveinal mesophyll, and its relationship to vein endings, in *Solidago canadensis* (*Asteraceae*). Can J Bot 67: 1429–1433.

Lersten NR (1990) Sieve tubes in foliar vein endings: review and quantitative survey of *Rudbeckia laciniata* (Asteraceae). Am J Bot 77: 1132–1141.

Lersten NR, Curtis JD (1993) Paraveinal mesophyll in *Calliandra tweedii* and *C. emarginata* (Leguminosa; Mimosoideae). Am J Bot 80: 561–568.

Li Y, Hagen G, Guilfoyle TJ (1992) Altered morphology in transgenic tobacco plants that overproduce cytokinins in specific tissues and organs. Dev Biol 153: 386–395.

Li H, Taylor EL, Taylor TN (1996) Permian vessel elements. Science 271: 188–189.

Lucas WJ, Groover A, Lichtenberger R, Furuta K, Yadav SR, Helariutta Y, He XQ, Fukuda H, Kang J, Brady SM, Patrick JW, Sperry J, Yoshida A, López-Millán AF,Grusak MA, Kachroo P (2013) The plant vascular system: evolution, development and functions. J Integr Plant Biol 55: 294–388.

Luxová M (1986) The hydraulic safety zone at the base of barley roots. Planta 169: 465–470.

Mattsson J, Sung ZR, Berleth T (1999) Responses of plant vascular systems to auxin transport inhibition. Development 126: 2979–2991.

Medford JL, Horgan R, El-Sawi Z, Klee HJ (1989). Alteration of endogenous cytokinins in transgenic plants using a chimeric isopentenyl transferase gene. Plant Cell 1: 403–413.

*Meents MJ, Watanabe Y, Samuels AL (2018) The cell biology of secondary cell wall biosynthesis. Ann Bot 121: 1107–1125.

Minocha SC (1984) The role of benzyladenine in the differentiation of tracheary elements in Jerusalem artichoke tuber explants cultured *in vitro*. J Exp Bot 35: 1003–1015.

Oda Y, Fukuda H (2012) Secondary cell wall patterning during xylem differentiation. Curr Opin Plant Biol 15: 38–44.

*Ohtani M, Akiyoshi N, Takenaka Y, Sano R, Demura T (2017) Evolution of plant conducting cells: perspectives from key regulators of vascular cell differentiation. J Exp Bot 68: 17–26.

Pesquet E, Tuominen H (2011) Ethylene stimulates tracheary element differentiation in *Zinnia elegans* cell cultures. New Phytol 190: 138–149.

Pesquet E, Zhang B, Gorzsas A, Puhakainen T, Serk H, Escamez S, Barbier O, Gerber L, Courtois-Moreau C, Alatalo E, Paulin L, Kangasjarvi J, Sundberg B, Goffner D, Tuominen H (2013) Noncell-autonomous postmortem lignification of tracheary elements in *Zinnia elegans*. Plant Cell 25: 1314–1328.

Phillip R (1987) Effects of sequential exposure to auxin and cytokinin in cultured explants of Jerusalem artichoke (*Helianthus tuberosus* L.). Ann Bot 59: 245–250.

*Regnault T, Davière JM, Wild M, Sakvarelidze-Achard L, Heintz D, Carrera Bergua E, Lopez Diaz I, Gong F, Hedden P, Achard P (2015) The gibberellin precursor GA12 acts as a long-distance growth signal in *Arabidopsis*. Nat Plants 1: 15073.

Rensing SA, Lang D, Zimmer AD, Terry A, Salamov A, Shapiro H, Nishiyama T, Perroud PF, Lindquist EA, Kamisugi Y, Tanahashi T, Sakakibara K, Fujita T,Oishi K, Shin-I T, Kuroki Y, Toyoda A, Suzuki Y, Hashimoto S, Yamaguchi K, Sugano S, Kohara Y, Fujiyama A, Anterola A, Aoki S, Ashton N, Barbazuk WB,Barker E, Bennetzen JL, Blankenship R, Cho SH, Dutcher SK, Estelle M, Fawcett JA, Gundlach H, Hanada K, Heyl A, Hicks KA, Hughes J, Lohr M, Mayer K,Melkozernov A, Murata T, Nelson DR, Pils B, Prigge M, Reiss B, Renner T, Rombauts S, Rushton PJ, Sanderfoot A, Schween G, Shiu SH, Stueber K,Theodoulou FL, Tu H, Van de Peer Y, Verrier PJ, Waters E, Wood A, Yang L, Cove D, Cuming AC, Hasebe M, Lucas S, Mishler BD, Reski R, Grigoriev IV,Quatrano RS, Boore JL (2008) The Physcomitrella genome reveals evolutionary insights into the conquest of land by plants. Science 319: 64–69.

Raven JA (2003) Long-distance transport in non-vascular plants. Plant Cell Environ 26: 73–85.

Raven PH, Evert RF, Eichhorn SE (2005) *Biology of Plants*. 7th edn, Freeman, New York.

Rier JP, Beslow DT (1967) Sucrose concentration and the differentiation of xylem in callus. Bot Gaz 128: 73–77.

Roberts LW, Gahan BP, Aloni R (1988) *Vascular Differentiation and Plant Growth Regulators*. Springer-Verlag, Berlin.

Sachs T (1972) The induction of fibre differentiation in peas. Ann Bot 36: 189–197.

*Sachs T (1981) The control of patterned differentiation of vascular tissues. Adv Bot Res 9: 151–262.

Saks Y, Aloni R (1985) Polar gradients of tracheid number and diameter during primary and secondary xylem development in young seedlings of *Pinus pinea* L. Ann Bot 56: 771–778.

Saks Y, Feigenbaum P, Aloni R (1984) Regulatory effect of cytokinin on secondary xylem fiber formation in an *in vivo* system. Plant Physiol 76: 638–642.

*Scarpella E, Helariutta Y (2010) Vascular pattern formation in plants. Curr Top Dev Biol 91: 221–265.

Schmitz K (1990) Algae. In: *Sieve Elements. Comparative Structure, Induction and Development*, HD Behnke, RD Sjolund (eds). Springer, Berlin, pp 1–18.

*Serk H, Gorzsás A, Tuominen H, Pesquet E (2015) Cooperative lignification of xylem tracheary elements. Plant Signal Behav 10: e1003753.

Sieburth LE (1999) Auxin is required for leaf vein pattern in *Arabidopsis*. Plant Physiol 121: 1179–1190.

Sircar SM, Chakraverty R (1960) The effect of gibberellic acid on jute (*corchorus capsularis* Linn). Sci Cult 26: 141–143.

Smith RA, Schuetz M, Roach M, Mansfield SD, Ellis B, Samuels L (2013) Neighboring parenchyma cells contribute to *Arabidopsis* xylem lignification, while lignification of interfascicular fibers is cell autonomous. Plant Cell 25: 3988–3999.

Stant MY (1961) The effect of gibberellic acid on fibre-cell length. Ann Bot 25: 453–462.

Stant MY (1963) The effect of gibberellic acid on cell width and the cell wall of some phloem fibres. Ann Bot 27: 185–196.

Taiz L, Zeiger E, Møller IM, Murphy A (2018) *Fundamentals of Plant Physiology*. Sinauer, Sunderland, MA.

Thompson NP, Jacobs WP (1966) Polarity of IAA effect on sieve-tube and xylem regeneration in *Coleus* and tomato stems. Plant Physiol 41: 673–682.

Tokunaga N, Uchimura N, Sato Y (2006) Involvement of gibberellin in tracheary element differentiation and lignification in *Zinnia elegans* xylogenic culture. Protoplasma 228: 179–187.

Torrey JG, Fosket DE, Hepler PK (1971) Xylem formation: a paradigm of cytodifferentiation in higher plants. Amer Sci 59: 338–352.

Tyree MT, Zimmermann MH (2002) *Xylem Structure and the Ascent of Sap*, 2nd edn. Springer, Berlin.

Uggla C, Moritz T, Sandberg G, Sundberg B (1996) Auxin as a positional signal in pattern formation in plants. Proc Nat Acad Sci USA 93: 9282–9286.

Wetmore RH, De Maggio AE, Rier JP (1964) Contemporary outlook on the differentiation of vascular tissue. Phytomorphology 14: 203–217.

Wetmore RH, Rier JP (1963) Experimental induction of vascular tissue in callus of angiosperms. Am J Bot 50: 418–430.

*Xu B, Ohtani M, Yamaguchi M, Toyooka K, Wakazaki M, Sato M, Kubo M, Nakano Y, Sano R, Hiwatashi Y, Murata T, Kurata T, Yoneda A, Kato K, Hasebe M, Demura T (2014) Contribution of NAC transcription factors to plant adaptation to land. Science 343: 1505–1508.

*Yamaguchi M, Goué N, Igarashi H, Ohtani M, Nakano Y, Mortimer JC, Nishikubo N, Kubo M, Katayama Y, Kakegawa K, Dupree P, Demura T (2010) VASCULAR-RELATED NAC-DOMAIN6 and VASCULAR-RELATED NAC-DOMAIN7 effectively induce transdifferentiation into xylem vessel elements under control of an induction system. Plant Physiol 153: 906–914.

Yoshida S, Kuriyama H, Fukuda H (2005) Inhibition of transdifferentiation into tracheary elements by polar auxin transport inhibitors through intracellular auxin depletion. Plant Cell Physiol 46: 2019–2028.

Zimmermann MH (1983) *Xylem Sructure and the Ascent of Sap*. Springer-Verlag, Berlin.

Apical Dominance and Vascularization

6

6.1 Organ Communication

The hormonal signals that induce vascular differentiation are the controlling signals that synchronize plant development, organ growth regulation, and feedback cross talks between the shoot and the root. These signals regulate organ development by promoting or inhibiting plant organ growth and therefore should be clarified in order to understand the regulation of vascular differentiation. There is a continuous positive hormonal feedback communication between the shoot apices and the root tips that synchronizes plant development; each plant pole sends its growth-promoting hormonal signal to the opposite side of the plant, informing the other plant pole about its activity and quantity. The major shoot signal is auxin produced in the apical bud and young leaves, while the basic root tip signals produced in the root cap are cytokinins. Auxin promotes the initiation and development of the roots, while cytokinins from the root tips promote the development and growth of the shoot organs.

The auxin flows polarly from the young leaves to root tips and induces continuous vascular tissues along the plant, forming the pathways for further signal flows, while cytokinins from the root tips are transported upward, promote bud, and leave development, cell division activity in the vascular system, activating the meristematic cambium and secondary vascular differentiation along the plant (Sachs 1981; Aloni 1995; Scarpella and Helariutta 2010).

Conversely, as will be clarified below, due to inside organ competition, identical organs may cause inhibition, when one of them becomes dominant and retards the others, which is well studied in shoots and known as shoot apical dominance.

© Springer Nature Switzerland AG 2021

R. Aloni, *Vascular Differentiation and Plant Hormones*,

https://doi.org/10.1007/978-3-030-53202-4_6

6.2   Regulation of Shoot Apical Dominance

Apical dominance is the control exerted by the apical bud over the outgrowth of the lateral buds (Thimann and Skoog 1933; Sachs 1991), which is most easily demonstrated by shoot decapitation that results in the prompt outgrowth of the axillary buds.

1. **Auxin** produced in the apical bud and its young leaves is the primary signal that inhibits the growth of the axillaries. This phenomenon of apical bud control is replaceable by auxin application after shoot decapitation (Thimann and Skoog 1933, 1934; Thimann et al. 1971). The auxin produced in the shoot apical bud induces well-developed vascular tissues which supply the apical bud, while the inhibited lateral buds cannot develop their supporting vascular tissues and therefore remain suppressed.

Understanding the control mechanism of apical dominance is important for agriculture implications to determine plant architecture by pruning trees for increasing fruit productivity and quality or for ornamental purposes. Apical dominance is an adaptive mechanism which gives the plant an advantage in competition for light, as it enables plants to concentrate their energy, nutrients, and growth to promote the optimal fast growth of the main stem and successfully compete with neighboring plants. On the other hand, the release from apical dominance is a plant survival mechanism, which enables recovery and regeneration of the shoot after a damage to the main stem.

In addition to the classic model that the auxin indole-3-acetic acid is produced in the shoot's apical bud and transported down the stem, where it inhibits axillary bud growth (Thimann and Skoog 1933, 1934), shoot apical dominance is also regulated by the following major signals: strigolactone, cytokinin, gibberellin, and sugar availability, and it is clarified below.

Shoot branching evolved independently in flowering plant sporophytes and moss gametophytes (the haploid growth phase). Lateral branches of the gametophytic shoots of the moss *Physcomitrella* arise by re-specification of epidermal cells into branch initials. Interestingly, like in sporophytic branching of higher plants, the ancient hormonal signals, namely, auxin, cytokinin, and strigolactone, regulate branch development in moss (Coudert et al. 2015). The size of the apical inhibition zone significantly increased by an auxin application, the branch number was strongly reduced, and branch initiation in the branching zone significantly decreased, suggesting that auxin can act as a global suppressor of shoot branching of mosses and higher plants. However, the apical dominant mechanisms are different, lateral bud inhibition in higher plants is regulated by PIN-mediated basipetal auxin transport, whereas in the moss *Physcomitrella*, the mechanism is less developed and a bi-directional transport occurs likely through plasmodesmata (Coudert et al. 2015).

2. **Strigolactone** is an important hormonal signal involved in promoting shoot apical dominance by preventing axillary bud development. The strigolactone is a

root hormone which moves upward to the shoot apex through the xylem (Kohlen et al. 2011), where it inhibits axillary bud development by downregulating the cytokinin biosynthetic genes (*IPT*). Auxin from the apical bud promotes strigo-lactone activity in nodes, where the strigolactone inhibits the activation of cyto-kinin producing genes (El-Showk et al. 2013). Additionally, the strigolactone can inhibit axillary growth by triggering rapid depletion of the auxin efflux pro-tein PIN1 from the plasma membrane of parenchyma cells, located between the axillary buds and the vascular system of the stem (Shinohara et al. 2013), namely, preventing polar auxin flow from the axillary buds to the stem, which prevents vascular development from the inhibited buds into the stem.

3. **Cytokinin** (CK) produced primary in root tips is known to promote the out-growth of axillary buds (Sachs and Thimann 1967; Thimann et al. 1971). Tanaka et al. (2006) demonstrated that auxin negatively regulates local CK biosynthesis in the nodes along the stem by controlling the expression level of the pea (*Pisum sativum*) gene adenosine phosphate-isopentenyltransferase (*PsIPT*), which encodes a key enzyme in CK biosynthesis. Expression of *PsIPT* was repressed by the application of the auxin IAA. These results indicate that in apical domi-nance, one role of auxin is to repress local biosynthesis of CK in the stem nodes. After decapitation, the CKs, which are thought to be derived from the root caps, can also be locally synthesized in the nodes along the stem, rather than only in the root tips. Müller et al. (2015) confirmed that shoot decapitation promotes cytokinin biosynthesis in the stem, suggesting that CK acts to overcome auxin-mediated bud inhibition, allowing buds to escape apical dominance under favor-able conditions, such as high nitrate availability, conditions known to stimulate CK production in the root, which is transported upward to the shoot (Geßler et al. 2004; Miyawaki et al. 2004) (see Sect. 3.3 and Chap. 4).

4. **Gibberellin** (GA) acts as a positive regulator in promoting shoot branching in the woody plant *Jatropha curcas* and other trees like papaya, indicating that the regu-latory control of shoot branching in some perennial woody plants may be more complicated. Ni et al. (2015) showed that GA and CK synergistically promote lateral bud outgrowth. Treatment with paclobutrazol, an inhibitor of de novo GA biosynthesis, significantly reduced the promotion of bud outgrowth by CK, sug-gesting that GA is required for CK-mediated axillary bud outgrowth (Ni et al. 2015).

5. **Sugar availability** regulates shoot apical dominance as has been proposed by the *auxin-directed nutrient hypothesis* (see Cline 1991), suggesting that auxin originating in the growing shoot apex directs nutrient transport to the actively growing apex supplied with well-developed vascular tissues, but away from the inactive lateral buds which do not grow out because of insufficient nutrients. This idea was supported by Mason et al. (2014) findings in *Pisum sativum* show-ing that sugar demand, not auxin, is the initial regulator of apical dominance, supporting the idea that apical dominance is controlled by the shoot tip's strong demand for sugars arriving via the phloem, which inhibits axillary bud outgrowth by substantially limiting the amount of sugar available for the axillary buds. After the loss of the shoot tip, sugars are rapidly redistributed over large distances

and accumulate in axillary buds within a timeframe that correlates with bud release. Barbier et al. (2015) confirmed the importance of sucrose as an early modulator of the hormonal mechanism controlling bud outgrowth in *Rosa hybrida* shoots.

Aloni et al. (2003, 2006b) found that from their early developmental stage, the leaf and flower primordia are loaded with conjugated auxin that can release the bioactive IAA when the tissues above (or in flowers beside, see Sect. 8.3) them mature or experimentally removed. As will be discussed in Chaps. 7 and 8 on leaves and flowers, the concept of shoot apical dominance can be applied to understand flower-organ development and organized vascularization in leaves and flowers, as will be clarified in the two following chapters (Chaps. 7 and 8).

6.3   Cytokinin-Dependent Root Apical Dominance

In tissue cultures, applications of elevated-auxin concentrations inhibit shoot-organ formation and promote root development, while high-cytokinin concentrations inhibit root formation and promote shoot development (Skoog and Miller 1965; Taiz and Zeiger 2006; Evert and Eichhorn 2013). The auxin from the apical bud and young leaves which inhibits shoot axillary buds is the promoting signal for root initiation and growth (Casimiro et al. 2001; Bhalerao et al. 2002). On the other hand, with a similar biological behavior, but in the opposite direction, the cytokinin produced in the main root tip inhibits the initiation of lateral roots (Rani Debi et al. 2005; Laplaze et al. 2007; Chang et al. 2013; Márquez et al. 2019) and is the promoting signal for shoot development (Aloni et al. 2006a). Therefore, the concept of apical dominance that originally was developed for understanding shoot development and architecture (Thimann and Skoog 1933, 1934) can properly be used for roots (Böttger 1974; Aloni et al. 2006a) as a general concept for both shoots and roots.

Actively growing primary roots of dicot plants may exhibit root apical dominance, where the primary root inhibits lateral root initiation (Zhang and Hasenstein 1999; Lloret and Casero 2002). The phenomenon of root apical dominance can be compared with the well-known shoot apical dominance; in both cases, the actively growing leader inhibits lateral identical organ initiation and development (Aloni et al. 2006a). In lettuce (*Lactuca sativa*) seedlings, the removal of the primary root tip stimulates rapid formation of lateral-root primordia, which normally are not produced by the intact lettuce root (Zhang and Hasenstein 1999). In maize (*Zea mays*), cytokinin application prevented lateral root initiation in the primary root initiation zone, and the inhibitory effect of CK occurs in the earliest stages of lateral-root development (Márquez et al. 2019).

Free bioactive CK can be visualized by the expression of *ARR5::GUS* (a CK-activated promoter sequence of an *Arabidopsis response regulator* fused to β-glucuronidase), which reflects the sites of the transcriptional activation of this CK-sensitive promoter. The construct reacts with free bioactive CK in a concentration-dependent manner (D'Agostino et al. 2000; Aloni et al. 2004, 2005). This construct shows that the cap of the primary root of a soil-grown *Arabidopsis* plant produces elevated concentrations of free bioactive CK (Fig. 6.1A), which is usually much higher than free CK concentrations of lateral roots (Fig. 6.1B, C) (Aloni et al. 2005). When primary root tips are excised, one or more lateral roots become dominant and their tip may start to produce higher concentrations of free CK (Forsyth and Van Staden 1981; Lloret and Casero 2002).

Root apical dominance may occur in wild-type plants with an actively growing primary root that inhibits lateral root initiation by the root-cap-synthesized CK, and their lateral roots develop further away from the tip of the main root. By contrast, a low CK content in CK-deficient transgenic plants (overexpressing the *CYTOKININ OXIDASES/DEHYDROGENASES (CKX)* genes that catalyze irreversible degradation of the cytokinins; Werner et al. 2001, 2003), or almost CK insensitivity [in the double and triple loss-of-function CK receptor, of the *ahk* (*Arabidopsis histidine kinase*) mutants, which are almost insensitive to cytokinin], result in the formation of lateral roots closer to the root tip and an increase in root branching (Schmülling 2002). Emphasize and extend the previously formulated concept of root apical dominance (Böttger 1974; Zhang and Hasenstein 1999; Lloret and Casero 2002) by focusing on CK mediation (Aloni et al. 2006a). The evidence that the root cap predominantly produces CK by expressing the *IPT* genes (Miyawaki et al. 2004; Takei et al. 2004; Sakakibara et al. 2006; Ruffel et al. 2011) and that the highest concentration of free CK in a root is found in the root tip (Aloni et al. 2004, 2005) justify a more precise term for this instance of apical dominance, namely, **cytokinin-dependent root apical dominance**. From an ecological point of view, this CK-dependent root apical dominance gives priority to the primary root in competition with its own lateral roots as well as neighboring root systems and enables the main root to reach water in deeper soil layers faster, which might be vital for plants before the dry season. CK regulates root architecture by balancing the promoting role of IAA on lateral root development (see Chap. 9). CK produced in the active root cap (see Chap. 2, Fig. 2.2B) of a primary root is the hormonal signal which enables maximum development of an actively growing primary root by retarding lateral root initiation (see Chap. 9, Fig. 9.2). This reduces the quantity of lateral roots, their development and requirements of which would be at the expense of the primary root growth.

Summary

- A continuous positive hormonal feedback communication occurs between the auxin-producing young leaves and the cytokinin-producing root tips that synchronize plant development. Auxin promotes the development of the root system, and cytokinin stimulates the growth of the shoot organs.
- Actively growing leader, of either the shoot apical bud or the primary root tip, inhibits lateral identical organ initiation and development, which prevents competition of similar organ. This behavior of the leader stimulated the developed of the apical dominance concept.

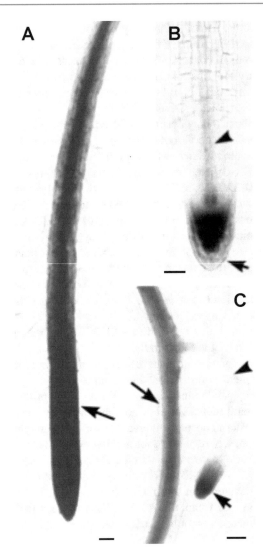

Fig. 6.1 *GUS* expression patterns visualizing free cytokinin (CK) in the root tip of CK-responsive *ARR5::GUS* transformant of *Arabidopsis thaliana* plants grown protected from wind, demonstrating the high-CK concentrations in the main root compared with much lower CK production in the lateral roots. (**A**) Main root of soil-grown wind-protected plant at 35 days after germination, with strong *ARR5::GUS* expression reflecting massive CK production and accumulation due to wind protection in the entire elongating zone (*arrow*), in the vascular cylinder, cortex, and epidermis. The concentration of *GUS* expression in the cortex decreases gradually. (**B** and **C**) Lateral roots and main root grown on MS basal medium in closed boxes. (**B**) Due to typically low CK production in a lateral root, *ARR5::GUS* expression was only observed in the root cap, with CK export restricted to the base of the central vascular tissue (*arrowhead*). The periphery of the root cap was almost free of *GUS* expression (*arrow*). (**C**) CK production in the tip of a lateral root (*short arrow*), and almost absent from its vascular cylinder, cortex and epidermis (*arrowhead*), while strong CK accumulation is observed along the entire axis of the main root (*large arrow*). Bars = 25 μm (**B**), 50 μm (**A**, **C**). (From Aloni et al. 2005)

- Shoot apical dominance is the control imposed by the apical bud to inhibit the outgrowth of lateral buds. Auxin produced in the apical bud is the primary signal that prevents the development of axillary buds.
- In addition to the primary role of auxin in inhibiting axillary buds, shoot apical dominance is also regulated by strigolactone, cytokinin, gibberellin, and sugar availability.
- Root apical dominance is regulated by the production of high concentrations of cytokinin in the root cap of the primary root, which inhibits the initiation and growth of lateral roots.

References and Recommended Readings[1]

Aloni R (1995) The induction of vascular tissues by auxin and cytokinin. In: *Plant hormones: physiology, biochemistry and molecular biology*, PJ Davies (ed). Kluwer Academic Publishers, Dordrecht, pp 531–546.

*Aloni, R, Aloni E, Langhans M, Ullrich CI (2006a) Role of cytokinin and auxin in shaping root architecture: regulating vascular differentiation, lateral root initiation, root apical dominance and root gravitropism. Ann Bot 97: 883–893.

Aloni R, Aloni E, Langhans M, Ullrich CI (2006b) Role of auxin in regulating *Arabidopsis* flower development. Planta 223: 315–328.

Aloni R, Langhans M, Aloni E, Dreieicher E, Ullrich CI (2005) Root-synthesized cytokinin in *Arabidopsis* is distributed in the shoot by the transpiration stream. J Exp Bot 56: 1535–1544.

Aloni R, Langhans M, Aloni E, Ullrich CI (2004) Role of cytokinin in the regulation of root gravitropism. Planta 220: 177–182.

Aloni R, Schwalm K, Langhans M, Ullrich CI (2003) Gradual shifts in sites of free-auxin production during leaf-primordium development and their role in vascular differentiation and leaf morphogenesis in *Arabidopsis*. Planta 216: 841–853.

Barbier F, Péron T, Lecerf M, Perez-Garcia MD, Barrière Q, Rolčík J, Boutet-Mercey S, Citerne S, Lemoine R, Porcheron B, Roman H, Leduc N, Le Gourrierec J, Bertheloot J, Sakr S (2015) Sucrose is an early modulator of the key hormonal mechanisms controlling bud outgrowth in *Rosa hybrida*. J Exp Bot 66: 2569–2582.

Bhalerao RP, Eklof J, Ljung K, Marchant A, Bennett M, Sandberg G (2002) Shoot-derived auxin is essential for early lateral root emergence in *Arabidopsis* seedlings. Plant J 29: 325–332.

Böttger M (1974) Apical dominance in roots of *Pisum sativum* L. Planta 121: 253–261.

Casimiro I, Marchant A, Bhalerao RP, Beeckman T, Dhooge S, Swarup R, Graham N, Inzé D, Sandberg G, Casero PJ, Bennett M (2001) Auxin transport promotes *Arabidopsis* lateral root initiation. Plant Cell 13: 843–852.

Chang L, Ramireddy E, Schmülling T (2013) Lateral root formation and growth of *Arabidopsis* is redundantly regulated by cytokinin metabolism and signalling genes. J Exp Bot 64: 5021–5032.

Cline MG (1991) Apical dominance. Bot Rev 57: 318–358.

Coudert Y, Palubicki W, Ljung K, Novak O, Leyser O, Harrison CJ (2015) Three ancient hormonal cues co-ordinate shoot branching in a moss. eLife 4: e06808.

D'Agostino IB, Deruère J, Kieber JJ (2000) Characterization of the response of the *Arabidopsis* response regulator gene family to cytokinin. Plant Physiol 124: 1706–1717.

*El-Showk S, Ruonala R, Helariutta Y (2013) Crossing paths: cytokinin signalling and crosstalk. Development 140: 1373–1383.

---

[1] Papers of particular interest for suggested reading have been highlighted (with *).

Evert RF, Eichhorn SE (2013) *Raven Biology of Plants*. 8th edn. Freeman, New York.

Forsyth C, Van Staden J (1981) The effect of root decapitation on lateral root formation and cytokinin production in *Pisum sativum*. Physiol Plant 51: 375–379.

Geßler A, Kopriva S, Rennenberg H (2004) Regulation of nitrate uptake at the whole-tree level: interaction between nitrogen compounds, cytokinins and carbon metabolism. Tree Physiol 24: 1313–1321.

Kohlen W, Charnikhova T, Liu Q, Bours R, Domagalska MA, Beguerie S, Verstappen F, Leyser O, Bouwmeester H, Ruyter-Spira C (2011) Strigolactones are transported through the xylem and play a key role in shoot architectural response to phosphate deficiency in nonarbuscular mycorrhizal host *Arabidopsis*. Plant Physiol 155: 974–987.

Laplaze L, Benkova E, Casimiro I, Maes L, Vanneste S, Swarup R, Weijers D, Calvo V, Parizot B, Herrera-Rodriguez MB, Offringa R, Graham N, Doumas P, Friml J, Bogusz D, Beeckman T, Bennett M (2007) Cytokinins act directly on lateral root founder cells to inhibit root initiation. Plant Cell 19: 3889–3900.

Lloret PG, Casero PJ (2002) Lateral root initiation. In: *Plant roots – the hidden half,* Y Waisel, A Eshel, U Kafkafi (eds). Marcel Dekker, New York, pp 127–155.

Márquez G, Alarcón MV, Salguero J (2019) Cytokinin inhibits lateral root development at the earliest stages of lateral root primordium initiation in maize primary root. J Plant Growth Regul 38: 83–92.

*Mason MG, Ross JJ, Babst BA, Wienclaw BN, Beveridge CA (2014) Sugar demand, not auxin, is the initial regulator of apical dominance. Proc Natl Acad Sci USA 111: 6092–6097.

Miyawaki K, Matsumoto-Kitano M, Kakimoto T (2004) Expression of cytokinin biosynthetic isopentenyltransferase genes in *Arabidopsis*: tissue specificity and regulation by auxin, cytokinin, and nitrate. Plant J 37: 128–138.

Müller D, Waldie T, Miyawaki K, To JP, Melnyk CW, Kieber JJ, Kakimoto T, Leyser O (2015) Cytokinin is required for escape but not release from auxin mediated apical dominance. Plant J 82: 874–886.

Ni J, Gao C, Chen MS, Pan BZ, Ye K, Xu ZF (2015) Gibberellin promotes shoot branching in the perennial woody plant *Jatropha curcas*. Plant Cell Physiol 56: 1655–1666.

Rani Debi B, Taketa S, Ichii M (2005) Cytokinin inhibits lateral root initiation but stimulates lateral root elongation in rice (*Oryza sativa*). J Plant Physiol 162: 507–515.

Ruffel S, Krouk G, Ristova D, Shasha D, Birnbaum KD, Coruzzi GM (2011) Nitrogen economics of root foraging: Transitive closure of the nitrate–cytokinin relay and distinct systemic signaling for N supply vs. demand. Proc Natl Acad Sci USA 108: 18524–18529.

Sachs T (1981) The control of patterned differentiation of vascular tissues. Adv Bot Res 9: 151–262.

Sachs T (1991) Hormones as correlative agents. In: *Pattern Formation in Plant Tissues,* T Sachs. Cambridge University Press, Cambridge, pp 52–69.

Sachs T, Thimann, KV (1967) The role of auxins and cytokinins in the release of buds from dominance. Am J Bot 54: 136–144.

Sakakibara H, Takei K, Hirose N (2006) Interactions between nitrogen and cytokinin in the regulation of metabolism and development. Trends Plant Sci 11: 440–448.

Scarpella E, Helariutta Y (2010) Vascular pattern formation in plants. Curr Top Dev Biol 91: 221–265.

*Shinohara N, Taylor C, Leyser O (2013) Strigolactone can promote or inhibit shoot branching by triggering rapid depletion of the auxin efflux protein PIN1 from the plasma membrane. PLOS Biology 11: e1001474.

Schmülling T (2002) New insights into the functions of cytokinins in plant development. J Plant Growth Regul 21: 40–49.

Skoog F, Miller CO (1965) Chemical regulation of growth and organ formation in plant tissues cultured *in vitro*. In: *Molecular and Cellular Aspects of Development*, E Bell (ed). Harper and Row. New York, pp 481–494.

Taiz L, Zeiger E (2006) *Plant Physiology*, 4th edn. Sinauer, Sunderland MA.

Takei K, Ueda N, Aoki K, Kuromori T, Hirayama T, Shinozaki K, Yamaya T, Sakakibara H (2004) *AtIPT3* is a key determinant of nitrate-dependent cytokinin biosynthesis in *Arabidopsis*. Plant Cell Physiol 45: 1053–1062.

Tanaka M, Takei K, Kojima M, Sakakibara H, Mori H (2006) Auxin controls local cytokinin biosynthesis in the nodal stem in apical dominance. Plant J 45: 1028–36.

Thimann KV, Sachs T, Mathur KN (1971) The mechanism of apical dominance in *Coleus*. Physiol Plant 24: 68–72.

Thimann KV, Skoog F (1933) Studies on the growth hormone of plants. III. The inhibiting action of growth substance on bud development. Proc Nat Acad Sci 19: 714–716.

Thimann KV, Skoog F (1934) On the inhibition of bud development and other functions of growth-substance in *Vicia faba*. Proc Roy Soc London B 114: 317–339.

*Werner T, Motyka V, Strnad M, Schmülling T (2001) Regulation of plant growth by cytokinin. Proc Nat Acad Sci USA 98: 10487–10492.

Werner T, Motyka V, Laucou V, Smets R, Van Onckelen HV, Schmülling T (2003) Cytokinin-deficient transgenic *Arabidopsis* plants show multiple developmental alterations indicating opposite functions of cytokinins in the regulation of shoot and root meristem activity. Plant Cell 15: 2532–2550.

Zhang NG, Hasenstein KH (1999) Initiation and elongation of lateral roots in *Lactuca sativa*. Inter J Plant Sci 160: 511–519.

Leaf Development and Vascular Differentiation

7

7.1 Auxin-Dependent Leaf Initiation and Phyllotaxis

Plants exhibit remarkable plasticity in their developmental program allowing them to produce new leaves and other organs during their growth, which are under hormonal regulation. Cytokinins produced in root tips and stem nodes promote apical bud growth and leaf development (Taiz et al. 2018). Leaf initiation depends on the establishment of dynamic gradients of upward polar auxin flow from the youngest leaves toward the shoot apex, where auxin maxima induce the initiation of leaf primordia. The IAA which originated in young growing leaves moves upward in the epidermis and through the outermost apical meristematic cell layer toward the shoot tip (Chap. 3, Fig. 3.2). This upward polar auxin flow regulates plant's phyllotaxis, namely, the orderly pattern of leaf initiation and position at the shoot apex (Reinhardt et al. 2003; Benková et al. 2003; Scheres and Xu 2006; Kuhlemeier 2017; Shi and Vernoux 2019). Existing leaf primordia act as sinks, redistributing IAA and creating its heterogeneous distribution at the shoot tip. Auxin accumulation occurs only at certain minimal distances from existing primordia, defining the position of future leaf primordia. The model accounts for the repetitive nature, as well as the regularity and stability of phyllotaxis, showing that the upward polar auxin flow functions at the shoot apex as a "pattern generator" (Reinhardt et al. 2003; Kuhlemeier 2017). The upward flow of auxin at the shoot tip through the epidermis can be detected and followed by the polar location of the auxin transport protein PIN1 on the upper cell membranes (Reinhardt et al. 2003; Benková et al. 2003; Kuhlemeier 2017). It has been suggested that the PIN1-dependent local auxin gradients represent a common module for the formation of all plant organs in both the shoot and root, regardless of their mature morphology or developmental origin (Benková et al. 2003).

© Springer Nature Switzerland AG 2021
R. Aloni, *Vascular Differentiation and Plant Hormones*,
https://doi.org/10.1007/978-3-030-53202-4_7

7.2   Hormonal Control of Venation Pattern Formation in Leaves

Leaves are usually flat organs that harvest the sun energy for photosynthesis. During the vegetative phase of plant growth, young growing leaves are the major sites of auxin production that induce vascular differentiation and regeneration along the plant (Jacobs 1952; Sachs 1981; Aloni 2001; Scarpella and Helariutta 2010; Taiz et al. 2018) and are characterized by endless morphology and vascular network patterns. Leaves induce their own vascular tissues by the auxin polar transport and auxin signaling (Sachs 1981; Aloni 2001; Aloni et al. 2003; Benková et al. 2003; Mattsson et al. 2003; Scarpella et al. 2006; Wenzel et al. 2007; Verna et al. 2015, 2019; Scarpella 2017; Linh et al. 2018; Biedroń and Banasiak 2018). The question that is dealt in this chapter is how leaves produce their vascular tissues in organized and orderly patterns characterized by plasticity, continuity, and polarity of their vein network. Understanding the processes and the regulation of vein pattern formation in leaves has been, therefore, a major challenge.

Although the leaves on the same shoot are identical genetically, each leaf has a different vein pattern, demonstrating their inherent developmental plasticity (Sachs 1975, 1989, 2002; Scarpella 2017). In many leaves, the vein network produced in their right side is different from the pattern developed on the left side (e.g., Figs. 7.5 and 7.12B). This asymmetric and unpredictable vein pattern is determined by random patterns of auxin production sites and flows, regulated by selection of various alternative auxin pathways during leaf development.

Nevertheless, there is a general regularity in vein pattern formation in leaves where high-auxin concentrations at the leaf periphery inhibit and delay minor auxin production sites in the lamina; which therefore affect the developmental timing of vein hierarchy formation, which is discussed below regarding the differentiation and time course of primary and secondary veins versus tertiary veins and freely ending veinlets.

The availability of a procambium, or a nearby differentiating vein, promotes the formation of loops (Mattsson et al. 2003; Scarpella et al. 2006; Wenzel et al. 2007; Sawchuk et al. 2007; Marcos and Berleth 2014; Biedroń and Banasiak 2018). New streams of auxin tend to merge into already existing differentiating veins that transport auxin more efficiently (Sachs 1981) and therefore attracts new auxin streams to form loops (Fig. 7.1). This loop pattern, which develops from the leaf tip downward, is typically observed along the differentiating midvein in *Arabidopsis* leaves (Mattsson et al. 2003; Scarpella et al. 2006; Biedroń and Banasiak 2018).

The continuity of vein networks in wild-type leaves and the entire vascular system along the plant axis results from the continuous movement of the polar auxin inducing signal, which therefore results in continual functioning veins and bundles from the leaf's margins to the root tips (Sachs 1981; Berleth and Sachs 2001; Aloni 2001, 2013; Scarpella and Helariutta 2010; Scarpella 2017; Biedroń and Banasiak 2018).

In a young growing leaf, the bioactive auxin produced along the leaf's periphery moves polarly toward the midvein and continues down to the petiole (Fig. 7.2).

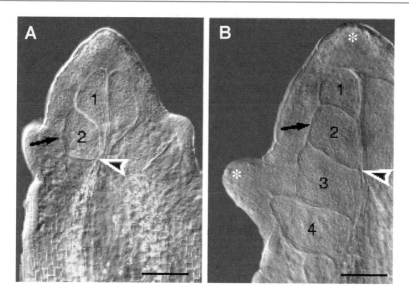

Fig. 7.1 Typical downward procambium development in wild-type *Arabidopsis thaliana* rosette leaf primordia, photographed under Nomarski illumination. (**A**) Very young primordium with two procambial loops, loop 2 (marked with *arrow*) developed after loop 1. (**B**) More developed primordium with four procambial loops numbered according to their downward (basipetal) developmental pattern. The *arrowhead* marks the analogous site in both photographs emphasizing the basipetal developmental pattern of the procambium induced by auxin. The sites of the inducing auxin maxima are marked by asterisks at the tip and lobe. The low-auxin concentration flows that induce the procambial loops tend to merge into the already formed procambium of the midvein, which is more polarized and transport efficiently the signal. Bar = 200 µm. (From Aloni 2004)

Along the plant axis, the IAA which moves polarly from the foliage to root tips induces a polar and continuous vascular system (Sachs 1981; Aloni and Zimmermann 1983; Berleth and Sachs 2001; Aloni 2001; Scarpella and Helariutta 2010; Scarpella 2017).

Most plants exhibit slight to considerable heteroblasty, where the leaves produced at different stages of shoot development vary in their morphology. In the model organism *Arabidopsis thaliana* (Meyerowitz 1989), the vegetative shoot is composed mainly of rosette leaves. Hence, understanding of leaf morphogenesis and vascular differentiation in leaves of this species is very important for correct evaluation of its mutants and their gene function. Early rosette leaves of *Arabidopsis* have rounded blades with smooth margins, while leaves produced in later stages of shoot development have slightly serrated edges (Poething 1997; Nikolov et al. 2019).

Leaves are characterized by complex networks of vascular tissues (Sachs 1975, 1981, 1989; Nelson and Dengler 1997; Candela et al. 1999; Biedroń and Banasiak 2018), fast vascular development, and early determination (Mattsson et al. 1999). Regeneration of both leaf shape (Sachs 1969) and leaf vascular differentiation is limited to early developmental stages (Sachs 1975, 1981; Mattsson et al. 1999; Sieburth 1999). In complex networks of leaves, there are strands in which the

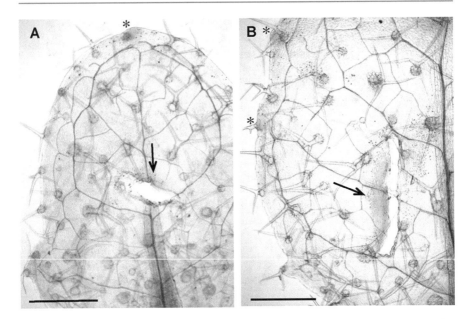

Fig. 7.2 The directions of polar auxin flow along the midvein and the secondary veins are detected by *DR5::GUS* expression beside a horizontal (**A**) and a longitudinal (**B**) cuts made in growing leaf primordia of the DR5::GUS-transformed *Arabidopsis thaliana*, revealing that the bioactive auxin flows from auxin maxima (detected by blue GUS staining) at the leaf's periphery, toward the midvein (**B**) and along the midvein (**A**) downward to the stem. The cuts stopped the auxin flow, which therefore accumulated at the cut surface. The *arrows* show the direction of the IAA flow. Auxin maxima are marked with asterisks. Bars = 1 mm (**A**), 0.65 mm (**B**)

individual cells do not have an obvious shoot-to-root polarity and therefore can be deemed not polar (Sachs 1975). It has been suggested that during the process of strand differentiation, the strand axis is determined first and along this axis auxin movement may occur in opposite directions (Sachs 1975, 1981, 1989; Avsian-Kretchmer et al. 2002; Mattsson et al. 2003; Scarpella et al. 2006). This phenomenon is a result of the early determination of parenchyma cells (between the veins) that lose the ability to redifferentiate; therefore any new auxin stream can induce a new vessel only from procambial initials inside the differentiating veins.

In wild-type plants, discontinuities in the path of xylem differentiation were detected during the process of vascular differentiation in their developing leaves (Esau 1965) (see Chap. 2, Fig. 2.21). In defective mutants, such vascular discontinuities may also be visualized in mature cotyledons and leaves (Przemeck et al. 1996; Candela et al. 1999; Carland et al. 1999; Deyholos et al. 2000; Hobbie et al. 2000; Koizumi et al. 2000). In leaves, vascular tissues become progressively confined to the leaf margin as the concentration of auxin transport inhibitors is increased, suggesting that the leaf vascular system depends on inductive signals from the margin of the leaf (Mattsson et al. 1999; Sieburth 1999). Staged application of auxin transport inhibitors has demonstrated that primary, secondary, and tertiary veins become unresponsive to further modulations of auxin transport at successive stages of early leaf development, indicating that the pattern of primary and secondary strands becomes fixed at the onset of lamina expansion (Mattsson et al. 1999; Sieburth 1999). The molecular and physiological mechanisms that govern and determine venation pattern formation in leaves were poorly understood (Telfer and Poething 1994; Nelson and Dengler 1997; Nelson 1998; Candela et al. 1999; Carland et al. 1999; Mattsson et al. 1999; Sieburth 1999; Deyholos et al. 2000); therefore, I suggested the below *leaf venation hypothesis* (Aloni 2001) as a working hypothesis for those interested in studying vascular differentiation in leaves.

7.2.1 The *Leaf Venation Hypothesis* and Evidence

In order to clarify where auxin is produced in dicotyledonous plant leaves, and how the production sites and movement of the IAA signal controls vascular differentiation in leaves, the following *leaf venation hypothesis* was proposed (Aloni 2001):

1. The expanding tissues of a leaf primordium are the sites where IAA is produced. Fast-growing regions of a leaf primordium are the major locations of high-auxin production. The term **"leaf apical dominance"** is proposed to describe, within a developing leaf, how the fast-growing tip of a primordium with its earlies high-auxin concentration suppresses the synthesis of auxin in the leaf tissues below it. Along the plant axis, this tip-inhibition effect may diminish with increasing distance of a leaf primordium from the root apices, the diminution due to declining levels of root-induced cytokinins with increasing distance from the root tips, which may enable a more complex morphology of leaves.

2. The rapidly elongating tip of a growing primordium is the first major site of high-auxin concentration in a leaf. The high-auxin concentration at the tip and expanding leaf periphery is drained basipetally into the stem. However, the leaf/stem junction is a local physiological barrier which slows down auxin flow through the junction due to its short cells (see Chap. 2, Fig. 2.27) causing a local increase in IAA concentration at the junction. Within a leaf, the junctions between the midvein (primary axial vein) and secondary veins might also slow auxin movement and cause local increasing levels of auxin concentrations at these sites. It is further proposed that during leaf development, there are gradual changes in primary high-auxin production sites along the leaf, which occur along the margins in a **downward "wave" pattern**, extending from the tip to the base of the primordium. Since the tip of a leaf primordium matures relatively fast, its auxin production decreases; then the auxin production increases in the upper lobes, which are not inhibited anymore by the tip, and they start to grow faster and become the major high-auxin synthesis sites. Later on, when the upper lobes mature, their auxin production decreases; therefore IAA production starts to increase in the lobes below them – a process that continues downward and diminishes at the base of the leaf.

3. As the primary production of high-auxin concentrations along the blade's margin decreases and accordingly their inhibition effect on the central lamina dimin-

ishes, secondary auxin synthesis sites within the lamina become relatively effective. I suggest that this relatively late auxin production in the central lamina occurs during blade expansion in the central region of lamina, possibly as a result of local intercalary expansion growth, where the inner parts of the leaf expand at different rates and in different directions (Avery 1933) – a process that has been termed anisotropic growth (Ashby 1948).

4. At very early stages of primordium development the auxin levels are low and therefore only the **basic framework** (main strands and loops) of procambium is induced first at the tip and progresses downward (Fig. 7.1). This provascular network design determines the later patterns of xylem and phloem differentiation, which will follow the procambium configuration during leaf morphogenesis.

5. A regular feature of xylem development is the formation of **discontinuous patterns** of vessels during the process of differentiation (see Chap. 2, Fig. 2.21). In other words, vessel differentiation proceeds normally in discontinuous fashion, entailing discrete vessel elements, which ultimately join into a continuous vessel. These discontinuities are formed along a differentiating vessel because the individual vessel elements have different speeds of maturation. A positive correlation is suggested between auxin level and the rate of vessel-element maturation. Differentiating vessel elements exposed to high-auxin concentrations, which are either near a site of auxin synthesis in the tip and lobes or above an obstacle to auxin flow (where auxin accumulates in junctions), differentiate faster than do the other intervening vessel elements.

6. Xylem maturation patterns are determined by auxin concentration and the ability of the differentiating cells to respond to the auxin signal. Fast leaf expansion results in high auxin production, which promotes xylem differentiation within the basic procambium network induced early in primordium development. Auxin accumulation at the base of a leaf primordium above the leaf/stem junction obstacle determines the upward (acropetal) pattern of xylem maturation in the midvein, although the general auxin polarity in leaves is from the margins downward towards the petiole (Fig. 7.2). Shortly after, the high-auxin levels produced at the tip and leaf margin induce xylem loops (within the already laid-down procambial pattern), which develop and mature downward in the basipetal direction, away from the auxin sources. High-auxin concentrations produced by the tip and lobes accelerate xylem differentiation in these locations at the leaf margin, which may result in discontinuous patterns of xylem differentiation in the lobes during early stages of leaf development (see Chap. 2, Fig. 2.21B, C). The high-auxin concentrations produced by the tip and lobes induce basipetal patterns of decreasing numbers of tracheary elements along the distal parts of the major veins, from the sites of auxin synthesis downward, resulting in the formation of *hydathodes* at the leaf tip and the lobes, which at leaf maturity allow guttation, namely, the exudation of drops of xylem sap at the tip and lobes of leaves (Fig. 7.3).

7. Phloem differentiation is induced by low-grade auxin stimulation, whereas xylem differentiation requires higher-auxin levels (Aloni 1980, 1987, 1995) (see

Fig. 7.3 Early morning guttation from hydathodes developed at the leaf margins of strawberries (*Fragaria ananassa*) leaves, showing exudation of drops of xylem sap (*arrows*) from young leaves and sepals

Chap. 5). In leaves, owing to proximity of the sites of auxin production to those of vascular differentiation, the differentiating cells are exposed to relatively high local auxin levels, which could therefore result in the differentiation of xylem in the absence of phloem in the hydathodes. Conversely, in locations where the terminal veins are induced by relatively low-auxin concentration streams, the sieve tubes may extend to the tip of the freely ending veinlets in the lamina. Intermediate sieve-tube configurations are probably induced by intermediate levels of auxin. These freely ending veinlets inside a leaf are induced by minor auxin synthesis sites produced in the lamina, only after the high-auxin production sites at the leaf periphery have stopped.

8. Due to fast development and early determination of tissues in leaves, there is a very rapid decrease in the ability of parenchyma cells in the lamina to regenerate. Consequently, vessel differentiation and regeneration in leaves is limited to existing strands, where new vessel elements differentiate mainly from meristematic cells in the veins (Fig. 7.1) rather than from parenchyma cells of the early determined lamina. Accordingly, although the general polarity of auxin movements in a leaf is from the margins toward the petiole (Fig. 7.2), late vessel differentiation, or regeneration, is restricted to already existing strands and might therefore occur along the same vein axis even in opposite polarities (Aloni 2001).

To test the *leaf-venation hypothesis* (Aloni 2001) and visualize the developmental patterns of free-auxin (IAA) production, movement, and accumulation in developing leaf primordia, Aloni et al. (2003) used the DR5::GUS-transformed *Arabidopsis thaliana*. The *DR5::GUS* expression was regarded to reflect sites of the bioactive auxin (Figs. 7.4, 7.5, 7.6 and 7.7), while immunolocalization with specific

Fig. 7.4 *DR5::GUS* gene expression in DR5::GUS-transformed *Arabidopsis thaliana* cleared in lactic acid after GUS staining, showing histochemical localization of GUS activity (blue GUS staining that marks free auxin) during early leaf primordium development. Incipient (low *GUS* expression) auxin activity is observed (where auxin maximum develops) in the tip (*arrowhead*) at the stage of acropetal (upward) xylem differentiation of the midvein (*arrow*) in a primordium of a rosette leaf. Bar = 250 μm. (From Aloni et al. 2003)

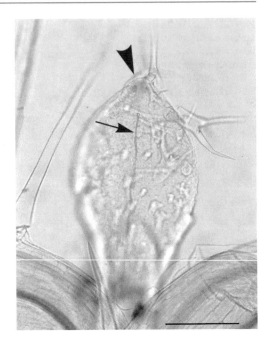

monoclonal antibodies indicated total auxin distribution (see Chap. 3, Figs. 3.4 and 3.5). Additionally, we used the mRNA expression of key enzymes involved in the synthesis, conjugate hydrolysis, accumulation, and polar transport of auxin (Aloni et al. 2003). The study shows that near the shoot apex, leaf's stipules were the earliest sites of high-IAA production, which possibly regulate the rate of shoot tip growth. During early stages of primordium development, leaf apical dominance was evident from strong GUS activity in the elongating tip (Fig. 7.4), likely suppressing the production of free auxin in the leaf tissues below it. **Differentiating hydathodes**, which develop in the tip and later gradually downward in the lobes, were apparently the primary sites of high-IAA production (Figs. 7.5 and 7.6), the latter supported by auxin-conjugate hydrolysis, auxin retention by the chalcone synthase-dependent action of flavonoids and also by the PIN1 – component of the carrier-mediated basipetal transport. **Developing trichomes** (see Chap. 1, Fig. 1.1B) and **mesophyll cells** (Fig. 7.7A) were secondary sites of free-auxin production. The study shows (Aloni et al. 2003) that during primordium development, there are gradual shifts in sites and concentrations of free-auxin production occurring first in the tip of a leaf primordium (Fig. 7.4), then progressing downward along the margins (Figs. 7.5A, B and 7.6), and finally appearing also in the central regions of the lamina (Fig. 7.7). These developmental patterns strongly support the *leaf venation hypothesis* (Aloni 2001) showing the expected developmental process of auxin production and vascular differentiation patterns proposed by the hypothesis.

Subcellular orientation of the auxin efflux-associated protein, PIN1, is polarly expressed in the cell membranes prior to pre-procambial formation, demonstrating the IAA flow directions and pathways in the primordium prior to procambium

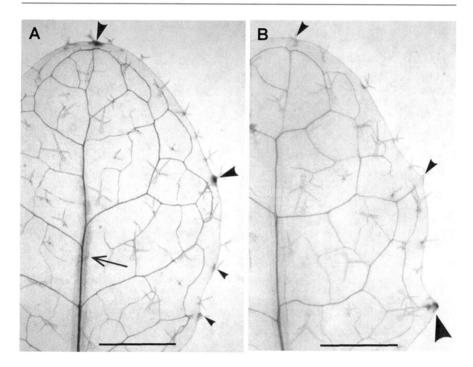

Fig. 7.5 Time course of a downward "wave" of auxin maxima developmental pattern along the periphery of young leaf primordia indicating "leaf apical dominance" detected by *DR5::GUS* gene expression in DR5::GUS-transformed *Arabidopsis*, depicting histochemical localization of GUS activity in a young primordium (**A**) and in an older primordium (**B**). (**A**) Early stage in basipetal progression of high-IAA production, showing strong *GUS* expression in the upper developing hydathodes (*large arrowheads*) and low activity in a lower hydathode (*small arrowhead*). Note the wide-band GUS activity (*arrow*) accumulated in the midvein. (**B**) Shows a later stage in the downward progression of free-auxin production, with low *GUS* expression (*small arrowheads*) in the upper hydathodes and strong GUS activity (*large arrowhead*) in the lower hydathode. Bar = 1 mm. (From Aloni et al. 2003)

formation. Integrated polarities in all emerging veins indicate IAA drainage toward pre-existing veins (as proposed by Sachs 1981), but veins could display divergent polarities until they become connected at both ends (Scarpella et al. 2006; Wenzel et al. 2007). Auxin from adjacent young growing leaves is directed to distinct "convergence points" at the leaf primordium's epidermis, from where the major veins are later produced (Scarpella et al. 2006; Wenzel et al. 2007). These "convergence points" promote lobe development, where local auxin maxima are produce (Aloni et al. 2003; Mattsson et al. 2003) and hydathodes will differentiate.

The visualization of auxin transport pathway in young leaf primordia with the subcellular location of PIN1 (Scarpella et al. 2006; Wenzel et al. 2007) by itself does not show possible changes in IAA concentrations along the auxin flow, while *DR5::GUS* expression can easily detect IAA maxima and auxin presence (Aloni et al. 2003; Mattsson et al. 2003). Only by combining the data of these two powerful

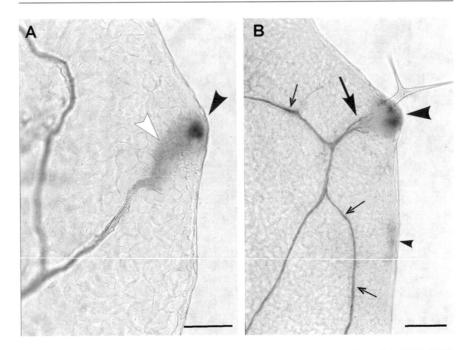

Fig. 7.6 Close up views of auxin maxima and their development into a hydathode in DR5::GUS-transformed *Arabidopsis thaliana*. (**A**) Demonstrating how, at the lobe, a center of strong *DR5::GUS* expression (*black arrowhead*) reflecting the high free-auxin concentration in a developing hydathode. From the auxin maximum, the IAA starts to move by diffusion, gradually becomes canalized to a narrow continuous stream (*white arrowhead*), flowing into a differentiating secondary vein, which is induced by this polar auxin movement. In the vein, the IAA flows more rapidly through the polarized and more efficient auxin-transporting cells, and therefore cannot be detected. (**B**) *GUS* expression at the lobe (*black arrowhead*) in a more developed hydathode with four freely ending vessels (*large arrow*) differentiating toward the margin. Note some weaker blue staining near the margin (*small arrowhead*) and low *GUS* expression within the veins (*small arrows*). Bar = 150 μm (**A**), 250 μm (**B**). (From Aloni et al. 2003)

techniques, together with auxin immunolocalization (Aloni et al. 2003), one can reach a comprehensive understanding of the process and the regulation of vein pattern formation in leaves. The auxin immunolocalization evidence shows that the youngest leaf primordium (see Chap. 3, Fig. 3.4) is loaded with very high amount of conjugated auxin (Aloni et al. 2003; Aloni et al. 2006) supporting the observations that the adjacent auxin-producing young leaves transport considerable quantities of IAA to emerging leaf primordia (Reinhardt et al. 2003; Benková et al. 2003; Scarpella et al. 2006; Scheres and Xu 2006) that accumulate this IAA as conjugated auxin. This mechanism is an economical process that establishes a conjugated reservoir of auxin in developing leaf primordia, which enables their rapid growth and differentiation, so that the development of leaf primordia is not depending solely, or be limited by their own rate of early auxin production.

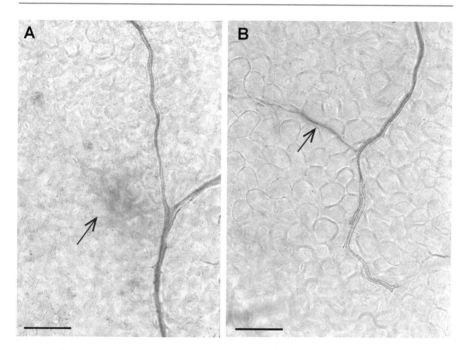

Fig. 7.7 Low-auxin production sites in the central region of the leaf's lamina, which induce the tertiary veins and the freely ending veinlets. These minor IAA production sites start to show DR5::GUS activity at the late stage of vascular differentiation in a leaf primordium, after the cessation of the auxin maxima along the primordium periphery. (**A**) *GUS* expression in the spongy parenchyma cells (*arrow*). (**B**) Reporter-gene activity (*arrow*) detecting IAA in a differentiating freely ending veinlet. Bar = 150 μm. (From Aloni et al. 2003)

Evidently, auxin is synthesized in the hydathodes at the periphery of leaf primordia (Aloni et al. 2003; Baylis et al. 2013). However, the auxin maxima in the tip hydathode is likely donated also by the adjacent young leaf, as analyzed by the acropetally oriented PIN1 pattern in the developing epidermal cells (see in Chap. 3, Fig. 3.3B) of emerging primordia (Reinhardt et al. 2003; Scarpella et al. 2006; Verna et al. 2015) and is drawn schematically with upward green line representing upward IAA flows in Fig. 7.8. The contribution of IAA from an adjacent auxin-producing young leaf to an emerging leaf primordium promotes the latter to induce its midvein in a very young developmental stage (Fig. 7.4), whereas the auxin maxima in the lower hydathodes are likely produced locally, either by hydrolysis of conjugated (bound) auxin (see Chap. 3, Figs. 3.4 and 3.5) or by local auxin synthesis (Aloni et al. 2003; Baylis et al. 2013).

In the leaves, because the mesophyll parenchyma cells between the veins become unresponsive to auxin during early stages of young leaf primordium, new conduits in the veins might differentiate also in opposite polarities (Sachs 1975, 1981, 1989). The early determination of the vein network in leaves also limits their vascular regeneration. I have made surgical experiments in *Arabidopsis* primordia to study vein pattern formation and discovered that a primordium longer than 1 mm is

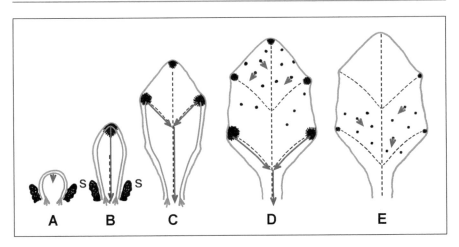

Fig. 7.8 Schematic diagrams showing the gradual changes in sites (*black spots locations*) and concentrations (*black spots size*) of high- versus low-IAA production during leaf primordium morphogenesis in *Arabidopsis*. The rising *green lines* illustrate the upward polar auxin flow through the differentiating epidermis during the early stage of primordium development (**A–C**), originating from nearby auxin-producing leaves. The *red arrows* show the experimentally confirmed directions of the downward (basipetal) vein-inducing polar IAA movement, descending from the differentiating hydathodes in the growing tip and lobes (**B–D**). Incisions (Fig. 7.2A) in leaf primordia show that although the midvein (*broken line,* in **B**) matures acropetally (Fig. 7.4), it is induced (by auxin accumulation above the short cells of the future abscission layer) by the basipetal (downward) polar IAA flow (*red arrow*) descending from the primordium tip (**B**). *Short red arrows* in the lamina originate from *small black spots* indicate random possible auxin flow directions from minor auxin production sites (**D, E**), which induce the tertiary veins and freely ending veinlets. The ontogeny of the midvein and secondary veins is illustrated by broken lines (marginal and minor veins are not shown). (**A**) Early high-auxin production occurs only in the stipules (s) of a very young leaf primordium, before free auxin is detectable in the tip. (**B**) Auxin maximum development in the tip of a fast elongating primordium induces acropetal midvein differentiation, illustrating "leaf apical dominance." (**C**) Auxin maxima development in the fast-expanding upper lobes induces the upper secondary veins and matures into hydathodes. (**D**) Auxin maxima development in the lower lobes inducing the lower secondary veins and mature into hydathodes; randomly distribution of minor auxin production sites start first in the upper lamina (**D**), and later also in the lower lamina (**E**), induce the tertiary veins and freely-ending veinlets, during later phase of primordium development

usually unable to regenerate its vascular system (Fig. 7.9A). Only shorter, very young leaf primordia showed vein regeneration after injury, with applied exogenous auxin that promoted vascular regeneration (Fig. 7.9B–D). These results reveal that veins in a leaf are induced and determined in a very early stage of primordium development. In addition, as the *Arabidopsis* rosette leaf primordia elongate considerably from about 1 mm to a few centimeters, the regenerated vessel elements produce helical or annular secondary wall thickenings (Fig. 7.10) which allowed them to elongate substantially during leaf primordium rapid growth. Instead of reticulate secondary wall thickenings typically found in regenerated vessel elements in wounded stems and tissue cultures (e.g., see Chap. 5, Fig. 5.11B).

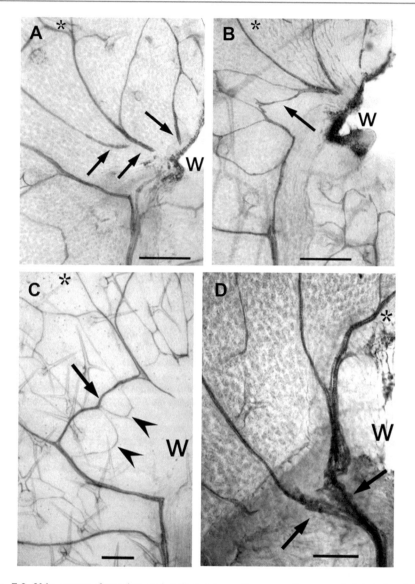

Fig. 7.9 Vein pattern formation and xylem regeneration around a wound (*w*) induced in *Arabidopsis thaliana* by injury and auxin application (1% IAA in lanolin) to leaf primordia shorter than 1 mm. The direction of auxin application site is marked by an *asterisk*. No xylem regeneration could be induced either without the application of exogenous auxin or in longer (older) primordia. After one regeneration week, the treated leaves were cleared in lactic acid and studied without staining. (**A**) There was typically no vein regeneration (*arrows*) around a wound, showing discontinuous patterns of veins, in about half of the wounded leaves. (**B**) Continuous pattern of xylem regeneration (*arrow*) occurred around a wound. (**C**) Two delicate loops (*arrowheads*) differentiated near the wound, attached to thicker regenerative vein (*arrow*) which became a major auxin transporting pathway, likely following procambial configuration. (**D**) Relatively rare typical of xylem regeneration in a leaf, showing continuous pattern of xylem regeneration around a wound, with many regenerative vessel elements that differentiated from parenchyma cells (*arrows*). Bars = 500 μm

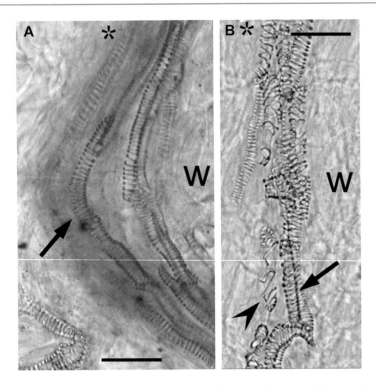

Fig. 7.10 Regenerative tracheary elements induced from parenchymatic cells around a wound (*w* – marks the wound direction) in leaves of *Arabidopsis thaliana*, by wounding and auxin application (1% IAA in lanolin) to leaf primordia shorter than 1 mm. The direction of the applied auxin site is marked by an *asterisk*. (**A**) Magnified view of a regenerative vein built of regenerated xylem elements characterized by helical wall thickenings (*arrow*). (**B**) Portion of a regenerative vein built of xylem elements with helical wall thickenings (*arrow*) and some annular wall thickenings (*arrowhead*) which allow considerable tissue elongation in a rapidly expending leaf primordium. Bars = 25 μm

Sawchuk et al. (2013) defined an additional regulation level of vein network formation in *Arabidopsis* leaves by two spatially separate auxin-transport pathways. In addition to the well-known polar IAA transport from cell-to-cell by the plasma-membrane localized PIN1, the other pathway occurs inside the cells involving auxin transport mediated by the evolutionarily older endoplasmic-reticulum-localized auxin transporters PIN5, PIN6, and PIN8, which are expressed at sites of vein formation. The study confirms that PIN1 in the plasma membrane controls vein network patterning by directing the canalized IAA flow during early stages of leaf primordium development, as was already found by Scarpella et al. (2006), and that PIN6 promotes vein network patterning redundantly with PIN1. While PIN5 inhibits that function of PIN6 in vein network patterning and that PIN8 inhibits the inhibitory function that PIN5 has on PIN6, as such, PIN8 ends up controlling PIN1-dependent vein network patterning redundantly with PIN6 (Sawchuk et al. 2013).

Verna et al. (2015) continued to analyze the relation and activity of these PINs, focusing on their role in vein connections in leaves. Their findings show that PIN1 inhibits the connection of veins into networks and that PIN6 acts redundantly with PIN1 in the inhibition of vein connections and that PIN8 acts redundantly with PIN6 in PIN1-dependent inhibition of vein connection. But PIN6 and PIN8 redundantly inhibit vein connections independently of PIN1, and PIN6 and PIN8 do so by inhibiting the vein formation-promoting function of PIN5 (Verna et al. 2015). These redundant activities of the evolutionarily old endoplasmic-reticulum-localized auxin transporters PIN5, PIN6, and PIN8 in regulating vein pattern formation in leaves have minor influences. Practically, as the activities of PIN5, PIN6, and PIN8 during vein formation in leaves are unnecessary and might have only minor effects on vein patterning; it is proper to focus on the plasma membrane location and polar activity of the evolutionarily advanced PIN1, which is the primary controlling auxin transporter that regulates and determines the polar auxin transport and canalization in leaves, as was found by Scarpella et al. (2006) and Wenzel et al. (2007).

7.2.2 Hydathodes Are Induced by Auxin Maxima that Suppress the Development of Lower Hydathodes

Mature hydathodes produced at the leaf tip and the lobes, can guttate (exudate) drops of xylem sap at the tip and lobes along the leaf margins (Fig. 7.3) usually during mornings, when the root pressure is high.

The hydathodes differentiate at the margin of leaf primordia in the sites of auxin maxima, which induce their unique vascular tissues (Aloni 2001; Aloni et al. 2003). The high-auxin concentration at the leaf's tip induces the primary vein and its largest hydathode (Fig. 7.11A). During primordium development the auxin maxima in the lobes induce the secondary veins descending from the lateral hydathodes. Figure 7.6A demonstrates how IAA from the local auxin maximum (detected by strong GUS staining) produced in a lobe starts flowing downward by diffusion, becomes canalized, and induces a secondary vein. During leaf growth, the auxin maximum will develop into a hydathode (Fig. 7.6B).

Aloni and Zimmermann (1983) predicted that near a site of high-auxin concentration, many tracheary elements will differentiate and their number would decrease with increasing distance from the auxin production site. Analyzing the pattern of vessel elements in large hydathodes indeed shows a significant basipetal decrease in the number of tracheary elements from the leaf margin toward the petiole (Aloni 2001). The greatest numbers of vessel elements in the tip hydathode are formed near the leaf's tip margin (Fig. 7.11A) indicating that this is where the auxin maximum occurred. Similar patterns of basipetal decrease in vessel element density along a leaf strand were experimentally induced by an exogenous auxin source applied along a cut in a very young cotyledon (Aloni 2001). Application of a high-auxin concentration to a longitudinal cut made in a cucumber cotyledon substantially increased vein density near the exogenous auxin source, thus mimicking the role of the rapidly elongating leaf tip and the lobes of a growing primordium as the major

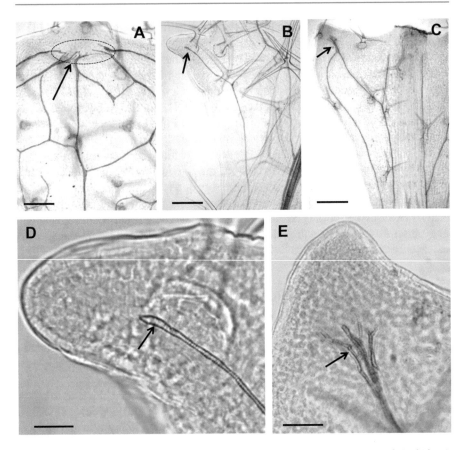

Fig. 7.11 Hydathode differentiation and lowest hydathode inhibition in leaves of *Arabidopsis thaliana*. Micrographs of leaves, cleared in lactic acid, showing the largest hydathode of the tip spread on a large region (*marked by a dotted ellipse* and *arrow*) that contains a few vessels arranged in a wide pattern (**A**), while the smallest hydathode of the lowest lobe contains only one vessel (**B**, **D**). Excision of the upper parts of the primordium, leaving only the lowest lobe (**C**, **E**), promoted the differentiation of four vessels in the lowest hydathode (**E**), demonstrating that when the growing tip and upper lobes were removed experimentally from the primordium, their auxin-maxima-inhibition effect was eliminated, allowing more vessel differentiation in the lowest hydathode. **D** and **E** are close-up views of B and C, respectively, in a different angle. Bars = 20 μm (**D**, **E**), 100 μm (**A**, **B**, **C**)

auxin synthesis sites in a leaf, causing a relatively high vessel-element density in the hydathodes of a dicotyledonous plant leaf (Aloni 2001).

   The first and second points of the *leaf venation hypothesis* (Aloni 2001) propose that the high-auxin concentration at the growing tip inhibits the development of the lobes and their hydathodes below the tip, termed "leaf apical dominance," which was confirmed by Aloni et al. (2003), showing that chalcone synthase, the key enzyme for biosynthesis of flavonoids, which are natural inhibitors of polar basipetal IAA transport and of IAA-degrading peroxidase, was strongly and preferentially expressed in the margin of *Arabidopsis* leaves, but was weakly expressed in the lamina of the primordia (Aloni et al. 2003).

To further clarify the possible inhibitory effect of the auxin maxima of the upper lobes on vascular differentiation in the lower lobes, I have removed all the upper region of young leaf primordia and analyzed the development of the hydathode in the remaining lowest lobe. The results are presented in Fig. 7.11, showing that while in an intact leaf the lowest hydathode contains only one single vessel (Fig. 7.11B, D); after the removal of the primordium's tissues above the lowest lobe, the hydathode in the lowest lobe produced 3–5 vessels (Fig. 7.11C, E), demonstrating that the high-auxin-producing upper lobes inhibit and minimalize the development of the lowest hydathodes. When the upper tissues of a leaf primordium were decapitated, the removal of the upper high-auxin-producing lobes enabled the development of the lowest hydathode, from usually one vessel in intact leaves to a few vessels after the removal of all the inhibiting upper auxin maxima.

7.2.3   Tertiary Veins and Freely Ending Veinlets Are Induced by Minor Auxin-Producing Sites

Leaves of dicots develop a reticulate network of veins; the primary vein (midvein) branches to secondary veins, which branch to tertiary veins, and can further branch to quaternary minor veins, and even to higher order of veins in species with large leaves, and to freely-ending veinlets (Nelson and Dengler 1997; Scarpella and Helariutta 2010; Biedroń and Banasiak 2018).

Point number 3 of the *leaf venation hypothesis* (Aloni 2001) proposes that "as the primary production of high-auxin concentrations along the blade's margin decreases and accordingly their inhibition effect on the central lamina diminishes, secondary auxin synthesis sites within the lamina become relatively effective". Accordingly, the delicate vein network inside the lamina can be produced only after the inhibition induced by the auxin maxima at the tip and lobes is terminated. The freely ending veinlets in the lamina mark the exact sites of minor IAA sources that induced them (e.g., see in Chap. 1, Fig. 1.1B; showing with *DR5::GUS* expression how trichromes produce IAA that induces their nearby freely-ending veinlets).

To clarify the mechanism that enables the differentiation of the tertiary veins and freely ending veinlets in the lamina, young DR5::GUS-transformed *Arabidopsis thaliana* plants were grown on a medium with the auxin efflux inhibitor 1-*N*-naphthylphthalamic acid (NPA). Analysis of the vascular tissues in their leaves showed that the endogenous high-auxin concentration was restricted to the leaf's periphery (detected by the green-blue GUS staining in Fig. 7.12C, D). The continuous high-auxin concentration at the leaf's margin induced more vessels in the midvein, but clearly inhibited the differentiation of the delicate vein network, of the tertiary veins and freely-ending veinlets inside the lamina (Fig. 7.12C, D). Likewise, application of IAA in lanolin to a young primordium that its upper part was removed prevented the differentiation of the tertiary veins and freely ending veinlets between the secondary veins (Fig. 7.12F). These results demonstrate that during primary and

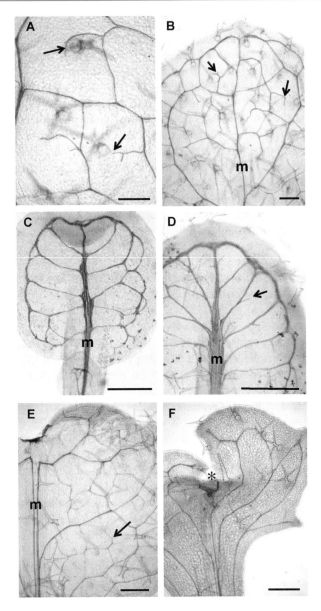

Fig. 7.12 Differentiation of tertiary veins and freely ending veinlet is inhibited by auxin maxima produced along the leaf primordium periphery. The DR5::GUS-transformed *Arabidopsis thaliana* was used in these experiments, showing intact control leaves (**A**, **B**); leaves of plants grown for 2 weeks on the auxin efflux inhibitor 1-*N*-naphthylphthalamic acid (20 μM NPA) (**C**, **D**); and leaves that their upper portion was excised (**E**, **F**), and high-auxin concentration was applied (**F**). The *arrows* mark freely ending veinlets. (**A**) Close-up view of freely-ending veinlets normally induced during the late phase of leaf morphogenesis by low-auxin producing trichomes in an intact leaf (see evidence in Chap. 1, Fig. 1.1B). (**B**) Low magnification shows the regular asymmetric pattern of freely ending veinlets in the lamina, with different vein patterns in the right versus the

secondary vein differentiation, the tertiary veins and freely ending veinlets are inhibited and retarded between the secondary veins by the auxin maxima of the tip and the lobes. Only after the cessation of the high-auxin concentrations in the lobes, the minor local auxin production sites in the lamina can induce the differentiation of the delicate tertiary veins and freely ending veinlets in the lamina (Fig. 7.12A, B, E). These results show that the general downward polar pattern of venation pattern formation in a leaf from the tip hydathode downward to the petiole is regulated by a "leaf apical dominance" mechanism induced by the auxin maxima of the tip and later in the upper lobes that retard vascular differentiation and maturation at the leaf base and lamina (Figs. 7.8 and 7.12).

Summary

- Leaf initiation depends on the establishment of dynamic gradients of upward polar auxin flows toward the shoot apex, where existing leaf primordia act as sinks and among them, the development of auxin maxima induce the initiation of new leaf primordia in an orderly pattern, namely, phyllotaxis.
- Young auxin-producing leaves donate the hormone to emerging leaf primordia that act as sinks and accumulate the IAA as conjugated auxin, establishing a conjugated reservoir of auxin in developing leaf primordia, from which free auxin can be released during leaf development.
- Polar flows of auxin during early primordium development induce the vein networks in leaves. Auxin induces asymmetric patterns of vascular veins in the same leaf, demonstrating the auxin-independent genetic control of vascular tissues. The vein patterns are determined by random patterns of auxin production sites and flows, regulated by selection of various alternative auxin pathways.
- Polar IAA transport in a growing leaf occurs from the margins towards the midvein and proceeds to the petiole, in continuous flows, therefore inducing continuous and functional vein pattern.
- Auxin maximum at the primordium tip induces the midvein. New auxin maxima develop along the primordium periphery, in a gradual downward pattern indicating a "leaf apical dominance," which is characterized by a gradual downward "wave" pattern of auxin maxima that induce the secondary veins and finally mature to hydathodes.
- Auxin maxima inhibit for a short time auxin production below them, repressing vascular differentiation in lower lobes and in the leaf's lamina.

Fig. 7.12 (continued) left sides of the midvein (m). (**C** and **D**) Show leaves that the endogenous auxin produced in their periphery remained continuously high (detected by green-blue DR5::GUS staining) along the margin, which prevented tertiary vein formation in the lamina. Note that the high-auxin produced in the primordium periphery induced more vessels in the midvein (m), compared with the thin midvein in the intact control (**B**). (**E**) Normal pattern of tertiary veins and freely-ending veinlets differentiated after the upper part of the leaf primordium was excised (**E** is the control of **F**). (**F**) Tertiary veins and freely ending veinlet differentiation was inhibited by the exogenous high-auxin (1% IAA in lanolin) applied on the cut surface (the auxin application site is marked with *asterisk*). Bars = 100 μm (**A**), 200 μm (**F**), 250 μm (**B, E**), 1 mm (**C, D**)

- Only after the inhibitory effect of the high-auxin concentrations along the leaf's periphery diminish, minor low-auxin synthesis sites within the lamina become relatively effective and induce the tertiary veins and freely ending veinlets.
- Due to early cell determination, there is normally no vascular regeneration in *Arabidopsis* leaves.

References and Recommended Readings[1]

Aloni R (1980) Role of auxin and sucrose in the differentiation of sieve and tracheary elements in plant tissue cultures. Planta 150: 255–263.

Aloni R (1987) Differentiation of vascular tissues. Annu Rev Plant Physiol 38: 179–204.

Aloni R (1995) The induction of vascular tissues by auxin and cytokinin. In: *Plant hormones: physiology, biochemistry and molecular biology*, PJ Davies (ed). Kluwer Academic Publishers, Dordrecht, pp 531–546.

*Aloni R (2001) Foliar and axial aspects of vascular differentiation: hypotheses and evidence. J Plant Growth Regul 20: 22–34.

Aloni R (2004) The induction of vascular tissue by auxin. In: *Plant Hormones: Biosynthesis, Signal Transduction, Action!* PJ Davies (ed). Kluwer Academic Publishers, Dordrecht, Boston, London, pp 471–492.

Aloni R (2013) The role of hormones in controlling vascular differentiation. In: *Cellular Aspects of Wood Formation*, J Fromm (ed). Springer-Verlag, Berlin, pp 99–139.

Aloni R, Aloni E, Langhans M, Ullrich CI (2006) Role of auxin in regulating *Arabidopsis* flower development. Planta 223: 315–328.

*Aloni R, Schwalm K, Langhans M, Ullrich CI (2003) Gradual shifts in sites of free-auxin production during leaf-primordium development and their role in vascular differentiation and leaf morphogenesis in *Arabidopsis*. Planta 216: 841–853.

Aloni R, Zimmermann MH (1983) The control of vessel size and density along the plant axis - a new hypothesis. Differentiation 24: 203–208.

Ashby E (1948) Studies in the morphogenesis of leaves: I. An essay on the leaf shape. New Phytol 47: 153–176.

Avery Jr GS (1933) Structure and development of tobacco leaf. Am J Bot 20: 565–592.

Avsian-Kretchmer O, Cheng JC, Chen L, Moctezuma E, Sung ZR (2002) Indole acetic acid distribution coincides with vascular differentiation pattern during Arabidopsis leaf ontogeny. Plant Physiol 130: 199–209.

*Baylis T, Cierlik I, Sundberg E, Mattsson J (2013) *SHORT INTERNODES/STYLISH* genes, regulators of auxin biosynthesis, are involved in leaf vein development in *Arabidopsis thaliana*. New Phytol 197: 737–750.

Benková E, Michniewicz M, Sauer M, Teichmann T, Seifertová D, Jürgens G, Friml J (2003) Local, efflux-dependent auxin gradients as a common module for plant organ formation. Cell 115: 591–602.

Berleth T, Sachs T (2001) Plant morphogenesis: long-distance coordination and local patterning. Curr Opin Plant Biol 4: 57–62.

*Biedroń M, Banasiak A (2018) Auxin-mediated regulation of vascular patterning in *Arabidopsis thaliana* leaves. Plant Cell Rep 37: 1215–1229.

Candela H, Martinez-Laborda A, Micol JL (1999) Venation pattern formation in Arabidopsis thaliana vegetative leaves. Dev Biol 205: 205–216.

---

[1] Papers of particular interest for suggested reading have been highlighted (with *).

Carland FM, Berg BL, FitzGerald JN, Jinamornphongs S, Nelson T, Keith B (1999) Genetic regulation of vascular tissue patterning in *Arabidopsis*. Plant Cell 11: 2123–2137.

Deyholos MK, Cordner G, Beebe D, Sieburth LE (2000) The *SCARFACE* gene is required for cotyledon and leaf vein patterning. Development 127: 3205–3213.

Esau K (1965) *Vascular differentiation in plants*. Holt, Rinehart and Winston, New York.

Hobbie L, McGovern M, Hurwitz LR, Pierro A, Yang Liu N, Bandyopadhyay A, Estelle M (2000) The axr6 mutants of *Arabidopsis thaliana* defective: a gene involved in auxin response and early development. Development 127: 23–32.

Jacobs WP (1952) The role of auxin in differentiation of xylem around a wound. Am J Bot 39: 301–309.

Koizumi K, Sugiyama M, Fukuda H (2000) A series of novel mutants of Arabidopsis thaliana that are defective in the formation of continuous vascular network: calling the auxin signal flow canalization hypothesis into question. Development 127: 3197– 3204.

Kuhlemeier C (2017) Phyllotaxis. Curr Biol 27: R882–887.

Linh NM, Verna C, Scarpella E (2018) Coordination of cell polarity and the patterning of leaf vein networks. Curr Opin Plant Biol 41: 116–124.

Marcos D, Berleth T (2014) Dynamic auxin transport patterns preceding vein formation revealed by live-imaging of Arabidopsis leaf primordia. Front Plant Sci 5: 235.

Mattsson J, Ckurshumova W, Berleth T (2003) Auxin signaling in *Arabidopsis* leaf vascular development. Plant Physiol 131: 1327–1339.

*Mattsson J, Sung ZR, Berleth T (1999) Responses of plant vascular systems to auxin transport inhibition. Development 126: 2979–2991.

Meyerowitz EM (1989) *Arabidopsis*, a useful weed. Cell 56: 263– 269.

Nelson T (1998) Polarity, vascularization and auxin. Trends Plant Sci 3: 245–246.

Nelson T, Dengler N (1997) Leaf vascular pattern formation. Plant Cell 9: 1121–1135.

Nikolov LA, Runions A, Gupta MD, Tsiantis M (2019) Leaf development and evolution. In: *Plant Development and Evolution*, U Grossniklaus (ed), Curr Top Develop Biol 131: 109–139.

Poething RS (1997) Leaf morphogenesis in flowering plants. Plant Cell 9: 1077–1087.

Przemeck GKH, Mattsson J, Hardtke CS, Sung ZR, Berleth T (1996) Studies on the role of the *Arabidopsis* gene *MONOPTEROS* in vascular development and plant cell axialization. Planta 200: 229–237.

*Reinhardt D, Pesce ER, Stieger P, Mandel T, Baltensperger K, Bennett M, Traas J, Friml J, Kuhlemeier C (2003) Regulation of phyllotaxis by polar auxin transport. Nature 426: 255–260.

Sachs T (1969) Regeneration experiments on the determination of the form of leaves. Isr J Bot 18: 21–30.

Sachs T (1975) The control of the differentiation of vascular networks. Ann Bot 39: 197–207.

*Sachs T (1981) The control of patterned differentiation of vascular tissues. Adv Bot Res 9: 151–262.

*Sachs T (1989) The development of vascular networks during leaf development. Curr Top Plant Biochem Physiol 8: 168–183.

Sachs T (2002) Consequences of the inherent developmental plasticity of organ and tissue relations. Evol Ecol 16: 243–265.

Sawchuk MG, Edgar A, Scarpella E (2013) Patterning of leaf vein networks by convergent auxin transport pathways. PLoS Genet 9: e1003294.

Sawchuk MG, Head P, Donner TJ, Scarpella E (2007) Time-lapse imaging of *Arabidopsis* leaf development shows dynamic patterns of procambium formation. New Phytol 176: 560–571.

Scarpella E (2017) The logic of plant vascular patterning. Polarity, continuity and plasticity in the formation of the veins and of their networks. Curr Opin Genet Dev 45: 34–43.

Scarpella E, Helariutta Y (2010) Vascular pattern formation in plants. Curr Top Dev Biol 91: 221–265.

*Scarpella E, Marcos D, Friml J, Berleth T (2006) Control of leaf vascular patterning by polar auxin transport. Genes Dev 20: 1015–1027.

Scheres B, Xu J (2006) Polar auxin transport and patterning: grow with the flow. Genes Dev 20: 922–926.

Shi B, Vernoux T (2019) Patterning at the shoot apical meristem and phyllotaxis. In: *Plant Development and Evolution*, U Grossniklaus (ed), Curr Top Develop Biol 131: 81–107.

Sieburth LE (1999) Auxin is required for leaf vein pattern in *Arabidopsis*. Plant Physiol 121: 1179–1190.

Taiz L, Zeiger E, Møller IM, Murphy A (2018) *Fundamentals of Plant Physiology*. Sinauer, Sunderland, MA.

Telfer A, Poething RS (1994) Leaf development in *Arabidopsis*. In: *Arabidopsis*, EM Meyerowitz, CR Somerville (eds). Cold Spring Harbor Laboratory Press: Cold Spring Harbor, pp 379–401.

Verna C, Sawchuk MG, Linh M, Scarpella E (2015) Control of vein network topology by auxin transport. BMC Biol 13: 94.

Verna C, Ravichandran SJ, Sawchuk MG, Linh NM, Scarpella E (2019) Coordination of tissue cell polarity by auxin transport and signaling. Elife 8: e51061.

*Wenzel CL, Schuetz M, Yu Q, Mattsson J (2007) Dynamics of MONOPTEROS and PIN-FORMED1 expression during leaf vein pattern formation in *Arabidopsis thaliana*. Plant J 49: 387–398.

Flower Biology and Vascular Differentiation

8

8.1 The FT Protein Is the Leaf-Inducing Signal for Flowering and Fiber Differentiation

The transition from the vegetative phase to a reproductive phase involves dramatic developmental changes. This transition is characterized by the induction of the floral meristem, which produces the flower, or flowers on an inflorescence.

A flower-inducing hormonal concept was established following experiments of partially darkened *Tropaeolum majus* and *Ipomoea purpurea* plants (Sachs 1865). The conclusion from these experiments was that leaves grown in light produce a flower-forming substance, which moved from the light-exposed leaves to the vegetative darkened shoot tip to induce flowering.

The floral-inducing stimulus was named "florigen" (flower-former) (Chailakhyan 1936) to define a specific leaf substance which induces flowering in the shoot tip. Grafting experiments between related species, but of a different photoperiodic response type (e.g., a short-day plant [SDP] and a long-day plant [LDP]), provided evidence for exchangeability of florigen among different species and light-inducing conditions. However, the florigen remained for a long time a physiological concept rather than a known chemical entity.

The hormone gibberellin can induce and promote flowering in many LDP plants (which would flower during late spring or early summer as the days begin to get longer) that grow as a rosette under short-day conditions. However, not all rosette plants can be induced to flower by GA (Zeevaart 1983, 2006).

In *Arabidopsis*, the mRNA of the *FLOWERING LOCUS T* (*FT*) gene was proposed as the long-sought "florigen" (Huang et al. 2005), suggesting that the *FT* mRNA is an important component of the elusive florigen signal that moves from a leaf to the shoot apex where it induces the floral meristem. However, other laboratories (Lifschitz et al. 2006; Lin et al. 2007; Jaeger and Wigge 2007) revealed that the mobile flower-inducing florigen signal is the FT protein (but not *FT* mRNA) which is the graft-transmissible protein that acts as the long-distance florigenic signal which is common in plants. In tomato, the *SFT* (*SINGLE-FLOWER TRUSS*)

© Springer Nature Switzerland AG 2021
R. Aloni, *Vascular Differentiation and Plant Hormones*,
https://doi.org/10.1007/978-3-030-53202-4_8

gene, which is the ortholog gene of *FT*, induces flowering in tomato and tobacco plants. The SFT protein signal was transmitted by grafting (see Sect. 10.2) from tomato (scion) to a non-flowering tobacco (stock), resulting in flowering of the tobacco (Lifschitz et al. 2006). In cucurbit species, a mass spectrometry analysis of florally induced heterograft of *Cucurbita moschata* scion revealed that from the stock of *Cucurbita maxima*, the FT protein crossed the graft union via the phloem translocation stream (Lin et al. 2007). These studies demonstrate that the FT protein is the mobile long-distance florigenic signal that travels via the phloem from the leaves to the shoot apex. FT is the only known protein that serves as a long-range developmental signal in plants (Jaeger and Wigge 2007). Thus the florigen may be consider as a universal proteinaceous hormonal signal that functions as a generic terminator of vegetative apical meristems to induce flowering (Shalit-Kaneh et al. 2019).

Florigen, which is encoded in tomato (*Solanum lycopersicum*) by the *SFT* gene, also regulates the vascular system in the stems of flowering plants by promoting fiber differentiation in their upper internodes below the developing fruits. RNA-sequencing analysis and complementary genetic and histological experiments revealed that florigen of endogenous, mobile, or induced origins accelerates the transcription network navigating of secondary cell wall biogenesis into the upper internodes of flowering stem, promoting differentiation of fibers in the upper internodes, which thereby adapt the shoot system to carry heavy tomato fruits, thus ensuing the stem's supporting needs during the reproductive phase (Shalit-Kaneh et al. 2019). As fibers are specifically induced by gibberellin (see Sects. 5.2.3 and 19.2), it is likely that florigen promotes a substantial increase in GA activity during the process of fiber differentiation in the flowering stems.

8.2   The ABC Model of Flower Development

Floral organs are initiated sequentially by the floral meristem. The flower of the model plant – *Arabidopsis thaliana* – and many other plant species produces four successive whorls, starting with the outermost whorl of **sepals** and progressing inward with a second whorl of the **petals**, a third whorl of the **stamens** (male), and finally a fourth whorl of the **carpels** (female). The remaining apical meristem, enclosed by the carpels, ultimately forms a placenta on which the ovules develop (Steeves and Sussex 1989; Evert and Eichhorn 2013).

The flower is a modified branch that its floral organs are homologous to foliage leaves. This old opinion (von Goethe 1790; Smyth 2005) has been confirmed by genetic studies, demonstrating that in the *Arabidopsis* ABC triple mutant: *apetala2, pistillata*, and *agamous1* (Bowman et al. 1991; Weigel and Meyerowitz 1994) and in the quadruple mutant: *ap1, ap2, ap3/pi*, and *ag* (Goto et al. 2001) the absence of floral organ identifying activities causes all floral organs to develop as leaves.

Based on the use of double and triple mutants of homeotic genes, the ABC model of flower development was proposed (Bowman et al. 1991). Accordingly, the action of floral organ identity genes in *Arabidopsis* named **A** (*APETALA2* gene function,

Fig. 8.1 Schematic representation of the ABC model proposed by Bowman et al. (1991) in a wild-type *Arabidopsis thaliana* flower, depicting how three classes of floral homeotic genes could specify the identity of each of the four whorls of floral organs. The function of *A* alone specifies sepal (se) identity in the first whorl (1); the combination of *A* + *B* function specifies petal (pe) identity in the second whorl (2); the combination of *B* + *C* function specifies stamen (st) identity in the third whorl (3); and *C* function alone specifies carpels (ca) in the fourth whorl (4)

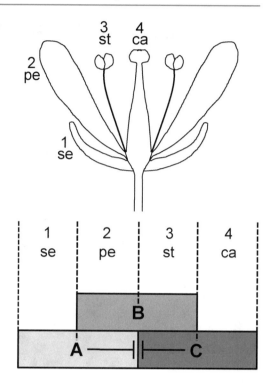

AP2), **B** (*APETALA3* and *PISTILLATA, AP3/PI*), and **C** (*AGAMOUS, AG*) was summarized in the following ABC model (Bowman et al. 1991; Coen and Meyerowitz, 1991): **A** function in whorl 1 defines the sepals, **A** + **B** function in whorl 2 induces the petals, **B** + **C** in whorl 3 specify the stamens, and **C** function in whorl 4 controls carpel identity. It was also proposed that **A** (*AP2*) and **C** (*AG*) antagonize each other's function; hence *AP2* seems to keep the *AG* gene inactive in the two outer whorls; conversely *AG* keeps *AP2* inactive in the two inner whorls (Fig. 8.1).

This basic ABC flower development model was confirmed in many studies and is widely used today as a framework for understanding floral development and flower evolution (Bowman et al. 2012; Ó'Maoiléidigh et al. 2014; Becker and Ehlers 2016; Stewart et al. 2016; Thomson and Wellmer 2019).

8.3   High-IAA-Producing Anthers Synchronize Flower Development

Unexpectedly, although the petal and stamen primordia in the flower of *Arabidopsis* appear simultaneously (early stage 5 of flower development), the stamens develop first (during stages 6–8) (Fig. 8.2B, C), whereas the petal primordia do not grow until stage 9 (Bowman et al. 1989) (Fig. 8.2D); a phenomenon which has wide occurrence among flowering plants (Endress 1994). However, the mechanism which controls this surprising developmental pattern was a mystery, which is clarified below.

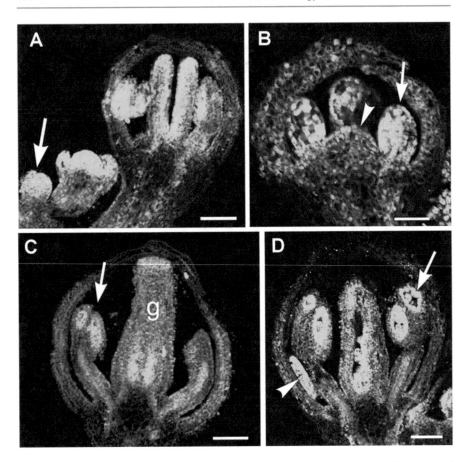

Fig. 8.2 Distribution of conjugated auxin detected by immunolocalization with rabbit polyclonal antibodies (green Alexa Fluor 488 fluorescence) in DR5::GUS-transformed *Arabidopsis*, viewed with confocal laser scanning microscope (CLSM), showing early development of the stamens and gynoecium (**A–C**), and late development of the petals (**D**). (**A**) Auxin distribution in three developmental stages of flower primordia. The youngest flower bud (*arrow*) before any visible organ initiation is already loaded with conjugated auxin, while the most developed flower (before petal growth) shows differential auxin patterns with highest auxin concentrations mainly in the anthers and gynoecium. (**B**) Very young stamen primordia (*arrow*) with the highest auxin concentrations, higher than in the promeristem (*arrowhead*) and sepals which already protect the stamens. (**C**) Elevated auxin in the stigma, developing ovules in the gynoecium (g) and the differentiating pollen sacs, before petal initiation. (**D**) Early stage of petal growth characterized by a short phase of high auxin concentration (*arrowhead*). The anthers (*arrow*) and gynoecium continue to maintain high-auxin concentrations during later developmental stages. Bars = 40 μm (**B**), 80 μm (**A, C, D**). (From Aloni et al. 2006)

Early indication that auxin is involved in flower formation derived from the *pin1* mutant that produces inflorescence which often does not produce flowers (Okada et al. 1991). The *PIN1* gene encodes the major transmembrane protein participating in the polar auxin efflux (Gälweiler et al. 1998; Wisniewska et al. 2006) proving a requirement for the polar auxin transport in early stages of flower initiation and development. By using confocal imaging of green fluorescent protein (GFP) reporter gene in living plants, Heisler et al. (2005) monitored the expression patterns of PIN1 and multiple other proteins and genes involved in the process of flower primordium development in *Arabidopsis*. Their beautiful images show that PIN1 undergoes cycles of activity during early primordial development in the inflorescence meristem, suggesting a role for auxin transport during flower primordium development (Heisler et al. 2005). This role of polar auxin transport was supported by studies showing that auxin coordinates specific gene activity during initiation and development of floral primordia (Krizek 2011; Yamaguchi et al. 2013, 2014).

As the main function of flowers is sexual reproduction, IAA is likely involved in the mechanism which controls the growth of the male gametophyte to the egg cell in the ovule. At the stage of pollen germination on the stigma, IAA reaches its highest content in the stigma of *Arabidopsis thaliana* (Aloni et al. 2006) and of *Nicotiana tabacum* (Chen and Zhao 2008) and in the transmitting tissue. After the pollen tubes entered the styles, the auxin signal increased in the part where pollen tubes would enter and then rapidly declined in the part where pollen tubes had penetrated (Chen and Zhao 2008). The role of auxin in the sexual processes of flower biology is supported by Aloni et al. (2006) findings that developing pollen grains accumulate high concentrations of auxin originating in the tapetum cells (Fig. 8.3B), which likely promotes pollen germination on the stigma and triggers the rapid intrusive growth of pollen tubes to the embryo sac. High amounts of free auxin in the stigma induce a wide xylem fan immediately beneath it (Fig. 8.4B, C). After fertilization, the developing embryos and seeds show elevated concentrations of auxin, which establish their axial polarity (Fig. 8.5D) (Aloni et al. 2006).

The youngest flower bud is loaded with conjugated auxin (see *arrow* in Fig. 8.2A), while the free bioactive auxin cannot be detected at this early stage (*arrow* in Fig. 8.3A). During flower morphogenesis, the levels of free auxin increase in the flower organs (probably, by hydrolysis of the conjugated auxin reservoir shown in Fig. 8.2 and possibly also by auxin biosynthesis shown by Robert et al. (2015). Free auxin appears in gradual patterns, always starting at the floral-organ tip (similar to the IAA pattern in foliage leaves, see Chap. 7, Fig. 7.4).

Anthers were found to be the major sites of very-high concentrations of free auxin (Figs. 8.3A, 8.4A and 8.5A), produced in their tapetum cells (Fig. 8.3B) that provides free auxin and nutrition to the developing pollen grains. Evidence for de novo local auxin biosynthesis during floral organ development was detected and confirmed by the *TRYPTOPHAN AMINOTRANSFERASE OF ARABIDOPSIS (TAA)* and *YUCCA (YUC) flavin monooxygenase* genes activity (Cheng et al. 2006; Robert et al. 2015), which are families of genes active in the TAA/YUC auxin biosynthesis pathway (Zhao 2018).

The extremely high-IAA concentrations produced in the anthers suppress the elongation of the petal primordia. In intact flowers, the petals produce very low-IAA concentrations (Aloni et al. 2006) and therefore probably cannot compete with the high-auxin-producing stamens. This anther inhibition of both the petals and nectar glands was confirmed experimentally by selective anther removal, after which the petals started to elongate while producing higher amounts of free auxin (Aloni et al.

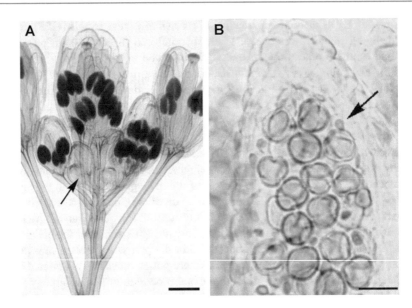

Fig. 8.3 Auxin-dependent *DR5::GUS* gene expression in the DR5::GUS transformed *Arabidopsis thaliana* (cleared in lactic acid) during early stages of inflorescence and flower morphogenesis, detecting free-auxin by the GUS staining. (**A**) Inflorescence with very weak (*arrow*) or without any GUS activity in the youngest flower buds (*arrow*), [which are loaded with conjugated auxin, detected in Fig. 8.2A]; the larger flower buds show strong *GUS* expression (dark blue-black staining) in the stamens. (**B**) Close-up view of a very young anther shows high GUS activity in the tapetum cells (*arrow*) that produce IAA and supply the pollen grains with auxin and nutrients. Bars = 25 μm (**B**), 250 μm (**A**). (From Aloni et al. 2006)

2006) and the nectaries became active (Fig. 8.5C). Naturally, the petals reach maximum size and the nectaries start to produce IAA and nectar just before anthesis (a known phenomenon widely occurring among flowering plants; see Endress 1994), at the time when the polar auxin flow from the anthers diminishes (Fig. 8.6). Aloni et al. (2006) findings demonstrate that the **high-IAA-producing anthers synchronize flower development** (Fig. 8.6) by retarding petal development and nectar gland activity, which start to function only shortly before pollination, only after the pollen grains mature and the flower becomes ready for pollination. By producing considerable free-auxin concentrations (Figs. 8.3A, 8.4A and 8.5A), the anthers inhibit the development of adjacent floral organs, emphasizing the validity of the apical dominance concept (discussed in Chap. 6) that can be applied for understanding various developmental behaviors of different plant organs.

Interestingly, even after fertilization, the nectaries (the nectar-secreting glandular organs) continue to produce free IAA and nectar for a relatively long period (Fig. 8.5D), up to seed maturation, probably because the hormonal repression by the stamens has ceased. These results demonstrate that IAA may play two opposing roles in flower development: free auxin produced at the tip of a floral organ promotes its own development and vascular differentiation, but can repress the growth or activity of a neighboring organ (Aloni et al. 2006).

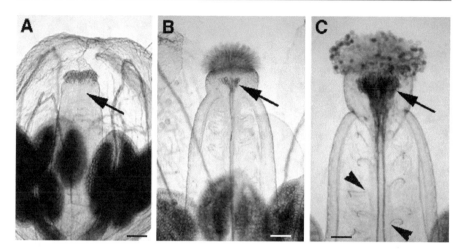

Fig. 8.4 Free-auxin production detected by *DR5::GUS* expression in transformed *Arabidopsis* (cleared with lactic acid), showing localization of GUS activity during flower morphogenesis, demonstrating how the bioactive auxin induces vascular differentiation in the gynoecium. (**A**) Young flower bud, prior to detectable IAA production in the stigma and therefore with no vascular development (*arrow*) in the gynoecium. At this developmental stage the stamens produce the highest free-auxin concentrations (*stained dark blue*), which likely delay free-auxin production in the stigma. (**B**) In a young primordium, free-auxin production (*detected by blue GUS expression*) in the stigma induces the central bundle and an early stage of xylem fan formation (*arrow*) immediately below the stigma. (**C**) Mature gynoecium with germinating pollen grains (*stained dark blue*). A typical well-developed wide fan of xylem (*arrow*) descending into two central bundles which were induced by the high-IAA-producing stigma (*stained blue*). The discontinuous short veinlets (*arrowheads*) induced by the ovules do not connect to the gynoecium's central bundles, because the bundles are well supplied with high-IAA concentration streams descending from the stigma, which prevents veinlet linkage. Bars = 100 μm. (From Aloni et al. 2006)

8.4   Vascular Differentiation in Floral Organs and Fruit Development

Flower organs have been derived from leaves and a flower is a modified branch. Genetic studies have shown that the floral organs are homologous to foliage leaves (Bowman et al. 1991; Weigel and Meyerowitz 1994; Goto et al. 2001). The sepals and the petals which protect the reproductive organs have leaf shapes, while the stamens and carpels have been considerably modified to produce the male and female gametes. Vascular differentiation in sepals is very similar to that of foliage leaves, usually with a hydathode in their tip. Petals develop relatively simple vasculatures (Figs. 8.7A–C and 8.8A). In all the floral organs, like in leaves, auxin maximum at the organ tip promotes early vascular differentiation near it (see Chap. 2, Fig. 2.22A, B) progressing in discontinuous patterns downward. This is evident in both young stamens that show initial vessel differentiation between the anthers progressing basipetally (Fig. 2.22A) and in the gynoecium, from the stigma downward (Fig. 2.22B).

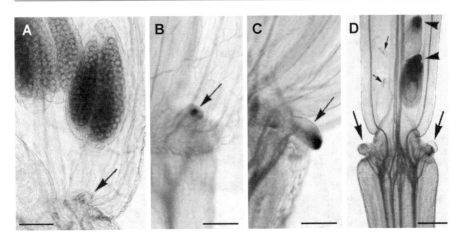

Fig. 8.5 Free auxin detected by *DR5::GUS* gene expression in transformed *Arabidopsis* during early stages of flower morphogenesis. The micrographs reveal the inhibitory mechanism of the stamens on nectar-gland activity, which during early flower development suppress nectar gland activity by the high-IAA concentrations produced by the stamens. (**A**) Young nectary (*arrow*) without *GUS* activity, likely inhibited by the high-IAA concentration produced in the anthers. (**B**) Incipient low *GUS* expression at the tip (*arrow*) of a nectary before anthesis (starting simultaneously with nectar secretion). (**C**) Early strong *GUS* expression (*arrow*) in a nectary of a younger flower promoted by four-stamen removal. Both flowers shown in **B** and **C** are from the same inflorescence. (**D**) GUS activity in nectaries (*arrows*) continues after pollination, likely due to the absence of repression by the anthers (that stopped auxin production), and is detectable below a developing silique with differentiating seeds showing strongest *GUS* expression where the embryonic root develops (*arrowheads*), while unfertilized ovules (*small arrows*) lack *GUS* expression. Bars = 100 μm (**B**, **C**), 200 μm (**A**, **D**). (From Aloni et al. 2006)

A unique type of discontinuous differentiation in the vascular system occurs at the base of the ovules, where the short veinlets that they induce do not connect to the central bundle of the gynoecium (Fig. 8.4C). This discontinuous veinlet pattern (see also in Chap. 2, Fig. 2.22C) results from the continuous high-auxin supply descending from the stigma (Fig. 8.4B, C) to the central bundle that prevents the connection of the veinlets induced by the ovules. This special phenomenon is explained by Sachs (1968) classical experiment in *Pisum sativum* seedling (shown in Chap. 3, Fig. 3.7C), which demonstrates that a central vascular bundle which is well supplied with high-auxin concentration stream prevents the connection of a new artificially induced strand to the pre-existing vascular bundle.

Petals of wild-type *Arabidopsis* are characterized by the formation of closed loops (Fig. 8.7A); forkedly open veins that lack distal meeting (Fig. 8.7B) and their combinations (Fig. 8.7C). Although FORKED genes were described and considered as essential for distal vein meeting in *Arabidopsis* (Steynen and Schultz 2003), combinations of the above close and open vein patterns (Fig. 8.7A–C) are frequently found on the same wild-type inflorescence and may occur in the same flower. A simple explanation to the formation of open branched veins is that the sites from where IAA is synthesized at their tip become active simultaneously, which therefore

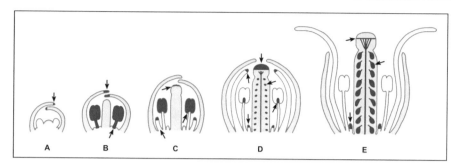

Fig. 8.6 Schematic diagrams showing the gradual changes in sites (*blue spot* locations) and concentrations (*blue symbol* sizes) of free-IAA production (detected by *DR5::GUS* expression) during *Arabidopsis* flower and early fruit development. *Arrows* mark sites of auxin production starting at the tip of floral organs during their development (**A–E**) and at the ovules and developing seeds in the gynoecium (**D, E**). The ontogeny of the gynoecium midvein, characterized by its wide fan xylem induced by free IAA descending from the stigma (**D, E**), and the short xylem veinlets induced by developing seeds are illustrated by *red lines* (**E**). (**A**) Young floral bud with incipient free-IAA production at the tip of the sepals (the bud is loaded with conjugated auxin). (**B**) Free-IAA production at the sepal tips and massive bioactive-auxin production in the stamens, demonstrating stamen dominance characterized by complete petal suppression. (**C**) Decreased auxin production in the stamens (DR5::GUS activity limited to the anthers) is followed by incipient auxin production in the growing petals and stigma. (**D**) High free-IAA production in the stigma; low-auxin production in the ovules, the nectaries, the petal tips, and stamen-filament tips. (**E**) Residual free-IAA production beneath the stigma, elevated auxin production in developing seeds, and continuous production in nectaries. (From Aloni et al. 2006)

prevents their meeting. While a loop is produced when the two IAA sites produce the auxin signals not at the same time, one after the other and therefore the second induced vessel becomes connected to the first formed one (Fig. 8.7D, E) and forms a loop. In a loop pattern, usually the upper, inside vessel (marked #1) of the loop is produced first, and the lower, outside vessel (#2) of the loop is produced later (Fig. 8.7D). Higher magnifications reveal that the loops are built of two vessels that overlap at the site of their connection; in such a pattern, the vessels cannot be produced at the same time, revealing that the vessels were produced one after the other. Spraying young flower buds with auxin can increase the number of vessels in a secondary vein (Fig. 8.7F); while auxin transport inhibitor is more effective in promoting additional vessels in the midvein and secondary veins (Fig. 8.8B).

Developing embryos and seeds are major sites of IAA production, which is accumulated at their basal side, where the auxin induces their roots (Figs. 8.5D and 8.9D). This auxin accumulation (due to polar auxin movement) at the root side is evident by the blue staining of *DR5::GUS* expression (the root side develops upward toward the stigma) detected in the DR5::GUS-transformed *Arabidopsis thaliana* (Fig. 8.9D).

The auxin produced in developing seeds promotes fruit development and growth (Nitsch 1952, 1953; Kumar et al. 2014; Giovannoni et al. 2017). The classical experiments done by Nitsch (1950) on fruit development of the garden strawberry (*Fragaria* × *ananassa*) hybrid, which is an aggregate accessory fruit originated

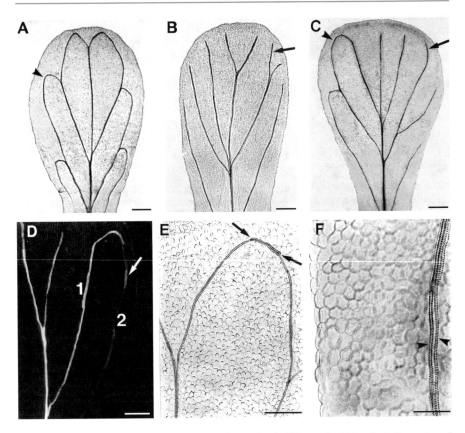

Fig. 8.7 Naturally occurring vascular patterns in petals of wild-type *Arabidopsis thaliana* cleared in lactic acid. (**A**) Pattern of closed loops (*arrowhead*). (**B**) Pattern of open branching veins with freely ending veinlets (*arrow*) that lack distal meetings. (**C**) Combined patterns of closed loops (*arrowhead*) and open branching (*arrow*) with freely ending veins. (**D**) Gradual development of a closed loop viewed under dark-field illumination, showing that the upper vein (#1) is already mature, while the lower vein (#2) is still differentiating (*arrow*) downwards. There are two freely ending veins at the petal's tip. (**E**) Close-up view on the top of a loop revealing the upper ends (*arrows*) of the two attached overlapping vessels that together built the loop. (**F**) Vein containing two vessels induced by one application of auxin (40 μM NAA spray) to a developing flower. Usually the petal veins contain only one vessel per vein, which is likely induced by low auxin stimulation. Bars = 100 μm (**E**, **F**), 125 μm (**D**), 200 μm (**A–C**)

from one flower, where the fleshy part is induced by auxin and derived from the receptacle that holds the ovaries (Fig. 8.9Aa). Each small achene (one-seeded fruit that does not open to release the seed) on the surface of the strawberry fruit develops from one of the fertilized ovaries of the flower. Each developing achene produces auxin that induces its vascular vein (*arrow* in Fig. 8.8C), which supports achene development and the fleshy fruit tissue around and below the achene. By selective removal of most of the fertilized ovaries of a pollinated flower, I have left only a few fertilized ovaries, resulting in two (Fig. 8.9A) or five (Fig. 8.9B) separate fleshy

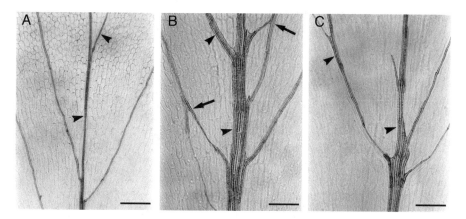

Fig. 8.8 Effects of influx auxin transport inhibitor (2,3,5-triiodobenzoic acid, TIBA) and auxin (naphthaleneacetic acid, NAA) spray applications (over 14-day period) on vascular differentiation in wild-type *Arabidopsis* petals. (**A**) Typical delicate vein system in the middle of an intact petal, consisting of two or three vessels in the midvein (*arrowhead*) and one in each secondary vein (*arrowhead*). (**B**) The auxin transport inhibitor (20 μM TIBA) induced a very wide midvein (comprised of 10–12 vessels), and wide secondary bundles, promoted new bypassing pathways resulting in additional regenerative bundles (*arrows*). (**C**) Repeated alternated-day treatments of 20 μM TIBA (on 1 day) and 20 μM NAA (on the next) caused an increase in vessel number at the base of the midvein and a moderate effect in the secondary veins. Bars = 125 μm. (From Aloni 2004)

small "fruit" below each untouched achene (located in the tip of each separate "tiny red fruit"), demonstrating how developing seeds induce around them their fleshy fruit tissue by the auxin they biosynthesis, confirming Nitsch (1950) elegant experiments.

Summary

- The floral-inducing signal, known as "florigen" (flower-former) is the FT protein, which is a mobile long-distance hormonal signal that travels via the phloem from the inducing leaves to the shoot apex, where it induced flowering. Florigen also promotes secondary cell wall biogenesis in the upper internodes of flowering stems, promoting differentiation of fibers, which adapt the shoot system to carry heavy fruits.
- Like vegetative leaves, young flower organ primordia start to show high-auxin production in their tip, resulting in discontinuous patterns of vessel differentiation, which appear first at the organ tip and extends downward.
- Anthers produce the highest-auxin concentrations that delays petal growth and nectary activity until the pollen grains are mature. This unique type of apical dominance shows how high-IAA-producing anthers synchronize flower development.
- A unique type of discontinuous vessel differentiation in the vascular system occurs in the gynoecium, at the base of the ovules, where the short veinlets that they induce do not connect to the central bundle of the gynoecium. This pattern

Fig. 8.9 Experimental modification of fruit development in garden strawberry (*Fragaria × ananassa*); and bioactive auxin accumulation in developing seeds of the DR5::GUS transformed *Arabidopsis thaliana* (the orientation of the silique is correct, with the seed's roots upward towards the stigma). (**A**) Selective removal of most of the fertilized ovaries in a strawberry flower (its white sepals remained untouched), resulted in the development of two separate miniature "fleshy red fruit" induced by the two remaining achenes (*arrowhead*). (**B**) Aggregation of five separate "tiny red fruits" each developed below its inducing achene (*arrowhead*), demonstrating that the developing auxin-producing achene induces the fleshy strawberry fruit tissue below it. (**C**) Longitudinal close-up view of a cleared intact strawberry fruit showing the achene (*dark brown*) embedded in the surface of the fruit with its supporting vascular bundle (*arrow*) that the auxin-synthesizing-achene induced, and the remaining basal part of its pistil (*arrowhead*). An early developmental stage of the pistil (*arrowhead*) on the surface of an intact growing white fruit (before it accumulates the red color) is shown in the small inserted photograph. (**D**) Auxin accumulation (due to polar auxin movement in the embryo) at the root side (evident by the green-blue staining of *DR5::GUS* gene expression) inducing the short veinlets during seed development in a silique of *Arabidopsis*. Bars = 200 μm (**C, D**), 500 μm (inserted photo), 10 mm (**A, B**). (**A–C**, R. Aloni unpublished; **D**, from Aloni et al. 2006)

is due to the high-auxin concentration streams occurring along the gynoecium's longitudinal bundles, which prevent veinlets connection.

- The high-auxin concentration descending from the stigma induces a wide fan of many vessel elements immediately below the stigma.
- Developing embryos and seeds produce auxin that induces their vascular tissues to support their growth and differentiation.

References and Recommended Readings[1]

Aloni R (2004) The induction of vascular tissue by auxin. In: *Plant Hormones: Biosynthesis, Signal Transduction, Action!* PJ Davies (ed). Kluwer Academic Publishers, Dordrecht, Boston, London, pp 471–492.

*Aloni R, Aloni E, Langhans M, Ullrich CI (2006) Role of auxin in regulating *Arabidopsis* flower development. Planta 223: 315–328.

Becker A, Ehlers K (2016) *Arabidopsis* flower development - of protein complexes, targets, and transport. Protoplasma 253: 219–230.

Bowman JL, Smyth DR, Meyerowitz EM (1989) Genes directing flower development in *Arabidopsis*. Plant Cell 1: 37–52.

*Bowman JL, Smyth DR, Meyerowitz EM (1991) Genetic interactions among floral homeotic genes of *Arabidopsis*. Development 112: 1–20.

Bowman JL, Smyth DR, Meyerowitz EM (2012) The ABC model of flower development: then and now. Development 139: 4095–4098.

Chailakhyan MKh (1936). New facts in support of the hormonal theory of plant development. *C R Acad Sci URSS* 13: 79–83.

Chen D, Zhao J (2008) Free IAA in stigmas and styles during pollen germination and pollen tube growth of *Nicotiana tabacum*. Physiol Plant 134: 202–215.

Cheng Y, Dai X, Zhao Y (2006) Auxin biosynthesis by the YUCCA flavin monooxygenase controls the formation of floral organs and vascular tissues in *Arabidopsis*. Genes Dev 20: 1790–1799.

Coen ES, Meyerowitz EM (1991) The war of the whorls: genetic interactions controlling flower development. Nature 353: 31–37.

Endress PK (1994) *Diversity and Evolutionary Biology of Tropical Flowers*. Cambridge University Press, Cambridge.

Evert RF, Eichhorn SE (2013) *Raven Biology of Plants*, 8th edn. Freeman, New York.

Gälweiler L, Guan C, Müller A, Wisman E, Mendgen K, Yephremov A, Palme K (1998) Regulation of polar auxin transport by AtPIN1 in *Arabidopsis* vascular tissue. Science 282: 2226–2230.

Giovannoni J, Nguyen C, Ampofo B, Zhong S, Fei Z (2017) The epigenome and transcriptional dynamics of fruit ripening. Annu Rev Plant Biol 68: 61–84.

Goethe JW von (1790). Versuch die Metamorphose der Pflanzen zu erklären. (Gotha: Carl Wilhelm Ettinger). [English translation: Arber A (1946). Goethe's botany. Chronica Botanica 10: 63–126.].

Goto K, Kyozuka J, Bowman JL (2001) Turning floral organs into leaves, leaves into floral organs. Curr Opin Genet Dev 11: 449–56.

*Heisler MG, Ohno C, Das P, Sieber P, Reddy GV, Long JA, Meyerowitz EM (2005) Patterns of auxin transport and gene expression during primordium development revealed by live imaging of the *Arabidopsis* inflorescence meristem. Curr Biol 5: 1899–1911.

Huang T, Böhlenius H, Eriksson S, Parcy F, Nilsson O (2005) The mRNA of the *Arabidopsis* gene *FT* moves from leaf to shoot apex and induces flowering. *Science* 309: 1694–1696.

---

[1] Papers of particular interest for suggested reading have been highlighted (with *).

Jaeger KE, Wigge PA (2007) FT protein acts as a long-range signal in *Arabidopsis* Curr Biol 17: 1050–1054.

Krizek BA (2011) Auxin regulation of *Arabidopsis* flower development involves members of the AINTEGUMENTA-LIKE/PHETHORA (AIL/PLT) family. J Exp Bot 62: 3311–3319.

Kumar R, Khurana A, Sharma AK (2014) Role of plant hormones and their interplay in development and riping of fleshy fruits. J Exp Bot 65: 4561–4575.

*Lifschitz E, Eviastar T, Rozman A, Shalit A, Goldschmidt A, Amsellem Z, Alvarez JP, Eshed Y (2006) The tomato *FT* ortholog triggers systemic signals that regulate growth and flowering and substitute for diverse environmental stimuli. Proc Natl Acad Sci USA 103: 6398–6403.

Lin MK, Belanger H, Lee YJ, Varkonyi-Gasic E, Taoka K, Miura E, Xoconostle-Cázares B, Gendler K, Jorgensen RA, Phinney B, Lough TJ, Lucas WJ (2007) FLOWERING LOCUS T protein may act as the long-distance florigenic signal in the cucurbits. Plant Cell 19: 1488–1506.

Nitsch JP (1950). Growth and morphogenesis of the strawberry as related to auxin. Am J Bot 37: 211–215.

Nitsch JP (1952) Plant hormones in the development of fruits. Q Rev Biol 27: 33–57.

*Nitsch JP (1953) The physiology of fruit growth. Annu Rev Plant Physiol 4: 199–236.

Okada, K.; Ueda, J.; Komaki, M.K.; Bell, C.J.; Shimura, Y (1991) Requirement of the auxin polar transport system in early stages of *Arabidopsis* floral bud formation. Plant Cell 3: 677–684.

Ó'Maoiléidigh DS, Graciet E, Wellmer F (2014) Gene networks controlling *Arabidopsis thaliana* flower development. New Phytol 201: 16–30.

*Robert HS, Crhak Khaitova L, Mroue S, Benková E (2015) The importance of localized auxin production for morphogenesis of reproductive organs and embryos in *Arabidopsis*. J Exp Bot 66: 5029–5042.

Sachs J (1865). Wirkung des lichtes auf die blütenbildung unter vermittlung der laubblätter. Bot Ztg 23: 117–121; 125–131; 133–139.

Sachs T (1968) On the determination of the pattern of vascular tissues in pea. Ann Bot 32: 781–790.

*Shalit-Kaneh A, Eviatar-Ribak T, Horev G, Suss N, Aloni R, Eshed Y, Lifschitz E (2019) The flowering hormone florigen accelerates secondary cell wall biogenesis to harmonize vascular maturation with reproductive development. Proc Natl Acad Sci USA 116: 16127–16136.

Smyth DR (2005) Morphogenesis of flowers - our evolving view. Plant Cell 17: 330–341.

Steeves TA, Sussex IM (1989) *Patterns in Plant Development*, 2nd edn. Cambridge University Press, Cambridge.

Stewart D, Graciet E, Wellmer F (2016) Molecular and regulatory mechanisms controlling floral organ development. FEBS J 283: 1823–1830.

Steynen QJ, Schultz EA (2003) The FORKED genes are essential for distal vein meeting in *Arabidopsis*. Development 130: 4695–4708.

Thomson B, Wellmer F (2019) Molecular regulation of flower development, (Ch 8) in: *Plant Development and Evolution*, U Grossniklaus (ed), Curr Top Develop Biol 131: 185–210.

Weigel D, Meyerowitz EM (1994) The ABCs of floral homeotic genes. Cell 78: 203–209.

Wisniewska J, Xu J, Seifertová D, Brewer PB, Ruzicka K, Blilou I, Rouquié D, Benková E, Scheres B, Friml J (2006) Polar PIN localization directs auxin flow in plants. Science 312: 883.

Yamaguchi N, Wu MF, Winter CM, Berns MC, Nole-Wilson S, Yamaguchi A, Coupland G, Krizek BA, Wagner D (2013) A molecular framework for auxin-mediated initiation of floral primordia. Dev Cell 24: 271–282.

Yamaguchi N, Wu MF, Winter CM, Wagner D (2014) *LEAFY* and polar auxin transport coordinately regulate *Arabidopsis* flower development. Plants 3: 251–265.

Zeevaart JAD (1983) Gibberellins and flowering. In: *The Biochemistry and Physiology of Gibberellins*, Vol 2, A Crozier (ed). Praeger Scientific, New York, pp 333–374.

Zeevaart JAD (2006) Florigen coming of age after 70 years. Plant Cell 18: 1783–1789.

Zhao Y (2018) Essential roles of local auxin biosynthesis in plant development and in adaptation to environmental changes. Ann Rev Plant Biol 69: 417–435.

Can a Differentiating Vessel Induce Lateral Root Initiation?

9

The previous chapters discussed the mechanisms how the shoot organs, the leaves, and flowers induce the initiation and differentiation of vessels by the polar auxin flow these organs produce (Sachs 1981; Aloni et al. 2003, 2006b Mattsson et al. 2003; Scarpella et al. 2006; Wenzel et al. 2007; Aloni 2010; Scarpella and Helariutta 2010). On the other hand, the basic question addressed in this chapter is whether there is a possibility of an opposite phenomenon in plants, namely, whether a differentiating vessel can induce the initiation of an organ?

The unexpected answer is positive; these organs are the lateral roots that are induced by differentiating vessels in the primary root. The mechanism of this common phenomenon, namely, how a differentiating vessel induces **lateral root initiation (LRi)** will be clarified below by the *lateral root initiation hypothesis* (Aloni et al. 2006a; Aloni 2013).

Structure and differentiation of the vascular systems in roots of dicots and monocots were discussed in Sect. 2.3. Here the focus is on the mechanism of lateral root initiation, aiming to clarify how lateral roots are induced by an inner signal originating in differentiating vessel elements and how the latter determine the location of a new lateral root. In dicotyledonous roots, LRi and growth starts at the protoxylem pole (Fig. 9.1), and auxin is a key promoting signal in LRi and development (Benková et al. 2003; Aloni et al. 2006a; De Smet et al. 2007; Dubrovsky et al. 2008; Lavenus et al. 2013; Marhavý et al. 2013; Du and Scheres 2018; Ilina et al. 2018). However, in spite of the increasing physiological and molecular information on lateral root formation (Himanen et al. 2002; Negi et al. 2008; Benková and Bielach 2010; Dubrovsky et al. 2011; De Smet 2012; Muraro et al. 2013; Bellini et al. 2014; Du and Scheres 2018; Ilina et al. 2018; Goh et al. 2019; Kiryushkin et al. 2019), the hormonal mechanism that controls LRi remains poorly understood.

Aloni et al. (2006a) proposed the *lateral root initiation hypothesis* as a working hypothesis that should be experimentally analyzed. The hypothesis suggests that **ethylene (C_2H_4) produced in a differentiating protoxylem vessel (DPV)** near the root tip is the signal that promotes the earliest stage of LRi at the xylem pole. When

© Springer Nature Switzerland AG 2021
R. Aloni, *Vascular Differentiation and Plant Hormones*,
https://doi.org/10.1007/978-3-030-53202-4_9

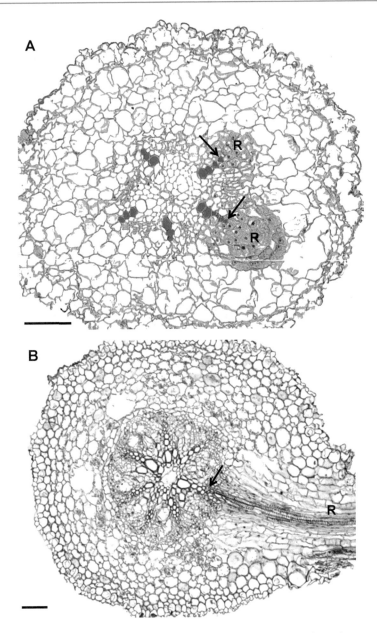

Fig. 9.1 Lateral root initiation and development occurs at the xylem pole, shown in cross sections of sweet potato (*Ipomoea batatas*). (**A**) Early stage of two young lateral roots (R) growing through the cortex of a root with five xylem arches (strands). Each lateral root starts from a protoxylem vessel (*arrow*). To mark the primary vessels, they were artificially filled with a red color. (**B**) More developed root region with a wider stele that enabled the development of seven alternating xylem and phloem stands. Note that the lateral root (R) developed at the xylem pole (*arrow*), the lignified vessels were stained red by safranin. Bars = 100 μm. (Courtesy of V. Singh and N. Firon)

we proposed this concept, it was mostly intuition, with no supporting evidence. Today, there is new published experimental evidence from diverse laboratories that strongly support the ***lateral root initiation hypothesis***, although no one has related the evidence to our concept:

(i) Supporting evidence is the finding of Pesquet and Tuominen (2011) that ethylene is produced in maturing vessel elements, indicating that a DPV at the root tip can produce a low amount of ethylene for a short duration.

(ii) Furthermore, Ivanchenko et al. (2008) have shown that the application of low concentration (0.04 μM **ACC**) of the ethylene precursor 1-aminocyclopropane-1-carboxylic acid (ACC) substantially promoted LRi; whereas high concentrations caused inhibition.

(iii) Moreover, the plant cytosolic enzyme 1-aminocyclopropane-1-carboxylate synthase (ACS) which catalyzes the rate-limiting step in the ethylene biosynthetic pathway is expressed near the root tip around DPVs in both the cell division zone (see: ACS2, ACS8, in Tsuchisaka and Theologis 2004, fig. 11) and the cell expansion zone (ACS5, ACS6, ACS7, also in fig. 11) marking ethylene production by DPVs (Tsuchisaka and Theologis 2004).

All these experimental results support the idea that low concentration of ethylene emission from DPVs is positively involved in LRi. One should realize that the phenomenon of LRi is controlled from the inside of the vascular cylinder by a stimulating signal arriving from the differentiating protoxylem vessels (De Smet 2012; Ilina et al. 2018). This is the reason why in dicotyledonous roots the LRi occurs at the xylem pole but never develops at the phloem pole (Dubrovsky et al. 2000). If the stimulation would have arrived from the cortex side, we would expect LRi also in the phloem pole. It has been shown experimentally that sieve-element differentiation is induced by very low-IAA concentrations and that there is a need for high-IAA concentration to induce a vessel (Aloni 1980, 1995, 2001) (see Chap. 5, Fig. 5.3). Accordingly, the high-IAA concentration streams which induce vessels can induce ethylene emission, while the very low-IAA concentration stream that induces sieve elements is too low to stimulate ethylene synthesis; therefore, the differentiating sieve tubes are not involved in LRi. Furthermore, the innermost layer of the cortex is the endodermis, which is compactly arranged and lacks air spaces among its cells. Consequently, the endodermis slows down the transport of ethylene emission from the differentiating protoxylem cells outward to the cortex, thus locally boosting ethylene concentration in the pericycle (which is located just inside the endodermis) from which the lateral roots initiate. Local ethylene accumulation in the pericycle stops the constant low-concentration polar auxin flow in the pericycle, resulting in local IAA accumulation above the ethylene obstacle forming local auxin maximum that can be clearly detected (see in: Ilina et al. 2018, their Fig. 1A, B visualizing auxin response maxima with *DR5* expression), which stimulate and induce LRi.

The *lateral root initiation hypothesis* (Aloni et al. 2006a; Aloni 2013) explains how ethylene, auxin, and cytokinin regulate LRi, above the elongation zone of dicotyledonous roots as follows. The IAA which promotes LRi descends from young leaves and moves polarly along the root in its vascular cylinder toward the root's tip in the two following canalized and distinct pathways: (1) for a short duration, a relatively high-IAA concentration stream flows through differentiating protoxylem vessels, during the process of vessel differentiation, and (2) a suggested low-IAA concentration polar flow which moves continuously via the pericycle (Fig. 9.2) that maintains its meristematic identity. The presence of the high-IAA concentration in the root's differentiating protoxylem vessels can be easily visualized with *DR5::GUS* expression (e.g., see in: De Smet 2012, Fig. 1b). While the suggested low-IAA concentration polar movement in the pericycle is too low to induce LRi and, therefore, needs to be locally concentrated (e.g., see concentrated auxin in the pericycle in: Fig. $1A_0$ of Benková et al. 2003; and Fig. 2, the 1st stage in Benková and Bielach 2010) just before LRi.

Logically, the only possible way to locally increase IAA concentration in the pericycle (observed in: Fig. $1A_0$ in Benková et al. 2003; and Fig. 2, the 1st stage in Benková and Bielach 2010) that induces the local auxin maxima (Ilina et al. 2018, Fig. 1A, B) in the pericycle is by obstructing the pericycle's low-concentration polar auxin flow, thus causing IAA accumulation in the pericycle, resulting in the auxin maxima (Ilina et al. 2018). Actually, the **accumulation of auxin maxima in the pericycle** confirms the existence of the low-concentration polar auxin flow in the pericycle, which maintains its meristematic nature to divide and initiate lateral roots.

Thus, the local protoxylem-induced ethylene is the trigger that determines the site of LRi at the xylem pole (Fig. 9.2) by locally stopping and concentrating the low-concentration polar IAA flow along the pericycle. The C_2H_4 is released from the differentiating protoxylem vessel elements and diffuses to the neighboring tissue. In the centrifugal direction, the C_2H_4 is locally accumulated in the pericycle (as its further centrifugal movement is slowed down by the densely packed endodermis). The local accumulation of ethylene in the pericycle obstructs the polar low-concentration IAA movement in the pericycle adjacent to the DPV. Therefore, immediately above this ethylene inhibition site, newly arriving IAA from young leaves is locally accumulated in the pericycle forming a local auxin maxim; this fast IAA buildup (detected by *DR5::GUS* expression: Benková et al. 2003, Fig. $1A_0$; Benková and Bielach 2010 the 1st stage; Ilina et al. 2018, Fig. 1A, B) stimulates cell divisions in the pericycle, inducing the founder cells of a new lateral root (Fig. 9.2).

Cytokinin (CK), which inhibits LRi (Laplaze et al. 2007; Márquez et al. 2019), originates in the root cap (Aloni et al. 2004) and moves upward through the root vascular cylinder (Aloni et al. 2005). The distance of LRi from the root cap is regulated by CK concentration. The high CK concentrations at the root cap antagonize IAA and inhibit LRi in the vicinity of the cap, which is crucial for enabling uninterrupted elongation of the root tip. Therefore, lateral roots initiate usually further away from the CK-synthesizing cap, occurring above the elongation zone, thus ensuring the elongation of a smooth primary root tip free from lateral roots. Above the elongation zone, where the concentrations of CK decrease, lateral roots can initiate.

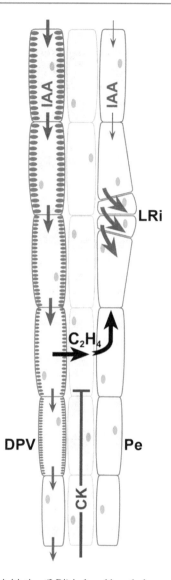

Fig. 9.2 Model of lateral root initiation (LRi) induced by ethylene and auxin (IAA) and inhibited by cytokinin (CK) at the tip of a young dicotyledonous root. *Arrows* indicate positive LRi regulation; *blunt-ended line* indicates negative regulation. The schematic diagram shows the three outermost cell columns of the vascular cylinder in a radial-longitudinal orientation at the xylem pole, with a differentiating protoxylem vessel (DPV) and the pericycle (Pe). Two pathways of parallel streams of polar IAA (marked with *red arrows*) are shown: in the left side is the high-IAA concentration stream that induces the protoxylem vessel (marked by the *gradual development of secondary wall thickenings*) and in the right side suggested low-IAA concentration stream maintaining the meristematic identity of the pericycle. During vessel differentiation, a local increase of IAA concentration in a differentiating protoxylem vessel element induces low-concentration ethylene (C_2H_4) synthesis. This C_2H_4 signal is released, and in the centrifugal direction (*black arrow*), it locally blocks the polar IAA movement in the adjacent pericycle cells, consequently boosting IAA concentration immediately above the blockage, thus forming a local auxin maximum. This elevated high-IAA concentration induces pericycle cell divisions and LRi. The CK (marked with a *blue line*) arriving from the root cap inhibits LRi in the vicinity of the cap. In *Arabidopsis* the protoxylem vessel is attached to the pericycle. (From Aloni 2013)

Summary

- The *lateral root initiation hypothesis* suggests that ethylene (C_2H_4) produced in a differentiating protoxylem vessel (DPV) near the root tip is the hormonal signal that promotes the earliest stage of lateral root initiation (LRi) at the xylem pole.
- Supporting evidence shows that: (1) ethylene is produced in maturing vessel elements, indicating that a DPV at the root tip can produce a low amount of ethylene for a short duration, (2) application of low concentration (0.04 μM) of the ethylene precursor ACC substantially promoted LRi, and (3) the plant cytosolic enzyme ACS, which catalyzes the rate-limiting step in the ethylene biosynthetic pathway is expressed near the root tip around DPVs marking ethylene production by differentiating protoxylem vessels.
- Vessels are induced by streams of high-auxin concentration (and sieve tubes by low-auxin concentration); therefore only differentiating vessels can induce ethylene emission, which is locally accumulated in the pericycle (because the centrifugal ethylene movement is stopped by the endodermis). This is the reason why lateral roots in dicotyledonous roots are induced at the xylem pole only.
- The downward polar auxin flow in the pericycle is stopped by the local ethylene accumulation (originating in DPV) in the pericycle, resulting in local IAA accumulation above the ethylene obstacle, forming local auxin maximum that can be easily detected (by *DR5* expression). From this site a lateral root is initiated.
- Cytokinin (CK), which inhibits LRi, is produced in the root cap, moves upward through the root vascular cylinder, and inhibits LRi in the vicinity of the cap.

References and Recommended Readings[1]

Aloni R (1980) Role of auxin and sucrose in the differentiation of sieve and tracheary elements in plant tissue cultures. Planta 150: 255–263.

Aloni R (1995) The induction of vascular tissues by auxin and cytokinin. In: *Plant Hormones: Physiology, Biochemistry and Molecular biology*, PJ Davies (ed). Kluwer, Dordrecht, pp 531–546.

Aloni R (2001) Foliar and axial aspects of vascular differentiation: hypotheses and evidence. J Plant Growth Regul 20: 22–34.

Aloni R (2010) The induction of vascular tissues by auxin. In: *Plant Hormones: Biosynthesis, Signal Transduction, Action!* PJ Davies (ed). Kluwer Academic Publishers, Dordrecht, pp 485–506.

*Aloni R (2013) Role of hormones in controlling vascular differentiation and the mechanism of lateral root initiation. Planta 238: 819–830.

*Aloni, R, Aloni E, Langhans M, Ullrich CI (2006a) Role of cytokinin and auxin in shaping root architecture: regulating vascular differentiation, lateral root initiation, root apical dominance and root gravitropism. Ann Bot 97: 883–893.

Aloni R, Aloni E, Langhans M, Ullrich CI (2006b) Role of auxin in regulating *Arabidopsis* flower development. Planta 223: 315–328.

---

[1] Papers of particular interest for suggested reading have been highlighted (with *).

Aloni R, Langhans M, Aloni E, Dreieicher E, Ullrich CI (2005) Root-synthesized cytokinin in *Arabidopsis* is distributed in the shoot by the transpiration stream. J Exp Bot 56: 1535–1544.

Aloni R, Langhans M, Aloni E, Ullrich CI (2004) Role of cytokinin in the regulation of root gravitropism. Planta 220: 177–182.

Aloni R, Schwalm K, Langhans M, Ullrich CI (2003) Gradual shifts in sites of free-auxin production during leaf-primordium development and their role in vascular differentiation and leaf morphogenesis in *Arabidopsis*. Planta 216: 841–853.

Bellini C, Pacurar DI, Perrone I (2014) Adventitious roots and lateral roots: similarities and differences. Annu Rev Plant Biol 65: 639–666.

Benková E, Bielach A (2010) Lateral root organogenesis - from cell to organ. Curr Opin in Plant Biol 13: 677–683.

Benková E, Michniewicz M, Sauer M, Teichmann T, Seifertová D, Jürgens G, Friml J (2003) Local, efflux-dependent auxin gradients as a common module for plant organ formation. Cell 115: 591–602.

De Smet I (2012) Lateral root initiation: one step at a time. New Phytol 193: 867–873.

De Smet I, Tetsumura T, De Rybel B, Frei dit Frey N, Laplaze L, Casimiro I, Swarup R, Naudts M, Vanneste S, Audenaert D, Inzé D, Bennett MJ, Beeckman T (2007) Auxin-dependent regulation of lateral root positioning in the basal meristem of *Arabidopsis*. Development 134: 681–690.

Du Y, Scheres B (2018) Lateral root formation and the multiple roles of auxin. J Exp Bot 69: 155–167.

Dubrovsky JG, Doerner PW, Colon-Carmona A, Rost TL (2000) Pericycle cell proliferation and lateral root initiation in *Arabidopsis*. Plant Physiol 124: 1648–1657.

Dubrovsky JG, Napsucialy-Mendivil S, Duclercq J, Cheng Y, Shishkova S, Ivanchenko MG, Friml J, Murphy AS, Benková E (2011) Auxin minimum defines a developmental window for lateral root initiation. New Phytol 191: 970–983.

Dubrovsky JG, Sauer M, Napsucialy-Mendivil S, Ivanchenko MG, Friml J, Shishkova S, Celenza J, Benková E (2008) Auxin acts as a local morphogenetic trigger to specify lateral root founder cells. Proc Natl Acad Sci USA 105: 8790–8794.

Goh T, Toyokura K, Yamaguchi N, Okamoto Y, Uehara T, Kaneko S, Takebayashi Y, Kasahara H, Ikeyama Y, Okushima Y, Nakajima K, Mimura T, Tasaka M, Fukaki H (2019) Lateral root initiation requires the sequential induction of transcription factors LBD16 and PUCHI in *Arabidopsis thaliana*. New Phytol 224: 749–760.

Himanen K, Boucheron E, Vanneste S, de Almeida Engler J, Inzé D, Beeckman T (2002) Auxin-mediated cell cycle activation during early lateral root initiation. Plant Cell 14: 2339–2351.

*Ilina EL, Kiryushkin AS, Semenova VA, Demchenko NP, Pawlowski K, Demchenko KN (2018) Lateral root initiation and formation within the parental root meristem of *Cucurbita pepo*: is auxin a key player? Ann Bot 122: 873–888.

*Ivanchenko MG, Muday GK, Dubrovsky JG (2008) Ethylene–auxin interactions regulate lateral root initiation and emergence in *Arabidopsis thaliana*. Plant J 55: 335–347.

Kiryushkin AS, Ilina EL, Puchkova VA, Guseva ED, Pawlowski K, Demchenko KN (2019) Lateral Root Initiation in the Parental Root Meristem of Cucurbits: Old Players in a New Position. Front Plant Sci 10: 365.

Laplaze L, Benkova E, Casimiro I, Maes L, Vanneste S, Swarup R, Weijers D, Calvo V, Parizot B, Herrera-Rodriguez MB, Offringa R, Graham N, Doumas P, Friml J, Bogusz D, Beeckman T, Bennett M (2007) Cytokinins act directly on lateral root founder cells to inhibit root initiation. Plant Cell 19: 3889–3900.

Lavenus J, Goh T, Roberts I, Guyomarc'h S, Lucas M, De Smet I, Fukaki H, Beeckman T, Bennett M, Laplaze L (2013) Lateral root development in *Arabidopsis*: fifty shades of auxin. Trends Plant Sci 18: 450–458.

Marhavý P, Vanstraelen M, De Rybel B, Zhaojun D, Bennett MJ, Beeckman T, Benková E (2013) Auxin reflux between the endodermis and pericycle promotes lateral root initiation. EMBO J 32: 149–158.

Márquez G, Alarcón MV, Salguero J (2019) Cytokinin inhibits lateral root development at the earliest stages of lateral root primordium initiation in maize primary root. J Plant Growth Regul 38: 83–92.

Mattsson J, Ckurshumova W, Berleth T (2003) Auxin signaling in *Arabidopsis* leaf vascular development. Plant Physiol 131: 1327–1339.

Muraro D, Byrne H, King J, Bennett M (2013) The role of auxin and cytokinin signalling in specifying the root architecture of *Arabidopsis thaliana*. J Theor Biol 317: 71–86.

Negi S, Ivanchenko MG, Muday GK (2008) Ethylene regulates lateral root formation and auxin transport in *Arabidopsis thaliana*. Plant J 55: 175–187.

*Pesquet E, Tuominen H (2011) Ethylene stimulates tracheary element differentiation in *Zinnia elegans* cell cultures. New Phytol 190: 138–149.

Scarpella E, Helariutta Y (2010) Vascular pattern formation in plants. Curr Top Dev Biol 91: 221–265.

Scarpella E, Marcos D, Friml J, Berleth T (2006) Control of leaf vascular patterning by polar auxin transport. Genes Dev 20: 1015–1027.

Sachs T (1981) The control of patterned differentiation of vascular tissues. Adv Bot Res 9: 151–262.

*Tsuchisaka A, Theologis A (2004) Unique and overlapping expression patterns among the Arabidopsis 1-amino-cyclopropane1-carboxylate synthase gene family members. Plant Physiol 136: 2982–3000.

Wenzel CL, Schuetz M, Yu Q, Mattsson J (2007) Dynamics of MONOPTEROS and PIN-FORMED1 expression during leaf vein pattern formation in *Arabidopsis thaliana*. Plant J 49: 387–398.

Vascular Regeneration and Grafting

10

10.1 Vascular Regeneration

Plants are characterized by a remarkable **developmental and regeneration plasticity**, which enable them to replace damaged organs by de novo organogenesis. Likewise, plant tissues have exceptional ability to undergo de-differentiation and form various cell types from parenchyma cells (Gaillochet and Lohmann 2015; Ikeuchi et al. 2016). The regeneration of organs is mainly regulated by the balance between auxin and cytokinin, which determines the type of the regenerating organs (Sang et al. 2018). High-auxin concentrations induce roots, while elevated cytokinin promotes shoot development (Ikeuchi et al. 2016; Sang et al. 2018).

Steward et al. (1958) and Vasil and Hildebrandt (1965) demonstrated that even a single cell taken from carrot or tobacco retain **totipotency**, namely, the capacity to regenerate a whole plant, demonstrating the amazing regenerative potential of plant somatic cells. At the tissue level, vascular regeneration might occur from stem cells of a meristem, like from the initials of the vascular cambium or from parenchyma cells which could de-differentiate, divide, and re-differentiate to phloem and xylem elements (Fig. 10.1A). Additionally, parenchyma cells taken from young leaves of *Zinnia elegans* (Fukuda and Komamine 1980) can transdifferentiate in liquid cell culture directly into tracheary elements without a need for cell division (see Sect. 5.2.4, Fig. 5.14).

We should realize that **naturally occurring regenerative differentiation** of vascular tissues is a normal process inside intact growing plants, which it is not induced by external wounding. Regenerative vascular elements re-differentiate from parenchyma cells at the junctions of the plant axis with lateral or adventitious roots, where regenerative vessel elements and sieve tube elements connect the vascular systems of new lateral roots to the main root or adventitious roots to the stem (Aloni and Plotkin 1985). Likewise, new vascular connection and bundle development between the stem and emerging axillary buds induce regenerative vascular elements from parenchyma cells in their junctions. Interestingly, in some plant species, like in *Luffa cylindrica*, the emergence of adventitious roots, which likely

© Springer Nature Switzerland AG 2021
R. Aloni, *Vascular Differentiation and Plant Hormones*,
https://doi.org/10.1007/978-3-030-53202-4_10

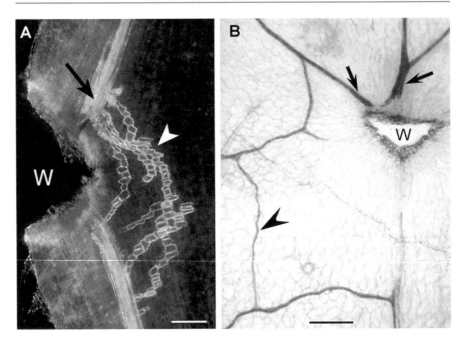

Fig. 10.1   Polar patterns of xylem regeneration (*arrowheads*) revealing the pathways of the inducing polar auxin transport around a wound (w). (**A**) In a decapitated young internode of *Cucumis
sativus* treated with auxin (0.1% IAA in lanolin for 7 days), which was applied to the upper side of
the internode immediately after wounding and removing the leaves and buds above it. Showing in
a longitudinal view (observed after clearing with lactic acid, staining with phloroglucinol, and
photographed in dark field) that the regenerated vessel elements (re-differentiated from parenchyma cells) formed above the wound (*arrow*) (where IAA was concentrated by the cut) differentiated close to the wound, while those below the wound differentiated at greater distances from the
wound. (**B**) Pattern of limited xylem regeneration restricted to veins, around a wound (w) made in
the midvein (primary vein) of a very young leaf primordium of the DR5::GUS-transformed
Arabidopsis thaliana (harvested after 7 days and cleared in lactic acid), revealing that the bioactive
auxin descending from the leaf's tip is detected above the wound (by green-blue staining of
DR5::GUS expression). The *arrows* mark the lower ends of the veins that were cut by the wound,
which became thicker due to the differentiation of additional regenerated vessel elements induced
in these veins by IAA accumulation. The absence of vessels immediately below the wound indicates that the wounding was made at the stage of early procambium development. The remote
xylem differentiation (*arrowhead*) occurred inside veins forming a continuous vein pattern around
the wound that was determined by the configuration of the procambium. Note that there is no
xylem regeneration from parenchyma cells around the wound in the leaf. Bars = 500 μm (**A**), 1 mm
(**B**). (By R. Aloni)

interrupts the longitudinal pathway of polar auxin movement in the periphery of the
vascular cylinder, induces naturally occurring vessel regeneration around the adventitious roots (Aloni and Baum 1991). Phloem anastomoses develop naturally
between vascular bundles along elongating internodes of many plant species (Aloni
and Sachs 1973; Aloni and Peterson 1990; Aloni and Barnett 1996). The phloem
anastomoses are absent in very young internodes, and their sieve elements and

companion cells re-differentiate from parenchyma cells between longitudinal vascular bundles during internode elongation (see Chap. 2, Fig. 2.18A, B) (Aloni and Barnett 1996). It is likely that these phloem anastomoses are the result of low-auxin concentration leaking streams from the elongating bundles that because of rapid internode elongation and stem growth oscillations, known as circumnutating growth, result in minor interruptions to auxin flow inside the bundles which induce IAA leaking. Additionally, transfusion tracheids are short tracheids that re-differentiate from parenchyma cells in conifer leaves. The transfusion tracheids are naturally occurring regenerative tracheids that can be induced experimentally by auxin application (Aloni et al. 2013).

The **regeneration of vascular tissues around an injury** is a recovery mechanism for survival, which heals the damaged organ, forming new vascular conduits that renew and restore water transport and assimilates translocation along the wounded organ. When a vascular bundle is cut, a bridge of both phloem and xylem elements differentiates around the wound (Simon 1908; von Kaan Albest 1934; Jost 1940; Sinnott and Bloch 1944; Camus 1949; Eschrich 1953). The regenerative wound vascular bypass develops in a polar pattern, starting from above the wound and progressing downward. Usually more regenerative vascular elements differentiate above the wound close to the wound surface, indicating that the wound retards and locally concentrates the moving signal arriving from buds and young leaves above the wound. However, patterns of regenerated vessels and sieve tubes may also differentiate from below the wound upward, and they are also induced by the young leaves above the wound (Camus 1949; Eschrich 1953; Aloni and Jacobs 1977a; Robbertse and McCully 1979).

In order to find the answer to the basic question whether (**i**) vascular regeneration around wounds includes a replacement of damaged tissues or (**ii**) only new sieve tubes and vessels which are normally formed are diverted around the wound, Benayoun et al. (1975) studied the induced vascular bridges of regenerative sieve tubes and vessels around wounds and also inside the wounded bundles in *Coleus blumei* and *Cucumis sativus* and found that the new bridge is part of new longitudinal conduits that do not form connections to damaged sieve tubes or vessels, so that the injured conduits continuity around the wounds is not restored, and the new bridges of regenerated sieve tubes and vessels around the wound are part of new continuous sieve tubes and vessels formed first along the bundles and later at the wound site where they are diverted around the wound. Behnke and Schulz (1980) confirmed this observation that regenerating sieve elements do not connect to pre-existing bundle of sieve elements and that they form a new vein of young sieve elements produced by cambial activity that build the bridge around the wound in *Coleus* stem.

The new regenerating sieve tubes and vessels differentiate in **fragmented patterns**; they first mature inside the pre-existing bundles and finally mature around the wound (Aloni and Jacobs 1977a,b; Aloni and Barnett 1996). This fragmented maturation pattern is a result of the different ability of the differentiating cells along the new auxin pathway to respond to auxin. The meristematic cambial cells inside the bundles differentiate first as they are already in a process of becoming the next sieve tubes and vessels, while the non-active cambial initials between the bundles and especially the parenchyma cells around the wound need more time to divide, de-differentiate, and then re-differentiate. Therefore, the parts of the new regenerating sieve tubes and regenerating vessels at the wound site are the last ones to mature, forming the local downward and upward regeneration patterns detected around wounds (marked by *arrows* in Fig. 10.2).

Young leaves induce the phloem and xylem regeneration around a wound, and this effect of the leaves can be replaced by external auxin (IAA) application (Jacobs 1952, 1970; LaMotte and Jacobs 1963; Thompson and Jacobs 1966) (Fig. 10.1A). When isolated non-elongating internodes of *Coleus* with cambial activity (the fifth internode counted from the apical bud) were not treated with any hormone, they showed some regenerated sieve tube elements, but did not regenerate any tracheary elements. With increasing concentrations of applied IAA added apically, the number of wound sieve tube strands increased. Higher-auxin concentrations induced also regenerated wound xylem strands, and their number increased with increasing IAA concentrations. These results demonstrate that low-auxin concentrations can induce phloem regeneration and higher IAA concentrations are needed to induce xylem regeneration (Thompson and Jacobs 1966).

In stems of *Coleus blumei* (see Chap. 2, Fig. 2.6), there are strands of phloem with no xylem. When these phloem-only strands are cut by a small wound, regenerative sieve elements with no xylem are produced around the injury (Houck and LaMotte 1977). Larger wounds that cut both phloem-only strands and collateral bundles caused more wound phloem strands than regenerated xylem strands (Fig. 10.2A, B). On the 4th day after wounding, the ratio of strands severed by the wound in the phloem to those in the xylem was 2.14, and the ratio of the regenerative sieve tubes to the regenerative vessels was 2.24. For both tracheary and sieve tube cells, the initial regeneration was strongly polar (mostly above the wound), as expected by the downward polar transport of the inducing auxin signal. The path of regenerated vessel was obviously related to that of the sieve tubes on the other side of the cambium (Fig. 10.2) (Aloni and Jacobs 1977b).

When phloem-only strands were cut by minor wounds, in stem stumps (with no leaves but with intact root system),1% IAA in lanolin completely restored phloem regeneration (with no xylem regeneration) to that of the intact plant level. IAA failed to restore phloem regeneration in excised internodes (with no roots), to the intact plant level. Zeatin, or zeatin riboside, applied to the bases of excised internodes in aqueous solution that their apical ends received IAA restored phloem regeneration to the level found in whole plants (Houck and LaMotte 1977). These results were confirmed with wounded collateral bundles (containing both phloem and xylem) with either kinetin or zeatin for phloem regeneration (Aloni et al. 1990) and for xylem regeneration around a wound in excised *Coleus* internodes (Baum et al. 1991), demonstrating the need for cytokinin, originating in root tips, to obtain the same level of vascular regeneration as in intact plants. Chen et al. (2019) studied regeneration of secondary vascular tissue after large-scale bark girdling in *Populus* trees under in vitro conditions; their results show that auxin was sufficient to induce

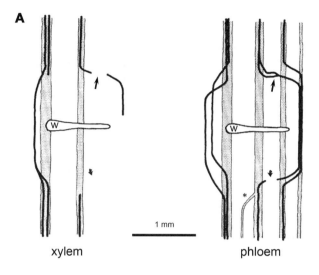

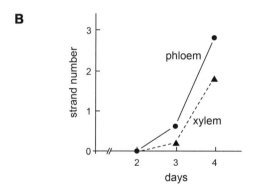

Fig. 10.2 (**A**) Diagrams comparing the regeneration of vessels (the heavy lines in the "xylem" diagram) with that of the sieve tubes (the heavy lines in the "phloem" diagram) from the same *Coleus blumei* wounded internode, grown in a greenhouse and collected 4 days after wounding. The longitudinal running stippled bands represent the pre-existing vascular bundles that contained differentiated vessels (in the xylem diagram) or sieve tubes (the phloem diagram). Both diagrams are at the same magnification (see 1 mm scale bar marked in lower center). The different *arrows* show equivalent locations in the two preparations, which were separated at the cambium. "W" designates the wound, and the small star (lower center of the phloem diagram) marks one end of a phloem anastomosis (that was not counted as phloem regeneration). The diagrams demonstrate that (i) there are more longitudinal phloem strands (on the right side two phloem-only strands, and on the left side two collateral bundles); (ii) there are more regenerated phloem strands than regenerated xylem strands; (iii) vascular regeneration follows the configuration of the pre-existing strands; (iv) vascular regeneration occurs in discontinuous patterns (*arrows*); (v) the regenerated conduits start differentiation inside the pre-existing phloem strands and bundles and continue regeneration in a fragmental fashion between the bundles; (vi) regeneration between the bundles is polar, occurring first above the wound; (vii) phloem regeneration is faster than the xylem regeneration; and (viii) regenerated vessel elements and wound sieve tube elements around the wound are the connecting parts of continuous new vessels and new sieve tubes along the bundles. (**B**) Time course of complete regenerated sieve tubes and regenerated vessels around a wound in the fifth internode of *Coleus blumei*. Each point is an average of five plants. (Aloni and Jacobs 1977b)

regeneration of phloem prior to continuous cambium restoration, while cytokinin promoted only the formation of new phloem, without cambium formation.

In dicotyledonous plants, parenchyma cells possess the ability to re-differentiate, whereas parenchyma cells in monocotyledons cannot divide and re-differentiate from a relatively early stage. However, in very young internodes of *Zea mays* and at the nodes, meristematic and young parenchyma cells regenerated around a wound and at sites of adventitious root initiation, to connect the vascular tissues of the emerging adventitious root to those of the stem (Aloni and Plotkin 1985). Within about 1 day after wounding, the injury induces cell divisions around the wound surface and at the cut vascular system, resulting in new cambial initials and paren-chyma cells that re-differentiate to become vascular elements. Regenerated sieve tube elements are the first to be detected around a wound. Time course experiments conducted on the fifth internode of *Coleus* (Thompson 1967) and on pea roots (Robbertse and McCully 1979) reveal that the first recognizable regenerated sieve tube elements can be detected 2 days after wounding, while regenerated vessel ele-ments appear 3 days after the injury (Thompson 1967; Robbertse and McCully 1979).

Not every plant organ can regenerate vascular tissues around a wound. While stems and roots respond to injury and produce regenerated phloem and xylem con-duits around the wound from parenchyma and cambium, leaves show a limited response (Fig. 10.1B) (see Chap. 7) and usually do not produce vascular regenera-tion from parenchyma cells around a wound (Fig. 10.3B). Petals that grow relatively fast for a short duration do not show vascular regeneration (Fig. 10.3C).

I started this chapter with the general declaration that plants are characterized by remarkable regeneration plasticity. This is true in young tissues of dicotyledonous

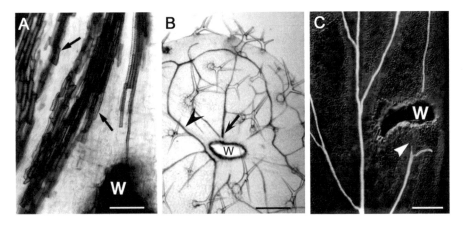

Fig. 10.3 Micrographs of typical responses around a wound (w) comparing the ability of very young organs of *Arabidopsis thaliana* to respond and regenerate vascular tissues (harvested after 7 days and cleared in lactic acid). (**A**) Well-developed massive xylem regeneration (*arrows*) was induced around a wound made in the inflorescence. (**B**) Functional vein bypass (*arrowhead*) beside the wound done in the leaf's midvein (*arrow*) connecting the tip veins with the lower part of the midvein. (**C**) No vascular regeneration response could be obtained after wounding a vein (*arrow-head*) in the petal (the petal was photographed in a dark field). Bars = 1 mm. (By R. Aloni)

plants that their parenchyma cells possess the ability to re-differentiate when they are young or from cambial initials in old organs. But **aging parenchyma cells gradually lose their plasticity** with time (Aloni and Barnett 1996). For example, in young internodes of *Cucurbita maxima*, although many phloem anastomoses were already present before wounding, the regenerated vessels re-differentiated from interfascicular parenchyma cells (Fig. 10.4A), and the anastomoses remained inactive. This result indicates that when parenchyma cells are young and can re-differentiate into vascular elements, they will serve as the preferred pathway for auxin. However, when the parenchyma cells become old, they gradually lose the ability to respond, and if the internodes do not develop cambium, the anastomoses become the main pathway for the polar auxin flow descending from the young leaves, resulting in xylem regeneration within the anastomoses, in patterns of the available phloem anastomoses around the wound (Fig. 10.4B) (Aloni and Barnett 1996; Aloni 1995).

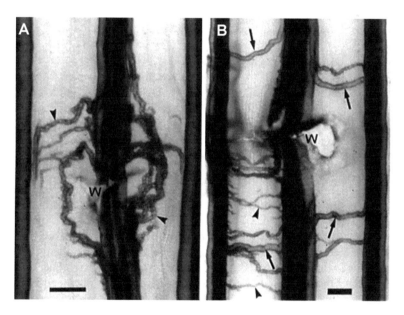

Fig. 10.4 Micrographs of typical regenerated vessels (*arrowheads*) and giant (very wide) regenerated vessels (*arrows*) around a wound (w) in young (**A**) and old (**B**) internodes of *Cucurbita maxima* sampled 7 days after wounding, showing the transition from the typical curved bypass of xylem regeneration from parenchyma cells immediately around the wound (**A**) to xylem regeneration mostly limited to phloem anastomoses further away from the wound (**B**). (**A**) In the third internode from the apical bud, many parenchyma cells have re-differentiated to produce regenerated vessels bypassing the wound and rejoining the damaged bundle. (**B**) In the sixth internode, there are no direct bridges of xylem formed around the wound. The differentiation of the regenerated xylem elements is exclusively associated with phloem anastomoses, leading to the wound being bypassed indirectly via neighboring, undamaged vascular bundles. Note that most of the regenerated xylem in the anastomoses consist of wide regenerated vessels (*arrows*). Delicate (narrow) regenerated vessels re-differentiated from parenchyma (*arrowheads*) are present (lower left side). Bars = 1 mm. (From Aloni and Barnett 1996)

The extremely wide regenerated vessel elements of the giant vessels in the phloem anastomoses (Fig. 10.4B) had undergone considerable diameter expansion growth during differentiation. It has been shown that slow vessel differentiation, which permits a long phase of vessel expansion, is induced by relatively low-auxin concentrations (Aloni and Zimmermann 1983) (see Chap. 13). Therefore, the differentiation of the giant regenerated vessels could only occur under relatively low-auxin concentration streams arriving to the old internodes that were located further away from the auxin-inducing young leaves (Aloni and Barnett 1996).

The pattern of secondary wall thickenings of regenerative vessel elements is determined by the growth rate of the differentiating tissue, likely regulated by auxin, gibberellin, and ethylene. When young internodes are injured, their regenerative tracheary elements produced vessel elements with spiral or reticulated secondary wall thickenings (Sinnott and Bloch 1944). Pitted secondary wall thickenings of tracheary elements differentiate in non-elongating internodes from cambial fusiform initials, as the reticulated and pitted cell walls limit tissue elongation. Wounding a very young leaf primordium promotes the differentiation of spiral or even annular secondary wall thickenings (see Chap. 7, Fig. 7.10) which allow substantial leaf elongation.

Wounding is used in many studies as a useful experimental research tool for studying the role of hormones in vascular differentiation and regeneration (Jacobs 1970; Sachs 1981). For example, wounding was used to reveal that gibberellin is the specific hormonal signal that induces and regulates fiber differentiation (Aloni 1976, 1979; Dayan et al. 2012) (see Chap. 2, Fig. 2.26; Sect. 5.2.3, Fig. 5.5).

10.2   Grafting

Grafting is a natural phenomenon, widely spread between roots of neighboring trees, either of the same species or might also occur among trees belonging to different species, demonstrating plants developmental plasticity and ability of plant tissues to adhere and reconnect their injured vascular tissues. Natural grafting among tree branches was also observed (Mudge et al. 2009). Cambial activity in roots and branches increases the width of organs which inflicts pressure on the bark of neighboring organs leading to bark impairment and cambial exposure. Cambial contact between neighboring organs results in the formation of a functional graft union. The new differentiating secondary phloem and xylem of these trees become united allowing water and nutrient movement between the grafted trees. In forests, this phenomenon of a connecting root network among trees establishes a **cooperative tree community** which could support each members of the group. Thus, nutrients from a large tree are transported to young trees growing in its shade that do not get enough light for photosynthesis. Without this supply the young trees might degenerate. Stumps of trees that lost their crowns in a storm might survive due to the continuous nutrient supply by neighboring trees via their root grafts (Bormann 1966). On the other hand, there is a serious risk of an easy pathogen transmission through root grafts, of viral, mycoplasmal (Epstein 1978), fungal, and bacterial diseases (Lopes et al. 2009).

Grafting is widely practiced in horticulture from ancient times (Darwin 1868) and is used in scientific research for studying the long-distance movement of molecules (Melnyk and Meyerowitz 2015; Wang et al. 2017). Grafting is an important tool in agriculture, providing a powerful technique for asexual propagation and the formation of a chimeric organism composed of two different plant genotypes (Liu 2018): the upper shoot portion, termed **scion**, and the root portion, **rootstock** or **stock**. Grafting keeps many agricultural selected plant genotypes which are propagated only by grafting and do not exist in natural habitats (Goldschmidt 2014). Grafting enables improving fruit productivity, e.g., by combining selected roots and shoots for generating chimeras that their scions are more vigorous and their rootstocks are more pathogen resistant and abiotic stress tolerant (Nanda and Melnyk 2018).

The reconnection of the cut vascular system is crucial and the limiting process during grafting of the scion to the vascular system of the rootstock, which enables transport of water and nutrients between the grafted parts. Both auxin and cytokinin play pivotal roles in the formation of vascular tissues at the graft union. The root and shoot do not contribute equally to the union. The scion supplies auxin and the rootstock cytokinin (Aloni et al. 2010; Yin et al. 2012; Goldschmidt 2014). Tissues above and below the graft rapidly developed an asymmetry of gene expression; many genes are highly expressed on one side than on the other (Kümpers and Bishopp 2015; Melnyk et al. 2018). This asymmetry correlated with sugar-responsive genes, promoting some accumulation of starch above the graft junction. Reconnected grafting restores gene symmetry (Melnyk et al. 2018).

After cutting and preparing the scion and rootstock, the ruptured cells on the cut surface collapse, and the surviving cells exposed on the surfaces of the opposing tissues adhere to each other. Cell proliferation above and below the graft junction fill the gaps and form a mass of pluripotent parenchymatic cells, namely, callus, composed of new formed parenchyma cells that can rapidly differentiate. Through these callus cells, auxin from the scion diffuses and induces the regeneration of phloem, which is followed by xylem through the callus which reconnects the original cut vascular tissues forming an active functional **graft union** (Kümpers and Bishopp 2015; Melnyk et al. 2015).

By studying the movement of fluorescent dyes and proteins through developing *Arabidopsis* grafts, the accurate **timing of phloem and xylem regeneration** and function at the graft junction between the scions and rootstocks were determined (Melnyk et al. 2015). A temporal separation between tissue attachment and callus formation, phloem connection, root growth, and xylem connection was observed. The first stage in graft formation was adhesion of the grafted tissues and cell proliferation forming a callus between the cut surfaces, during the first and second days. Reconnection of **phloem** was evident 3 days after grafting (DAG), and roots start to regrow about 5 DAG (promoted by auxin descending from the scion). While regenerated vessels reconnecting the **xylem** was observed 7 DAG, the regenerated vessels connected the xylem of the scion and rootstock, allowing efficient water transport throughout the new grafted plant (Melnyk et al. 2015; Kümpers and Bishopp 2015), likely forming continuous new vessels extending from the scion into the rootstock. This clear separation pattern of phloem differentiation a few days before the xylem indicates that initial diffusion of low-concentration auxin streams from the scion to rootstock through the callus tissue was enough to induce the regenerative phloem. While there was a need for a longer time to establish high-concentration polar auxin streams that induced the regenerative vessels, which therefore differentiated much later (for more information on the role of auxin in phloem versus xylem differentiation, see Sect. 5.1), it is likely that similar to vascular regeneration around a wound (Fig. 10.2), the regenerated sieve tubes and vessels in grafting started their differentiation first in the preexisting vascular bundles (of the scion) away from the graft union and finally differentiated also at the graft junction. In both cases phloem differentiated faster than the xylem. The gap of a few days of temporal separation between phloem to xylem regeneration in grafting is likely due to the involvement of callus parenchyma cells that during the early grafting stage are less polar than organized tissues.

Interestingly, grafting might create new species. The asexual formation of novel plants by grafting was suggested by Charles Darwin who coined the term "graft hybridization," namely, the formation of hybrid plants by grafting, without a sexual process (Darwin 1868). Graft hybridization of plants is a useful and efficient agricultural breeding technique for improving fruit trees. Graft hybridization may be explained by horizontal gene transfer and DNA transformation. Additionally, the long-distance transport of mRNA and small RNAs through the graft union might influence the development of graft hybrids (Wang et al. 2017; Liu 2018). Recent studies have demonstrated cell-to-cell movement of mitochondria through a graft junction (Gurdon et al. 2016), and even the entire nuclear genomes was transferred through a graft union between different species of *Nicotiana* (Fuentes et al. 2014). Evidently, this entire nuclear transfer created a new allopolyploid plant species by the grafting. The new species is fertile and produces fertile progeny. This result indicates that natural grafting provides a potential asexual mechanism of speciation, which may be used for generating novel polyploid crop species that produces fertile progeny (Fuentes et al. 2014; Hare 2014). These findings stimulated the development of a new successful method of "cell grafting" by mixing population of cells from two parents growing in vitro as callus and wounding the callus tissue, thus promoting the transfer of plastid or nuclear DNA between cells of different plant species. This promising method provides alternative pathway to sexual hybridization that can be a powerful tool for crop improvement (Sidorov et al. 2018).

Summary

- Regenerative differentiation of vascular tissues by re-differentiation from parenchyma cells occurs naturally and regularly inside intact growing plants, e.g., at the junctions of the plant axis with lateral roots and branches during their initiation, in the course of phloem anastomoses development between vascular bundles, and in the differentiation of short tracheids (transfusion tracheids) that re-differentiate from parenchyma cells in conifer leaves.

- Regeneration of vascular tissues around an injury is a recovery mechanism for survival, which heals the damaged organ, forming new vascular conduits that renew and restore water transport and assimilates translocation along the wounded organ.
- When a vascular bundle is cut, a bridge of regenerative vascular elements differentiates around the wound. This bridge is part of new longitudinal conduits, which do not form connections to the damaged sieve tubes and vessels, so that the injured conduits continuity around the wounds is not restored; and the new bridges of regenerated sieve tubes and vessels around the wound are part of new continuous sieve tubes and vessels formed first along the bundles and later at the wound site, where they are diverted around the wound.
- Auxin originating in young leaves is the primary hormonal signal that induces vascular regeneration around wounds along its canalized pathways from (pro) cambial initials (inside the cut strands) and parenchyma (or cambium) cells around the wound. Cytokinins from the root tips promote the regeneration of both phloem and xylem.
- The first recognizable regenerative sieve tube elements were detected 2 days after wounding, while regenerative vessel elements appeared 3 days after the injury in both *Coleus blumei* internodes and *Pisum sativum* roots.
- In the same order but in different timing after grafting, reconnection of phloem by new regenerative sieve tubes was evident 3 days after grafting, while regenerative vessels reconnecting the xylem was observed 7 days after grafting in *Arabidopsis thaliana* hypocotyls.
- With time, aging parenchyma cells gradually lose their plasticity to re-differentiate. In *Cucurbita maxima*, along mature internodes with low or no cambial activity, regenerative vessels can differentiate after wounding in phloem anastomoses. This regenerative process of vessel differentiation away from the auxin-producing young leaves proceeds relatively slowly, allowing vessel widening, which results in extremely wide elements, forming giant regenerative vessels along the phloem anastomoses.
- Grafting is a natural phenomenon, widely spread between roots of neighboring trees, either of the same species or might also occur among trees belonging to different species. In forests, this phenomenon of a connecting root network among trees establishes a cooperative tree community which could support each members of the group.
- Grafting is widely practiced in horticulture from ancient times, providing a powerful technique for asexual propagation and the formation of a chimeric organism composed of two different plant genotypes, combining selected shoots and roots for generating chimeras that their scions are more vigorous and their rootstocks are more pathogen resistant or abiotic stress tolerant.
- "Graft hybridization," namely, the formation of hybrid plants by grafting, without a sexual process, may be explained by horizontal gene transfer and DNA transformation. Cell-to-cell movement of mitochondria through a graft junction

was evident, and even the entire nuclear genomes was transferred through a graft union between different species of *Nicotiana*, forming a new species that is fertile and produces fertile progeny.
- A promising new method of "cell grafting" is accomplished by mixing populations of cells from two parents growing in vitro as callus and wounding these mixed callus tissues, thus promoting plastid or nuclear DNA transfer between cells of different plant species. Providing an alternative way to sexual hybridization that can be a powerful tool for crop improvement.

References and Recommended Reading[1]

Aloni R (1976) Regeneration of phloem fibres around a wound: a new experimental system for studying the physiology of fibre differentiation. Ann Bot 40: 395–397.

Aloni R (1979) Role of auxin and gibberellin in differentiation of primary phloem fibers. Plant Physiol 63: 609–614.

*Aloni R (1995) The induction of vascular tissues by auxin and cytokinin. In: *Plant Hormones: Physiology, Biochemistry and Molecular biology*, PJ Davies (ed). Kluwer, Dordrecht, pp. 531–546.

*Aloni R, Barnett JR (1996) The development of phloem anastomoses between vascular bundles and their role in xylem regeneration after wounding in *Cucurbita* and *Dahlia*. Planta 198: 595–603.

Aloni R, Baum SF (1991) Naturally occurring regenerative differentiation of xylem around adventitious roots of *Luffa cylindrica* seedlings. Ann Bot 67: 379–382.

*Aloni R, Baum SF, Peterson CA (1990) The role of cytokinin in sieve tube regeneration and callose production in wounded *Coleus* internodes. Plant Physiol 93: 982–989.

Aloni B, Cohen R, Karni L, Aktas H, M. Edelstein (2010) Hormonal signaling in rootstock-scion interactions. Sci Hortic 127: 119–126.

*Aloni R, Foster A, Mattsson J (2013) Transfusion tracheids in the conifer leaves of *Thuja plicata* (Cupressaceae) are derived from parenchyma and their differentiation is induced by auxin. Am J Bot 100: 1949–1956.

Aloni R, Jacobs WP (1977a) Polarity of tracheary regeneration in young internodes of *Coleus* (Labiatae). Am J Bot 64: 395–403.

Aloni R, Jacobs WP (1977b) The time course of sieve tube and vessel regeneration and their relation to phloem anastomoses in mature internodes of *Coleus*. Am J Bot 64: 615–621.

Aloni R, Peterson CA (1990) The functional significance of phloem anastomoses in stems of *Dahlia pinnata* Cav. Planta 182: 583–590.

Aloni R, Plotkin T (1985) Wound induced and naturally occurring regenerative differentiation of xylem in *Zea mays* L. Planta 163: 126–132.

Aloni R, Sachs T (1973) The three-dimensional structure of primary phloem systems. Planta 113: 343–353.

Aloni R, Zimmermann MH (1983) The control of vessel size and density along the plant axis - a new hypothesis. Differentiation 24: 203–208.

Baum SF, Aloni R, Peterson CA (1991) The role of cytokinin in vessel regeneration in wounded *Coleus* internodes. Ann Bot 67: 543–548.

Benayoun J, Aloni R, Sachs T (1975) Regeneration around wounds and the control of vascular differentiation. Ann Bot 39: 447–454.

---

[1] Papers of particular interest for suggested reading have been highlighted (with *).

Behnke HD, Schulz A (1980) Fine structure, pattern of division, and course of wound phloem in *Coleus blumei*. Planta 150: 357–365.

Bormann FH (1966). The structure, function, and ecological significance of root grafts in *Pinus strobus* L. Ecol Monogr 36: 1–26.

Camus G (1949) Recherches sur le rôle des bourgeons dans les phénomènes de morphogénèse. Rev Cytol Biol Vég 11: 1–195.

Chen JJ, Wang LY, Immanen J, Nieminen K, Spicer R, Helariutta Y, Zhang J, He XQ (2019) Differential regulation of auxin and cytokinin during the secondary vascular tissue regeneration in *Populus* trees. New Phytol 224: 188–201.

Darwin C (1868) *The Variation of Animals and Plants under Domestication*. John Murray, London.

Dayan J, Voronin N, Gong F, Sun T-p, Hedden P, Fromm H, Aloni R (2012) Leaf-induced gibberellin signaling is essential for internode elongation, cambial activity, and fiber differentiation in tobacco stems. Plant Cell 24: 66–79.

Epstein AH (1978) Root graft transmission of tree pathogens. Ann Rev Phytopathol 16: 181–192.

Eschrich W (1953) Beiträge zur Kenntnis der Wundsiebröhrenentwicklung bei *Impatiens holsti*. Planta 43: 37–74.

*Fuentes I, Stegemann S, Golczyk H, Karcher D, Bock R (2014) Horizontal genome transfer as an asexual path to the formation of new species. Nature 511: 232–235.

Fukuda H Komamine A (1980) Establishment of an experimental system for the study of tracheary element differentiation from single cells isolated from the mesophyll of *Zinnia elegans*. Plant Physiol 65: 57–60.

Gaillochet C, Lohmann JU (2015) The never-ending story: from pluripotency to plant developmental plasticity. Development 142: 2237–2249.

*Goldschmidt EE (2014) Plant grafting: new mechanisms, evolutionary implications. Front Plant Sci 5: 727.

Gurdon C, Svab Z, Feng Y, Kumar D, Maliga P (2016) Cell-to-cell movement of mitochondria in plants. Proc Natl Acad Sci USA 113: 3395–3400.

Hare P (2014) New plant species through grafting. Nat Biotechnol 32: 887.

Houck DF, LaMotte CE (1977) Primary phloem regeneration without concomitant xylem regeneration: its hormone control in *Coleus*. Am J Bot 64: 799–809.

Ikeuchi M, Ogawa Y, Iwase A, Sugimoto K (2016) Plant regeneration: cellular origins and molecular mechanisms. Development 143: 1442–1451.

Jacobs WP (1952) The role of auxin in differentiation of xylem around a wound. Am J Bot 39: 301–309.

Jacobs WP (1970) Regeneration and differentiation of sieve tube elements. Int Rev Cytol 28: 239–273.

Jost L (1940) Zur Physiologie der Gefäßbildung. Z Bot 35: 114–148.

von Kaan Albest A (1934) Anatomische und physiologische Untersuchungen über die Entstehung von Siebröhrenverbindungen. Z Bot 27: 1–94.

*Kümpers BM, Bishopp A (2015) Plant grafting: making the right connections. Curr Biol 25: R411–413.

LaMotte CE, Jacobs WP (1963) A role of auxin in phloem regeneration in *Coleus* internodes. Dev Biol 8: 80–98.

*Liu Y (2018) Darwin's pangenesis and graft hybridization. Adv Genet 102: 27–66.

Lopes SA, Bertolini E, Frare GF, Martins EC, Wulff NA, Teixeira DC, Fernandes NG, Cambra M (2009) Graft transmission efficiencies and multiplication of 'Candidatus Liberi bacter americanus' and 'ca. Liberi bacter asiaticus' in citrus plants. Phytopathology 99: 301–306.

Melnyk CW, Meyerowitz EM (2015) Plant grafting. Curr Biol 25: R183–188.

*Melnyk CW, Schuster C, Leyser O, Meyerowitz EM (2015) A developmental framework for graft formation and vascular reconnection in *Arabidopsis thaliana*. Curr Biol 25: 1306–1318.

Melnyk CW, Gabel A, Hardcastle TJ, Robinson S, Miyashima S, Grosse I, Meyerowitz EM (2018) Transcriptome dynamics at *Arabidopsis* graft junctions reveal an intertissue recognition mechanism that activates vascular regeneration. Proc Natl Acad Sci USA 115: E2447–E2456.

Mudge K, Janick J, Scofield S, Goldschmidt EE (2009) A history of grafting. Hortic Rev 35: 437–493.

Nanda AK, Melnyk CW (2018) The role of plant hormones during grafting. J Plant Res 131: 49–58.

Robbertse PJ, McCully ME (1979) Regeneration of vascular tissue in wounded pea roots. Planta 145: 167–173.

Sachs T (1981) The control of patterned differentiation of vascular tissues. Adv Bot Res 9: 151–262.

Sang YL, Cheng ZJ, Zhang XS (2018) Plant stem cells and de novo organogenesis. New Phytol 218: 1334–1339.

*Sidorov V, Armstrong C, Ream T, Ye X, Saltarikos A (2018) "Cell grafting": a new approach for transferring cytoplasmic or nuclear genome between plants. Plant Cell Rep 37: 1077–1089.

Simon S (1908) Experimentale Untersuchungen über die Entstehung von Gefäßverbindungen. Ber dt Bot Ges 68: 227–232.

Sinnott EW, Bloch R (1944) Visible expression of cytoplasmic pattern in the differentiation of xylem strands. Proc Nat Acad Sci USA 30: 388–392.

Steward FC, Mapes MO, Mears K (1958) Growth and organized development in cultured cells. II. Organization in cultures grown from freely suspended cells. Am J Bot 45: 705–708.

Thompson NP (1967) The time course of sieve tube and xylem cell regeneration and their anatomical orientation in *Coleus* stems. Am J Bot 54: 588–595.

Thompson NP, Jacobs WP (1966) Polarity of IAA effect on sieve-tube and xylem regeneration in *Coleus* and tomato stems. Plant Physiol 41: 673–682.

Vasil V, Hildebrandt AC (1965) Differentiation of tobacco plants from single, isolated cells in microcultures. Science 150: 889–892.

Wang J, Jiang L, Wu R (2017) Plant grafting: how genetic exchange promotes vascular reconnection. New Phytol 214: 56–65.

Yin H, Yan B, Sun J, Jia P, Zhang Z, Yan X, Chai J, Ren Z, Zheng G, Liu H (2012) Graft-union development: a delicate process that involves cell-cell communication between scion and stock for local auxin accumulation. J Exp Bot 63: 4219–4232.

Regulation of Cambium Activity

11

11.1 Hormonal Control of Cambial Activity and Secondary Vascular Differentiation

The continued widening of the plant axis is a result of cell division activity of the lateral vascular meristem, namely, the cambium (Larson 1994; Nieminen et al. 2015; Shi et al. 2019) and the differentiation of secondary vascular tissues. The cambium produces new cells that build the secondary xylem (the wood) and secondary phloem tissues which increase the thickness of stems and roots (see Chap. 2, Figs. 2.3, 2.4 and 2.5). The mechanisms that regulate cambial activity enable perennial trees to become the oldest organisms on earth by promoting their continuous lateral growth, replacing old non-functional vascular tissues with new ones, permitting almost unlimited life time. The cambium and the secondary vascular tissues are regulated by the following hormonal signals.

1. **Auxin** is the primary hormonal signal that regulates cambial activity. The pioneering study of Jost (1893) on the relationships between leaf development and vascularization in plants revealed that the activity of the cambium depends on some influence arriving from leaves, especially from growing leaves. This leaf stimulation travels only downward toward the root, and it is different from the movement of nutrient supply. In an attempt to replace the influence of the leaves on vascularization, Snow (1935) demonstrated that the application of auxin to the upper end of decapitated stems of sunflower (*Helianthus annuus*) seedlings induced fascicular cambium initiation and cell division activity in their vascular bundles and also the formation of interfascicular cambium between the bundles. Much later, precise quantitative measurements of auxin concentrations collected along the cambium region of the pine trees *Pinus sylvestris and P. contorta* revealed that the downward polar movement of auxin along the stem occurs in a wave-like pattern (see Sect. 3.1.5). The quantitative results showed oscillating IAA concentrations along the vascular cambium, evidently demonstrating that auxin moved downward in a wave-like pattern along the cambium region of the

© Springer Nature Switzerland AG 2021
R. Aloni, *Vascular Differentiation and Plant Hormones*,
https://doi.org/10.1007/978-3-030-53202-4_11

pine stems (Wodzicki et al. 1984, 1987). This downward wavy pattern of polar IAA transport can provide morphogenetic information along the plant axis, which can inform the cambium initials and their differentiating phloem and xylem derivatives about their location along a decreasing auxin gradient from the auxin-producing young leaves downward to the roots (Wodzicki et al. 1984, 1987; Zajaczkowski et al. 1984). By using microscale mass-spectrometry technique coupled with cryosectioning, Uggla et al. (1996, 1998) visualized that auxin moves preferably through the vascular cambium. Their results show a radial concentration gradient of endogenous indole-3-acetic acid (IAA) across the cambial meristem and the differentiating derivatives in *Pinus sylvestris* trees, indicating that the peak of IAA in the cambium provides positional information along the plant axis and that the auxin signal behaves like a plant morphogen (see Chap. 13). These results were confirmed in *Arabidopsis thaliana* (Aloni 2013a, b) also shown in Chap. 3, Fig. 3.11A, B and drawn in Fig. 3.12ii. Morris et al. (1969) calculated the velocity of polar auxin movement along stems of intact pea (*Pisum sativum*) seedlings and found a speed of 11 mm per hour. Morris and Thomas (1978) demonstrated that the slow polar movement of radioactive-labeled IAA occurred preferably in the cambium, which gives the meristem its identity. In the above experiments, the slow polar auxin movement was detected in the vascular cambium after the exogenous auxin was applied to the apical bud or to young leaves (Morris et al. 1969; Morris and Thomas 1978). But when auxin was applied to mature leaves, a completely different pathway of rapid nonpolar (up and down) auxin movement was found through the phloem in the sieve tubes, detected by using aphids feeding on the stem's sieve tubes (Morris et al. 1973; Goldsmith et al. 1974). The major role of the polar auxin movement in promoting cambial identity and cell division activity was confirmed in many studies and is well established (Savidge 1983; Uggla et al. 1996, 1998; Sundberg et al. 2000; Aloni 2004, 2007, 2013a, b; Spicer et al. 2013; Bhalerao and Fischer 2017; Fischer et al. 2019; Agustí and Blázquez 2020).

In stems of different angiosperm species, patterns of secondary vessels and sieve tubes indicate that the polar auxin moves in preferable locations along the cambium circumference. In these sites radial patterns of vessels and sieve tubes are formed, likely marking the preferable spots of polar auxin streams in the cambium (Fig. 11.1A, B), while between them along the cambial circumference there are the sites with fibers indicating preferable gibberellin flows (between the *arrows* in Fig. 11.1A). The occurrence of continuous radial files of vessels indicate that for a relatively long period, the location of preferable polar IAA stream is maintained (Fig. 11.1B), showing that along the circumference of the cambium, the distribution of IAA is not even and is high in some locations and low in others (Fig. 11.1B).

Numerous beautiful and complex patterns of secondary phloem and secondary xylem anatomy (e.g., Fig. 11.2) characterize liana species (Angyalossy et al. 2015), which are likely regulated by streams of preferable polar movements of IAA in different concentrations combined with other hormonal signals, in specific locations along the cambium circumference, indicating stable flows of hormonal signals along the cambium, which result in the production of diverse patterns of secondary

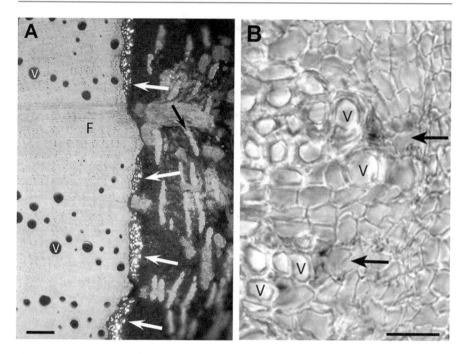

Fig. 11.1 Patterns of vascular differentiation (**A**, **B**) and bioactive auxin distribution in the cambium (**B**). (**A**) Cross section in the vascular tissues of the oak *Quercus calliprinos* (stained with 0.01% aniline blue and observed with epifluorescence microscope) showing a pattern of radial distribution of vessels (*black rounded spots marked by white* V) and their radially located adjacent sieve tubes (*white spots marked by white arrows*) indicating that the conduits were induced by preferably polar IAA movements in specific cambial cells between the conduits. Xylem fibers (F) and phloem fibers (*black arrow*) differentiated in the regions between the conduits likely induced by longitudinal flows of GA in the cambium. (**B**) Cross section in the vascular tissues of the DR5::GUS transformed *Arabidopsis thaliana* presented in the same orientation as the oak (*xylem on left side*), showing free-IAA distribution (marked by *DR5::GUS expression,* which forms the blue spots, marked by *arrows*) located preferably adjacent to the differentiating vessels (V) and absent from the cambium located adjacent of xylem areas without vessels, demonstrating that along the cambium circumference there are preferable longitudinal streams of polar IAA that specifically induce radial patterns of vessels and sieve tubes. Bar = 50 μm (**B**), 200 μm (**A**). (From Aloni 2007)

phloem and xylem. The cambial variant can produce xylem furrowed by phloem arcs (wedges) likely induced by polar streams of auxin at different concentrations (Fig. 11.2C). The hormonal regulation of the cambium in lianas is an interesting research field for further investigations.

Stem cell behavior is determined by their niche (Zhou et al. 2015), namely, that stem cells in embryonic tissues (meristems) are not the source of patterning information and that they are regulated by already differentiated tissues around them. In order to determine how the induction and activity of cambial stem cells are coordinated by organizer cells that direct the adjacent stem cells to undergo programmed cell division and differentiation, Smetana et al. (2019) used lineage tracing and molecular genetic studies in the roots of *Arabidopsis thaliana* and found that

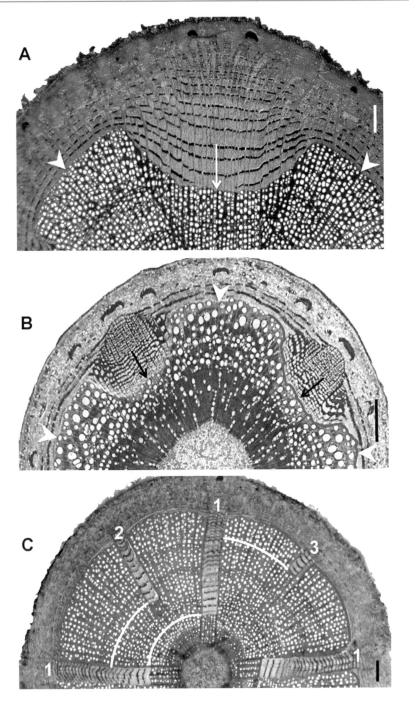

Fig. 11.2 Naturally occurring patterns of secondary phloem and secondary xylem produced by variant cambium in three liana species: *Perianthomega vellozoi* (**A**), *Fridericia speciose* (**B**), and

high-auxin-induced-differentiating xylem-identity cells direct adjacent cells in the root vascular cylinder to divide and function as cambial stem cells. Thus, these differentiating xylem-identity cells constitute organizers that promote the initiation of cambial stem cells. A local maximum of the auxin hormone, and consequent expression of CLASS III HOMEODOMAIN-LEUCINE ZIPPER (HD-ZIP III) transcription factors, promotes xylem identity of the organizer cells. Additionally, Smetana et al. (2019) suggest that the organizer also maintains phloem identity. Consistent with this dual function of the organizer cells, xylem and phloem originate from a single, bifacial cambial stem cell in each radial cell file. Smetana et al. (2019) propose that a local high level of auxin signaling within the root vascular tissue is sufficient to establish an organizer cell that promotes stem-cell divisions and identity in the adjacent cells. Further suggesting a dynamic nature of the organizer; differentiation of the organizer into a vessel element leads to formation of a new organizer in the adjacent cambial stem cell, thus ensuring the maintenance of the vascular cambium in the *Arabidopsis* root.

2. **Gibberellins** are major promoting hormonal signals of cambial activity. Bradley and Crane (1957) found that application of gibberellin to intact young branches of *Prunus armeniaca* stimulated cell division in the cambial zone and a substantial increase in xylem production. Wareing (1958) applied auxin and gibberellin to disbudded shoots of *Populus robusta* and found that the application of IAA produced some cambial cell divisions and the differentiation of small amounts of xylem while the application of GA also caused cambial division with no xylem differentiation. When both IAA and GA were applied simultaneously, there was a marked synergism between the two hormones, both in the stimulation of cambial division activity and in the production of xylem tissue. These results were

---

Fig. 11.2 (continued) *Mansoa onohualcoides* (**C**), demonstrating different activity of the cambium along the circumference and consequently formation of more, or less, of each secondary vascular tissue. *Arrows* mark cambial sites that produce mainly phloem, while *arrowheads* mark cambial locations which produced mainly xylem (in **A** and **B**). In some sites the location of the cambium is marked by a *red line* (in **B** and **C**). (**A**) In *P. vellozoi* the cambium alternates, forming wide xylem with thin phloem layers (*arrowheads*) or a cambial region that produced wide phloem layers on the expense of almost no xylem production (*arrow*). (**B**) *F. speciose* shows a cambial variant that produced mainly xylem with gradually enlarging vessels (*arrowheads*) and among them sites with more secondary phloem production (*arrows*) on the expense of xylem formation. (**C**) In *M. onohualcoides*, the cambial variant produced xylem furrowed by phloem arcs/wedges. Starting from four (the lower one is outside the figure frame) equidistant (*white arch*) phloem wedges (near the pith), each of the first-formed wedges is marked by number 1. During cambial growth, more space becomes available on the cambial circumference for a new inductive hormonal stream of a phloem signal that induced a new phloem wedge marked by number 2. Additional cambial growth produced more space on the cambial circumference to form a new hormonal stream that produced a new phloem wedge marked by number 3. The *white arches* mark similar distances on the cambium circumference at the time when a new wedge was formed. Bar = 2 mm (**B**), 3 mm (**A**), 4 mm (**C**). (Courtesy of Marcelo R. Pace)

confirmed in different experimental systems (Hess and Sachs 1972; Israelsson et al. 2005; Dayan et al. 2012; Aloni 2013a; Sorce et al. 2013; Fischer et al. 2019). Both auxin and gibberellin were also identified conclusively by mass spectrometry in the cambial region of some Pinaceae trees (Little and Savidge 1987) confirming that auxin is the primary signal to initiate cambium activity (Snow 1935; Mazur et al. 2014) and that GA promotes cell divisions in the cambium, but GA by itself (without an auxin background) does not induce xylem differentiation (Wareing 1958; Hess and Sachs 1972; Aloni 2013a). In order to induce tracheid or fiber differentiation, GA requires the presence of IAA (Aloni 1979, 2013a; Kalev and Aloni 1998) (see Sect. 5.2.3). The effect of GA in inducing cambial activity and fiber differentiation is nonpolar; GA induces cambial activity and fiber differentiation in both the basipetal and acropetal directions (Fig. 11.3). The highest concentrations of auxin were found in the actively dividing cambial cells (Uggla et al. 1996, 1998; Aloni 2013a, b) (see Chap. 3, Fig. 3.11A, B), while the bioactive gibberellins show maximum levels in the differentiating xylem cells indicating that the main role of GA during wood formation is to regulate early stages of tracheid and fiber differentiation, including their elongation (Israelsson et al. 2005; Fischer et al. 2019).

3. **Cytokinins** have an important role in promoting cambial cell division activity, promoting radial growth that results in stem and root thickening. The pioneering discovery that cytokinin is a factor necessary for cell divisions in tissue cultures (Skoog and Miller 1957) was confirmed in different plant tissues demonstrating that CKs are major controlling signals that promote cell division activity in the cambium (Aloni et al. 1990; Baum et al. 1991; Nieminen et al. 2008; Matsumoto-Kitano et al. 2008; Ursache et al. 2013; Immanen et al. 2016). Cambial initials and derivatives affected by applied CKs became more sensitive to the auxin stimulation (Baum et al. 1991). The increased sensitivity of the cytokinin-affected cambial initials enables them to respond to very low-auxin streams in the cambium (Aloni 1993) (see Sect. 19.3).

Fig. 11.3 (continued) The cross sections shown in the micrographs were done in the middle of internode number 3 (*marked in the experimental scheme by a blue bar; while the hormonal application sites are marked with arrows*), leaving two younger internodes above internode 3 (*as an endogenous source of low IAA production; because GA does not function without an auxin background*). The results were analyzed 3 weeks after treatments. (**A**) Effect of applied lanolin paste (*without gibberellin*), showing the differentiation of relatively large isolated slightly deformed vessels (V) indicating some endogenous auxin production by the two youngest internodes during the experiment (*the deformed thin-wall vessels resulted from shortage of nutrients because all the leaves were removed* – see Sect. 2.9, Fig. 2.32A, B). Note, that in the absence of GA stimulation, there is no cambial activity or any fiber differentiation. (**B**) Effect of 0.8% GA_3 in lanolin applied on the youngest internode, inducing cambial C activity, secondary fibers (SF) in the xylem, and primary phloem (PF) fibers inside and outside the xylem. (**C**) effect of 0.8% GA_3 in lanolin applied on a mature internode, two internodes below internode number 3, showing cambial activity and secondary fiber formation in the above xylem. The results demonstrate that the effect of gibberellin on cambial activity and fiber differentiation is nonpolar; the GA_3 promoted cambial activity and fiber differentiation when applied either from above (**B**) or from below (**C**) the studied internode. All micrographs are at the same magnification, bar = 50 μm. (Jonathan Dayan and R. Aloni, unpublished)

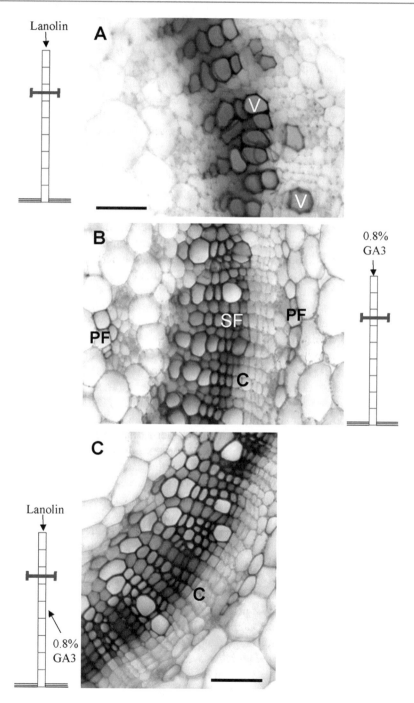

Fig. 11.3 The nonpolar effect of gibberellin (GA₃) on cambium activity and fiber differentiation in young decapitated stems of tobacco (*Nicotiana tabacum*) from which all the leaves were excised.

Cytokinins are produced in the root cap (Aloni et al. 2004) and transported upward through the xylem conduits (Aloni et al. 2005; Sakakibara et al. 2006; Ko et al. 2014). However, although the CKs from the root cap are transported upward in xylem conduits, Immanen et al. (2016) found that CK signaling and biosynthesis genes peak in the developing phloem cells at the maximum CK content, while most of the auxin response genes had maximal expression in the middle of the cambial zone, coinciding with the peak of auxin content. To explore the functional significance of CK signaling for cambial activity, transgenic *Populus tremula* × *P. tremuloides* trees with an elevated CK biosynthesis level were generated. These transgenic trees displayed stimulated cambial cell division activity resulting in dramatically increased (up to 80% in dry weight) production of the lignocellulosic trunk biomass, confirming that CKs are major regulators of cambial activity and secondary vascular tissues. Remarkably, in addition to the observed elevated cambial CK content and signaling level, the cambial IAA concentration and auxin-responsive gene expression were also increased in these transgenic trees (Immanen et al. 2016).

In young stem internodes and tissue cultures, the phloem sieve elements are induced by low-auxin stimulation (see Chaps. 2 and 5; Fig. 5.3), and in young stem internodes, the phloem differentiates before the xylem (see Chaps. 2 and 5; Fig. 2.6), while the vessels are induced by high-auxin stimulation. In small vascular bundles, only phloem is induced and may be followed by fascicular cambium formation with no xylem (e.g., see Chap. 3, the small bundle on left side in Fig. 3.10A). Furthermore, the interfascicular cambium produced from parenchyma cells between the vascular bundles is not connected to any high-auxin-induced differentiating vessels or low-auxin-induced differentiating sieve tubes. Therefore, the recently described mechanism for *Arabidopsis* roots (Smetana et al. 2019), where cambial stem-cell initiation and maintenance are regulated by high-auxin-induced differentiating vessels that direct adjacent cells in the root vascular cylinder to divide and function as cambial stem cells, is likely different or modified in the shoot. Therefore, another regulatory mechanism induces and maintains cambial stem cell in the interfascicular cambium between bundles and in the phloem only bundles of the young shoot, possibly involving the mobile PEAR transcription factors, which are active in differentiating phloem (Miyashima et al. 2019). The initiation of cambial radial growth occurs in the *Arabidopsis* root around early protophloem-sieve-element cell files. In this domain, cytokinin signaling promotes the expression of a pair of mobile transcription factors – PHLOEM EARLY DOF 1 (PEAR1) and PHLOEM EARLY DOF 2 (PEAR2) – and their four homologues (DOF6, TMO6, OBP2, and HCA2), which were collectively named PEAR proteins. The PEAR proteins form a short-range concentration gradient that peaks at protophloem sieve elements and activates gene expression that promotes cambial cell division activity. The expression and function of PEAR proteins are antagonized by the HD-ZIP III proteins (Miyashima et al. 2019) which are promoted by high-level auxin signaling in the differentiating vessel elements (Smetana et al. 2019). It is also possible that both lateral meristems, the cambium and phellogen, are directly and primarily determined by the polar auxin moving streams that maintain their meristematic identity and activity. Gibberellin might also be involved as well as ethylene that induces the vascular rays.

Overexpressing transgenic line of the *increased cambial activity (ICA)* gene (encoding a putative pectin methyltransferase), which could function as a modulator for the meristematic activity of the cambium in the inflorescence of *Arabidopsis*, showed accelerated stem elongation and radial thickening. The authors propose that the expression of *ICA* increases cambial activity by regulating CK and GA homeostasis, eventually leading to stem elongation and radial growth in the inflorescence stem (Kim et al. 2016).

4. **Ethylene** is the fourth major endogenous hormonal stimulator of cell division in the cambial meristem (Savidge 1988; Andersson-Gunnerås et al. 2003; Love et al. 2009); its concentrations increases dramatically in response to wounding and stress (Zhu and Lee 2015). Naturally, the ethylene is synthesized in the xylem (Eklund 1990; Ingemarsson et al. 1991), specifically in maturing tracheary elements (Pesquet and Tuominen 2011). Ethylene is the specific signal that induces and regulates the ray initials (see Chap. 15) in the cambium (Lev-Yadun and Aloni 1995; Aloni et al. 2000) and promotes reaction wood formation (Andersson-Gunnerås et al. 2003; Love et al. 2009; Seyfferth et al. 2019) (see Chap. 18).

5. In addition to the above mentioned four major hormonal signals, studies on *Arabidopsis* showed that **jasmonate** stimulates cambial activity following environmental and mechanical stresses (Sehr et al. 2010). Likewise, **strigolactone** that promotes main stem growth by inhibiting lateral bud development (see Chap. 4 and Sect. 6.2) interacts with auxin signaling to promote cambial activity (Agusti et al. 2011).

11.2 Cambial Dormancy

Plants respond to extreme conditions by stopping cambial activity. Cytokinin which promotes cell division activity in the cambium appeared in *Populus* stems as a clear ring of the cytokinin detected by *ARR5::GUS* reporter under comfortable growth conditions, whereas drought resulted in a drastic loss of this CK activity, which can stop secondary vascular production. Similarly, young leaves of trees that were well watered showed strong CK activity in their veins but low staining (activity) under drought stress, accompanied by diminished leaf expansion (Paul et al. 2017). However, low CK concentrations do not induce dormancy.

Abscisic acid (ABA), the universal stress hormone of higher plants, has a central role in plant developmental plasticity. ABA is involved in slowing down and stopping cell divisions in the cambium and wood formation in trees toward their winter dormancy by retarding and ending their cambium activity (Hou et al. 2006). ABA is the long-distance stress signal produced in the meristematic cells of the root tip (Koiwai et al. 2004), specifically in their meristematic endodermal cells (Duan et al. 2013; Bloch et al. 2019) when the soil is drying. ABA is transported upward through the xylem from roots to shoot to regulate the closure of stomata under stress and retard cambial activity before dormancy (Taiz et al. 2018). However, ABA can also be produced in the phloem of the shoot and even in the leaves by their stomata (Koiwai et al. 2004), making the shoot responsive to environmental stimuli, like seasonal day length changes. As the days get shorter toward winter, ABA promotes the **blockage of the plasmodesmata** (Tylewicz et al. 2018) which initiates dormancy of deciduous trees that lose their leaves toward winter. Cambial dormancy protects the meristem during harsh environmental periods. Dormancy sets in by communication shutdown, when symplastic intercellular movement through plasmodesmata is blocked. The blockage of the plasmodesmata makes the dormant cambial meristem nonresponsive to growth signals during an occasional sunny and warm day during winter. Therefore, the dormant trees stay dormant until spring (Tylewicz et al. 2018). The release from dormancy to hormonal responsiveness needs to be studied and clarified.

Nevertheless, cambium dormancy can occur in evergreen species during regular environmental conditions. For example, in the shrubby *Cordiera concolor*, the cambium enters into dormancy for a long time during the rainy season, up to 9 months, while the conducting phloem remains active. Cambial activity was positively related to day length, and although it occurred in the rainy season, the period of its onset and termination was not concurrent with the beginning and end of the rainy season (de Lara and Marcati 2016).

11.3 Cambial Activity Reflects the Social Status of a Forest Tree

A study on cambium dynamics and wood formation in a 40-year-old *Abies alba* plantation near Nancy, France, has shown that the timings, duration, and rate of tracheid production change according to the social status (relative size and vitality) of a tree in the forest (Rathgeber et al. 2011). The study demonstrates clear gradients of cambial activity related to the crown area and the height of the trees. Cambial activity started earlier, stopped later, and therefore lasted longer in dominant trees than in intermediate and suppressed ones. Cambial activity was more intense in dominant trees than in the smaller trees. It was estimated that about 75% of tree-ring width variability was attributable to the rate of cell production and only 25% to extend cambial duration. Interestingly, growth duration was correlated to tree height, while growth rate was correlated to crown area (Rathgeber et al. 2011).

Vigorous crowns produce more auxin in their young leaves and more bioactive gibberellins in their mature leaves. The synergistic effects of these two hormones upgrade cambial activity and enhance tracheid production. Together with the expected elevated hormonal production, a larger crown also provides higher sugar contents, as was found in the outer wood of the most productive poplar clones (Deslauriers et al. 2009). Sugars from photosynthesis act as both an energy source and as signaling molecules promoting IAA synthesis (Lilley et al. 2012; Sairanen et al. 2012). This positive regulation of IAA synthesis by the availability of free sugar means that under favorable light conditions more IAA can be produced in the higher trees promoting more growth and wood formation in response to upgraded environmental conditions. These results suggest that gradients in cambial activity and of wood formation are strongly related to tree size and vigor. It is likely that the dominant trees are genetically superior, and therefore their seeds should be collected for future plantations.

11.4 Woodiness

Early angiosperms were minimally woody. Increase in woodiness and changes in wood histology characterize the vascular tissues in the transition to shrubs, trees, and lianas in various clades. Herbs have been derived from variously woody ancestors by a gradual decrease and even absence (in monocots) of cambial activity. However, there is no understanding in the literature how more woodiness or less woodiness has been regulated during plant evolution (Spicer and Groover 2010; Carlquist 2013). Here I clarify that modifications in the hormonal stimulation of gibberellin and auxin have been the driving signals for more or less woodiness. Gibberellin, in the presence of auxin, promotes cambial activity, stimulating long tracheids and is the specific signal inducing fiber differentiation (Aloni 1979; Kalev and Aloni 1998; Dayan et al. 2012; Aloni 2013a, b). Increase of mature leaf biomass boosts gibberellin production which increases cambial activity, tracheid, and fiber production, thus boosting woodiness. The evolutionary trend of gradual changes in growth form and increase in plant size from early angiosperms to large trees occurred with the increase in foliage biomass, increasing the production of both auxin from young leaves and mainly gibberellin from mature leaves that synergistically increased woodiness. The important observation that in woodier species fusiform cambial initials become longer over time (Carlquist 2013) clearly indicates an increase in gibberellin stimulation from their mature leaves.

The pattern of decreasing woodiness in herbs occurred during the trend of decreasing foliage biomass and shortening plant life cycle which gives herbs a great advantage of producing many short generations that enable better and rapid genetic adaptation to changing environmental conditions. The annual herbs are adapted to extreme habitats, where their dormant seeds avoid the extreme conditions until the favorable growth season returns. The relatively small herbs do not produce enough hormonal (GA and IAA) stimulation for the development of cambium and wood and therefore show decreased or absence of woodiness.

Summary

- Polar auxin movement, which originates in buds and young leaves, along the cambium is the primary controlling signal of cambium activity. Auxin behaves like a morphogen that moves downward in a speed of 11 mm per hour (in *Pisum sativum*) and in an oscillating IAA concentrations forming a wave-like pattern (in pine trees) and provides a gradient of positional information to the cells along the cambium and its derivatives.

.

- The polar auxin moves in preferable locations along the cambium circumference, forming radial patterns of vessels and sieve tubes. In lianas, different streams of auxin concentrations in combinations with other signals produce beautiful anatomical patterns of cambial variants.
- Gibberellin produced in maturing leaves, in the presence of auxin, promotes cambial activity. GA specifically induces fiber differentiation. The effect of GA is nonpolar and promotes differentiation in both the basipetal and acropetal directions.
- Cytokines from the root cap and stem nodes are major controlling signals that promote cell division activity in the cambium. CKs also promote wood production.
- Ethylene synthesized in the xylem, specifically in maturing tracheary elements, promotes wood production and induces and regulates cambial ray initials. Ethylene also promotes reaction wood formation and response to wounding.
- Jasmonate stimulates cambial activity following environmental and mechanical stresses.
- Strigolactone that stimulates main stem growth by inhibiting lateral bud development interacts with auxin signaling to promote cambial activity.
- The hormonal signals that regulate cambial activity operate in synergism and crosstalk that need further research clarification to reveal their molecular mechanism.
- Abscisic acid slows down and stops cell divisions in the cambium when the soil is drying. Toward winter, the ABA promotes the blockage of the plasmodesmata which initiates cambial dormancy that protects the meristem during harsh environmental periods.
- The hormonal mechanisms that regulate the cambium enable perennials to continue their growth for unlimited years. Only the wood and secondary phloem produced by the cambium in the few recent years remain alive and active, while the xylem of the inner old annual years stop functioning and die and the old phloem becomes separated by cork cambium (phellogen) activity.
- Timings, duration, and rate of tracheid production in a conifer tree change according to the social status (relative size and vitality) of a tree in the forest. In dominant trees, cambial activity started earlier, stopped later, and therefore lasted longer, resulting in more wood production in dominant trees than in suppressed ones.
- Increase in hormonal stimulation with increasing plant size has been the driving stimulation for cambial activity resulting in more woodiness. Increase of mature leaf biomass boosts gibberellin production which promotes cambial activity, tracheid, and fiber production, thus boosting woodiness.
- Herbs have been derived from variously woody ancestors by a gradual decrease and even absence of cambial activity.

References and Recommended Reading[1]

*Agustí J, Blázquez MA (2020) Plant vascular development: mechanisms and environmental regulation. Cell Mol Life Sci 77: 3711–3728.

Agusti J, Herold S, Schwarz M, Sanchez P, Ljung K, Dun EA, Brewer PB, Beveridge CA, Sieberer T, Sehr EM, Greb T (2011) Strigolactone signaling is required for auxin-dependent stimulation of secondary growth in plants. Proc Natl Acad Sci USA 108: 20242–20247.

Aloni R (1979) Role of auxin and gibberellin in differentiation of primary phloem fibers. Plant Physiol 63: 609–614.

Aloni R (1993) The role of cytokinin in organised differentiation of vascular tissues. Aust J Plant Physiol 20: 601–608.

Aloni R, Feigenbaum P, Kalev N, Rozovsky S (2000) Hormonal control of vascular differentiation in plants: the physiological basis of cambium ontogeny and xylem evolution. In: *Cell and Molecular Biology of Wood Formation*, RA Savidge, JR Barnett, R Napier (eds). Oxford, BIOS Scientific Publishers, pp. 223–236.

Aloni R (2004) The induction of vascular tissue by auxin. In: *Plant Hormones: Biosynthesis, Signal Transduction, Action!* PJ Davies (ed). Kluwer Academic Publishers, Dordrecht, Boston, London, pp 471–492.

Aloni R (2007) Phytohormonal mechanisms that control wood quality formation in young and mature trees. In: *The Compromised Wood Workshop 2007*, K Entwistle, P Harris, J Walker (eds). The Wood Technology Research Centre, University of Canterbury, Christchurch, New Zealand, pp. 1–22.

*Aloni R (2013a) The role of hormones in controlling vascular differentiation. In: *Cellular Aspects of Wood Formation*, J Fromm (ed). Springer-Verlag, Berlin, pp. 99–139.

Aloni R (2013b) Role of hormones in controlling vascular differentiation and the mechanism of lateral root initiation. Planta 238: 819–830.

Aloni R, Baum SF, Peterson CA (1990) The role of cytokinin in sieve tube regeneration and callose production in wounded *Coleus* internodes. Plant Physiol 93: 982–989.

Aloni R, Langhans M, Aloni E, Dreieicher E, Ullrich CI (2005) Root-synthesized cytokinin in *Arabidopsis* is distributed in the shoot by the transpiration stream. J Exp Bot 56: 1535–1544.

Aloni R, Langhans M, Aloni E, Ullrich CI (2004) Role of cytokinin in the regulation of root gravitropism. Planta 220: 177–182.

Andersson-Gunnerås S, Hellgren JM, Björklund S, Regan S, Moritz T, Sundberg B (2003) Asymmetric expression of a poplar ACC oxidase controls ethylene production during gravitational induction of tension wood. Plant J 34: 339–349.

*Angyalossy V, Pace MR, Lima AC (2015) Liana anatomy: a broad perspective on structural evolution of the vascular system. In: *Ecology of Lianas*, SA Schnitzer, F Bongers, RJ Burnham, FE Putz (eds). Wiley & Sons, Hoboken, New Jersey, pp. 253–287.

Baum SF, Aloni R, Peterson CA (1991) The role of cytokinin in vessel regeneration in wounded *Coleus* internodes. Ann Bot 67: 543–548.

*Bhalerao RP, Fischer U (2017) Environmental and hormonal control of cambial stem cell dynamics. J Exp Bot 68: 79–87.

Bloch D, Puli MR, Mosquna A, Yalovsky S (2019) Abiotic stress modulates root patterning via ABA-regulated microRNA expression in the endodermis initials. Development 146: dev177097.

Bradley MV, Crane JC (1957) Gibberellin-stimulated cambial activity in stems of apricot spur shoots. Science 136: 973–974.

Carlquist S (2013) More woodiness/less woodiness: evolutionary avenues, ontogenetic mechanisms. Int J Plant Sci 174: 964–991.

---

[1] Papers of particular interest for suggested reading have been highlighted (with *)

Dayan J, Voronin N, Gong F, Sun T-p, Hedden P, Hillel Fromm H, Aloni R (2012) Leaf-induced gibberellin signaling is essential for internode elongation, cambial activity, and fiber differentiation in tobacco stems. Plant Cell 24: 66–79.

de Lara NOT, Marcati CR (2016) Cambial dormancy lasts 9 months in a tropical evergreen species. Trees 30: 1331–1339.

Deslauriers A, Giovannelli A, Rossi S, Castro G, Fragnelli G, Traversi L (2009) Intra-annual cambial activity and carbon availability in stem of poplar. Tree Physiol 29: 1223–1235.

Duan L, Dietrich D, Ng CH, Chan PMY, Bhalerao R, Bennett MJ, Dinneny JR (2013) Endodermal ABA signaling promotes lateral root quiescence during salt stress in Arabidopsis seedlings. Plant Cell 25: 324–341.

Eklund L (1990) Endogenous levels of oxygen, carbon dioxide and ethylene in stems of Norway spruce trees during one growing season. Trees 4: 150–154.

*Fischer U, Kucukoglu M, Helariutta Y, Bhalerao RP (2019) The dynamics of cambial stem cell activity. Annu Rev Plant Biol 70: 293–319.

Goldsmith MHM, Cataldo DA, Karn J, Brenneman T, Trip P (1974) The nonpolar transport of auxin in the phloem of intact *Coleus* plants. Planta 116: 301–317.

Hess T, Sachs T (1972) The influence of a mature leaf on xylem differentiation. New Phytol 71: 903–914.

Hou H-W, Zhou Y-T, Mwange K-N, Li W-F, He X-Q, Cui K-M (2006) ABP1 expression regulated by IAA and ABA is associated with the cambium periodicity in *Eucommia ulmoides* Oliv. J Exp Bot 57: 3857–3867.

Ingemarsson BSM, Lundqvist E, Eliasson L (1991) Seasonal variation in ethylene concentration in wood of *Pinus sylvestris* L. Tree Physiol 8: 273–279.

*Immanen J, Nieminen K, Smolander OP, Kojima M, Alonso Serra J, Koskinen P, Zhang J, Elo A, Mähönen AP, Street N, Bhalerao RP, Paulin L, Auvinen P, Sakakibara H, Helariutta Y (2016) Cytokinin and auxin display distinct but interconnected distribution and signaling profiles to stimulate cambial activity. Curr Biol 26: 1990–1997.

Israelsson M, Sundberg B, Moritz T (**2005**) Tissue-specific localization of gibberellins and expression of gibberellin-biosynthetic and signaling genes in wood-forming tissues in aspen. Plant J 44: 494–504.

Jost L (1893) Über Beziehungen zwischen der Blattentwicklung und der Gefässbildung in den Pflanzen. Bot Zeit 51: 89.

Kalev N, Aloni R (1998) Role of auxin and gibberellin in regenerative differentiation of tracheids in *Pinus pinea* L. seedlings. New Phytol 138: 461–468.

Kim H, Kojima M, Choi D, Park S, Matsui M, Sakakibara H, Hwang I (2016) Overexpression of *INCREASED CAMBIAL ACTIVITY*, a putative methyltransferase, increases cambial activity and plant growth. J Integr Plant Biol 58: 874–889.

Ko D, Kang J, Kiba T, Park J, Kojima M, Do J, Kim KY, Kwon M, Endler A, Song WY, Martinoia E, Sakakibara H, Lee Y (2014) Arabidopsis ABCG14 is essential for the root-to-shoot translocation of cytokinin. Proc Natl Acad Sci USA. 111: 7150–7155.

Koiwai H, Nakaminami K, Seo M, Mitsuhashi W, Toyomasu T, Koshiba T (2004) Tissue-specific localization of an abscisic acid biosynthetic enzyme, AAO3, in *Arabidopsis*. Plant Physiol 134: 1697–1707.

Larson PR (1994) *The vascular cambium: Development and Structure*. Springer Verlag, Berlin, Heidelberg.

Lev-Yadun S, Aloni R (1995) Differentiation of the ray system in woody plants. Bot Rev. 61: 45–84.

Lilley JL, Gee CW, Sairanen I, Ljung K, Nemhauser JL (2012) An endogenous carbon-sensing pathway triggers increased auxin flux and hypocotyl elongation. Plant Physiol 160: 2261–2270.

Little CHA, Savidge RA (1987) The role of plant growth regulators in forest tree cambial growth. Plant Growth Regul 6: 137–169.

Love J, Björklund S, Vahala J, Hertzberg M, Kangasjärvi J, Sundberg B (2009) Ethylene is an endogenous stimulator of cell division in the cambial meristem of *Populus*. Proc Natl Acad Sci USA 106: 5984–5989.

Matsumoto-Kitano M, Kusumoto T, Tarkowski P, Kinoshita-Tsujimura K, Václavíková K, Miyawaki K, and Kakimoto T (2008) Cytokinins are central regulators of cambial activity. Proc Natl Acad Sci USA 105: 20027–20031.

Mazur E, Kurczyńska EU, Friml J (2014) Cellular events during interfascicular cambium ontogenesis in inflorescence stems of *Arabidopsis*. Protoplasma 251: 1125–1139.

*Miyashima S, Roszak P, Sevilem I, Toyokura K, Blob B, Heo JO, Mellor N, Help-Rinta-Rahko H, Otero S, Smet W, Boekschoten M, Hooiveld G, Hashimoto K, Smetana O, Siligato R, Wallner ES, Mähönen AP, Kondo Y, Melnyk CW, Greb T, Nakajima K, Sozzani R, Bishopp A, De Rybel B, Helariutta Y (2019) Mobile PEAR transcription factors integrate positional cues to prime cambial growth. Nature 565: 490–494.

Morris DA, Briant RE, Thomson PG (1969) The transport and metabolism of ^{14}C-labelled indoleacetic acid in intact pea seedlings. Planta 89: 178–197.

Morris DA, Kadir GO, Barry AJ (1973) Auxin transport in intact pea seedlings (*Pisum sativum* L.): the inhibition of transport by 2,3,5-triiodobenzoic acid. Planta 110: 173–182.

Morris DA, Thomas AG (1978) A microautoradiographic study of auxin transport in the stem of intact pea seedlings (*Pisum sativum* L.). J Exp Bot 29: 147–157.

*Nieminen K, Blomster T, Helariutta Y, Mähönen AP (2015) Vascular cambium development. *Arabidopsis Book* 13: e0177.

Nieminen K, Immanen J, Laxell M, Kauppinen L, Tarkowski P, Dolezal K, Tähtiharju S, Elo A, Decourteix M, Ljung K, Bhalerao R, Keinonen K, Albert VA, Helariutta Y (2008) Cytokinin signaling regulates cambial development in poplar. Proc Natl Acad Sci USA 105: 20032–20037.

Paul S, Wildhagen H, Janz D, Polle A (2017) Drought effects on the tissue- and cell-specific cytokinin activity in poplar. AoB Plants 10: plx067.

Pesquet E, Tuominen H (2011) Ethylene stimulates tracheary element differentiation in *Zinnia elegans* cell cultures. New Phytol 190: 138–149.

Rathgeber CB, Rossi S, Bontemps JD (2011) Cambial activity related to tree size in a mature silver-fir plantation. Ann Bot 108: 429–438.

Sairanen I, Novák O, Pěnčík A, Ikeda Y, Jones B, Sandberg G, Ljung K (2012) Soluble carbohydrates regulate auxin biosynthesis via PIF proteins in *Arabidopsis*. Plant Cell 24: 4907–4916.

Sakakibara H, Takei K, Hirose N (2006) Interactions between nitrogen and cytokinin in the regulation of metabolism and development. Trends Plant Sci 11: 440–448.

Savidge RA (1983) The role of plant hormones in higher plant cellular differentiation. II. Experiments with the vascular cambium, and sclereid and tracheid differentiation in the pine, *Pinus contorta*. Histochem J 15: 447–466.

Savidge RA (1988) Auxin and ethylene regulation of diameter growth in trees. Tree Physiology 4: 401–414.

Sehr EM, Agusti J, Lehner R, Farmer EE, Schwarz M, Greb T (2010) Analysis of secondary growth in the *Arabidopsis* shoot reveals a positive role of jasmonate signalling in cambium formation. Plant J 63: 811–822.

Seyfferth C, Wessels BA, Gorzsás A, Love JW, Rüggeberg M, Delhomme N, Vain T, Antos K, Tuominen H, Sundberg B, Felten J (2019) Ethylene signaling is required for fully functional tension wood in hybrid Aspen. Front Plant Sci 10: 1101.

Shi D, Lebovka I, López-Salmerón V, Sanchez P, Greb T (2019) Bifacial cambium stem cells generate xylem and phloem during radial plant growth. Development 146: dev171355.

Skoog F, Miller CO (1957) Chemical regulation of growth and organ formation in plant tissues cultured in vitro. Symp Soc Exp Biol 11: 118–130.

*Smetana O, Mäkilä R, Lyu M, Amiryousefi A, Sánchez Rodríguez F, Wu MF, Solé-Gil A, Leal Gavarrón M, Siligato R, Miyashima S, Roszak P, Blomster T, Reed JW, Broholm S, Mähönen AP (2019) High levels of auxin signalling define the stem-cell organizer of the vascular cambium. Nature 565: 485–89.

Snow R (1935) Activation of cambial growth by pure hormones. New Phytol 34: 347–360.

Sorce C, Giovannelli A, Sebastiani L, Anfodillo T (2013) Hormonal signals involved in the regulation of cambial activity, xylogenesis and vessel patterning in trees. Plant Cell Rep 32: 885–898.

Spicer R, Groover A (2010) Evolution of development of vascular cambia and secondary growth. New Phytol 186: 577–592.

Spicer R, Tisdale-Orr T, Talavera C (2013) Auxin-responsive DR5 promoter coupled with transport assays suggest separate but linked routes of auxin transport during woody stem development in *Populus*. PLoS One 8: e72499.

Sundberg B, Uggla C, Tuominen H (2000) Cambial growth and auxin gradients. In: *Cell and Molecular Biology of Wood Formation*. RA Savidge, JR Barnett, R Napier, (eds), BIOS Scientific Publishers, Oxford, pp. 169–188.

Taiz L, Zeiger E, Møller IM, Murphy A (2018) *Fundamentals of Plant Physiology*. Sinauer, Sunderland, MA.

*Tylewicz S, Petterle A, Marttila S, Miskolczi P, Azeez A, Singh RK, Immanen J, Mähler N, Hvidsten TR, Eklund DM, Bowman JL, Helariutta Y, Bhalerao RP (2018) Photoperiodic control of seasonal growth is mediated by ABA acting on cell-cell communication. Science 360: 212–215.

Uggla C, Mellerowicz EJ, Sundberg B (1998) Indole-3-acetic acid controls cambial growth in Scots pine by positional signaling. Plant Physiol 117: 113–121.

Uggla C, Moritz T, Sandberg G, Sundberg B (1996) Auxin as a positional signal in pattern formation in plants. Proc Nat Acad Sci USA 93: 9282–9286.

Ursache R, Nieminen K, Helariutta Y (2013) Genetic and hormonal regulation of cambial development. Physiol Plant 147: 36–45.

Wareing PF (1958) Interaction between indole-acetic acid and gibberellic acid in cambial activity. Nature 181: 1744–1745.

*Wodzicki TJ, Abe H, Wodzicki AB, Pharis RP, Cohen JD (1987) Investigations on the nature of the auxin-wave in the cambial region of pine stems: validation of IAA as the auxin component by the *Avena* coleoptile curvature assay and by gas chromatography-mass spectrometry-selected ion monitoring. Plant Physiol 84: 135–143.

Wodzicki TJ, Knegt E, Wodzicki AB, Bruinsma J (1984) Is indolyl-3-acetic acid involved in the wave-like pattern of auxin efflux from *Pinus silvestris* stem segments? Physiol Plant 44: 122–126.

Zajaczkowski S, Wodzicki TJ, Romberger JA (1984) Auxin waves and plant morphogenesis. In: *Encyclopedia of Plant Physiology (New Series). Hormonal Regulation of Development, II, Vol 10*, TK Scott (ed). Springer-Verlag, Berlin, pp. 244–262.

Zhou Y, Liu X, Engstrom EM, Nimchuk ZL, Pruneda-Paz JL, Tarr PT, Yan A, Kay SA, Meyerowitz EM (2015) Control of plant stem cell function by conserved interacting transcriptional regulators. Nature 517: 377–380.

Zhu Z, Lee B (2015) Friends or foes: new insights in jasmonate and ethylene co-actions. Plant Cell Physiol 56: 414–420.

Regulation of Juvenile-Adult Transition and Rejuvenations

12

Woody perennials usually have two distinct growth phases: juvenile and adult. In the juvenile period, plants grow vigorously competing for light against other surrounding plants during their early years. Gradually their vegetative growth declines and fruit production begins. Similarly, inside a tree, there is a transition from **juvenile wood** (core wood) to **adult wood** (outer wood), which occurs gradually during tree growth and is likely regulated by hormonal signals from both the leaves and root tips. The juvenile wood in a tree occupies the center of the stem and therefore is also called core wood, which is produced during the first 5–20 annual growth rings (depends on tree species) is characterized by short cells, thin cell walls, and low specific gravity, and the outcome is low strength and low wood quality for industry (Zobel and van Buijtenen 1989; Moore and Cown 2017; Fischer et al. 2019), which may therefore result in the development of reaction wood that may be induced by wind. The adult features of different cell type along the tree axis could occur in different rates (e.g., large cells, thick cell walls and high specific gravity) and therefore various adult wood features might mature independently.

There is a limited understanding about the hormonal mechanisms that control the juvenile-adult transition in trees. Likewise the molecular mechanisms underlying these phase changes are poorly understood. Knowledge about developmental cambial plasticity depends solely on early descriptive work (Fischer et al. 2019), and therefore an experimental approach is needed for better understanding of the juvenile-adult transition and rejuvenation. In addition, there is confusion when wood formation is explained by the vague concept of **"cambial age."** Unfortunately, the use of this vague term does not reveal the inducing mechanisms, the controlling signaling, or hormonal regulation of any specific wood pattern. Therefore, it is better not to use the terms: "cambial age" or **"older cambia"** used, for example, by Rodriguez-Zaccaro et al. (2019), as well as the following approach: "The nature of the wood that is formed depends on the age of the vascular cambia: young cambia produce juvenile wood, and old cambia produce mature wood" (Zobel and Sprague 1998), which is also cited in these terms by Fischer et al. (2019), as if the cambium operates by itself and has a built-in program that regulates its activity and products.

© Springer Nature Switzerland AG 2021
R. Aloni, *Vascular Differentiation and Plant Hormones*,
https://doi.org/10.1007/978-3-030-53202-4_12

The changes from juvenile wood to mature wood or the downward increase in vessel and tracheid width along the plant (Aloni and Zimmermann 1983; Anfodillo et al. 2012) and other xylem cell patterns along a tree are induced and regulated by auxin and possibly also by other hormones and peptide signaling originating in leaves and root tips as well as by their gradients and carbohydrate levels, but not by "cambium age." As the cambium in young and mature internodes produces the specific xylem derivatives following the different hormonal signals, it receives at the different sites along the plant axis, e.g., auxin, gibberellin, cytokinin, ethylene, brassinosteroids, jasmonic acid, and strigolactone (Aloni 2007, 2013; Sorce et al. 2013; Miyashima et al. 2013; Oles et al. 2017; Campbell and Turner 2017; Fischer et al. 2019).

A molecular study on the transition from juvenile to adult cambium and wood was carried out on the evergreen conifer tree *Cunninghamia lanceolata* grown in southern China, due to its commercial importance as a timber tree characterized by rapid growth (Xu et al. 2016). Samples of cambium with wood were anatomically and molecularly analyzed at three stages, namely, juvenile, transition, and adult (3-, 13-, and 35-year-old trees, respectively) for transcriptome-wide analysis. The cambium zone at the juvenile stage consisted of about 7 layers of cells, at the transition zone increase to about 10 layers, and remarkably decreased to only about 3 cambial cell layers in the adult stage, demonstrating a substantial decrease in cambial activity likely due to a significant decrease in the concentration of the regulating hormonal signals at the adult phase, likely caused by growing distances from the hormonal producing sources (leaves and root tips). Aloni and Zimmermann (1983) proposed that low auxin concentrations, away from young leaves, permits a slow cell enlarging process of vascular elements that continue to expand and increase their size until secondary wall deposition stops the spreading process, a suggestion that was confirmed by a few studies (see Chap. 13). Accordingly, the increase in cell size of the cambial fusiform initials and their derivatives typical to mature wood is likely a result of low auxin stimulation at the sites of adult wood formation, which occurs far away from the auxin-producing young leaves.

Some putative genes involved in plant hormone biosynthesis were differentially regulated at the three stages along the trunk of *C. lanceolata*. Some transcripts responsive to both auxin and ABA signals were detected. Auxin-responsive proteins and an auxin-induced protein were upregulated at the juvenile and transition stages, whereas these were expressed at slightly lower levels at the adult stage, likely indicating a decreased auxin influence with increasing distance from the young leaves. A brassinosteroid-regulated protein was upregulated during the switch from the juvenile to transition stage, but downregulated at the adult stage (Xu et al. 2016). However, possible modifications in significant gene expression related to gibberellin and cytokinin regulation were not studied, which are likely involved in the graduate transition from juvenile to the adult phase.

Auxin is the primary regulating signal of wood formation (Aloni 2001, 2007, 2015). Near the young auxin-producing leaves, the high-IAA concentrations induce rapid cell differentiation resulting in short and narrow tracheids, fibers, vessel, and sieve elements, which become gradually larger downward with the increasing distance from the IAA sources (Aloni and Zimmermann 1983). Gibberellin from mature leaves (Hess and Sachs 1972; Dayan et al. 2012) promotes vascular cell elongation and, therefore, also contributes to the gradual increase in cell dimensions of the cambial fusiform initials and their derivatives in the secondary vascular tissues. Additionally, the mature leaves are the major source of photosynthetic products that build the secondary cell walls, e.g., carbohydrates, needed for the substantial increased deposition of thick secondary cell walls typical to mature wood. The distance from young leaves to a given site in the cambium and differentiating wood cells increases gradually during stem elongation, and this gradual increasing distance (which gradually decreases the effects of young leaves) likely promotes the gradual phase change from juvenile to adult wood. Thus, the secondary xylem in the recent most peripheral annual rings at the base of the trunk become gradually adult, while the core wood continuously produced in the upper twigs of the crown has juvenile features.

Gibberellin is a key signal in controlling wood development and maturation. Conifers and woody angiosperms might respond differently to the GA hormonal signal regarding their transition from juvenility to maturity. Similar to their opposing mechanisms evolved in responses to gravity: the compression wood in conifers *versus* tension wood of angiosperms (see Sect. 2.8 and Chap. 18). Likewise, gibberellin is a key controlling signal in juvenile-adult transition, which regulate differently the transition phase in conifers *versus* angiosperms. Thus, **in conifers**, GA promotes the transition to the adult phase. Exogenous GA application can promote early transition from juvenile to adult by inducing the reproductive phase and cone production in young conifer trees (Pharis et al. 1980, 1987; Bockstette and Thomas 2019).

Conversely, **in woody angiosperms** including many fruit trees, gibberellins delay the adult phase, by promoting vegetative growth and inhibition of flowering (Goldschmidt and Samach 2004), and when bioactive GAs are applied to mature woody angiosperms, they may induce rejuvenation (Taiz et al. 2018).

The elegant study on English ivy (*Hedera helix*) demonstrated that *H. helix* roots produce gibberellin-like substances that promote juvenile shoot growth (Frydman and Wareing 1973). During the early years of the juvenile phase, *H. helix* is a liana that grows slowly, and in later years the stem begins to elongate considerably and can attached itself by adventitious roots to cover high walls and trees. When the stems are young, the "root signals" retard the juvenile-adult phase transition. During the early 5–15 years, the stem elongates substantially, and consequently, with the increasing distance of the shoot apical meristem (SAM) from the root tips, the concentrations of the root signals decrease at the shoot apex, promoting apical bud maturation, which results in the development of a shrub-like adult shoot. Apical buds from juvenile tissue have higher levels of GAs than buds from adult tissue. Application of various **bioactive gibberellins** to adult *H. helix* shoots induce rejuvenation (Frydman and Wareing 1974), namely, a transition of adult shoots to the juvenile phase with typical juvenile features. Wareing and Frydman (1976) concluded that as the distance between the roots and the shoot apex increases, the amount of root originating GAs arriving at the shoot tip decreases, resulting in a loss of juvenility.

To uncover the possible role of signals from the root tips on the **juvenile-adult transition** and possible **rejuvenation** in an angiosperm tree, I have conducted grafting experiments on *Eucalyptus globulus* trees. By grafting flowering adult branches on a juvenile fast-growing rootstocks (the juvenile phase in this species lasts up to about 2 years, which is influenced by environmental conditions), the distances between the root tips and the grafted adult buds were substantially reduced, thus increasing the influence (signal strength) and the effect of the root-tip signals on the adult buds and the cambium in the grafted mature branches (Fig. 12.1).

In these experiments, well-developed healthy 20-year-old adult branches were grafted (see Sect. 10.2) on 6-month-old juvenile rootstocks. The results demonstrate that the juvenile roots can induce **rejuvenation** of the adult buds and branches. Some of the adult buds in the grafted adult branches started to show typical juvenile characteristics in leaf structure and color, branch phyllotaxis, and juvenile wood characterized by narrow vessels and thin secondary walled fibers. On well-developed rapid successful graft unions, on which the grafted adult branches grow vigorously, all the new leaves showed typical adult morphology (Fig. 12.1A), produced flowers and adult wood with relatively wider vessels and fibers. Conversely, when the apical bud and the youngest internodes of the grafted branch were incidentally broken, or degenerated (because of slow graft-union establishment), the remaining lateral adult buds developed into juvenile looking leaves, of both their shape, color, juvenile phyllotaxis, with no flowers (Fig. 12.1B), and typically juvenile wood (of relatively narrow vessels and fibers).

In addition, various in-between grafted shoots developed. Usually, only the lowest bud developed to become a juvenile branch (marked by *arrowhead* in Fig. 12.1C) although it originated from an adult bud of a 20-year-old grafted branch, demonstrating the strongest root influence on the lowest bud, which is the closest to the juvenile root. Rarely, instead of one adult apical bud, which was damaged during the grafting, a few apical buds developed, and they started to produce juvenile looking leaves for a few weeks and with time have gradually returned to produce adult leaves (marked by the *upper arrow* in Fig. 12.1D). These results indicate that the roots produce a signal, or signals, which can be transported through a graft union. This root signaling promotes juvenile features in the adult shoot. The root originating signals likely retard and delay the normally occurring juvenile-adult transition in intact woody angiosperms. Therefore, to improve wood quality by promoting early development of adult wood, rapid root growth should be encouraged, which will increase the distance of the root tips producing signals from the stem and thus reduce the juvenile-promoting effects of the root tips. **Gibberellins** (Frydman and Wareing 1973, 1974; Wareing and Frydman 1976) and **cytokinins** (Aloni et al. 2005, 2006) are the potential root-tip hormonal candidates that could promote the juvenile-adult wood transition. However additional signals from the root tips might be involved in extending the juvenile behavior of the shoot's apical bud.

On the other hand, mature leaves promote wood maturity predominantly by their photosynthetic products, e.g., carbohydrates, which massively contributing to thick

Fig. 12.1 Photographs showing the effects of a 6-month-old juvenile rootstock on 20-year-old adult grafted branch of *Eucalyptus globulus*, 4 months after grafting. The graft union is marked with a *red ring of a broken line*. (**A**) Showing well-developed grafted shoot with adult leaves (*arrow*) only. (**B**) Only juvenile (rounded and gray) looking leaves (*arrowhead*) developed from a lateral adult bud after the upper branch internodes died. (**C**) On a well-developed grafted branch with many adult leaves (*arrow*), the lowest adult bud developed into a juvenile branch (*arrowhead*). (**D**) Gradual transition of leaves on three branches (originated from the grafted adult branch that its apical bud was damaged during grafting). Note the two original adult leaves of the adult branch (*lower arrow*) and then the development of juvenile leaves (*arrowhead*) and the gradual transition to adult leaves (*upper arrow*). The scale bars are in cm. (From Aloni 2007)

secondary wall deposition in adult wood cells, as well as by their hormonal signaling of the bioactive gibberellins (GA_1 and GA_4) and their long-distance transported precursors, both GA_{12} and GA_{20} (Ragni et al. 2011; Dayan et al. 2012; Regnault et al. 2015; Tal et al. 2016; Binenbaum et al. 2018). The GA from mature leaves promotes lignification and regulates lignin composition and structure (Aloni et al. 1990). More research on this neglected topic is clearly needed from a practical perspective for promoting the early transition to adult secondary xylem, namely, to high-quality adult wood of forest trees used by industry.

The above study on *Eucalyptus* trees also demonstrates that a grafted cambium in adult flowering branches can modify its wood production and start to produce a new layer of juvenile wood, regulated by strong juvenile promoting signaling from the nearby young root and the experimentally induced juvenile leaves. These results support the idea that the cambium produces wood in response to the inductive signals that the cambium receives with no correlation to the cambium's chronological age. In the case of rejuvenation, the cambium of the 20-year-old adult branches became juvenile and started to produce juvenile wood, following the strong stimulation originated in the 6-month-old young juvenile root.

The mechanisms of juvenile-adult transition are not clearly understood. We can expect different mechanisms in conifers *versus* angiosperms due to their opposite responses to gibberellin. The transition of juvenile-to-mature wood and pith-to-bark-profile of wood density (Mutz et al. 2004) are probably controlled by downward-decreasing gradients of auxin concentrations from the young leaves (Aloni and Zimmermann 1983; Aloni 2001, 2007) combined with the gradually increasing contributions of mature leaves that supply carbohydrates and gibberellin that promotes lignification (Aloni 2007), as well as the involvement of the other hormonal signals mentioned above, which should be clarified.

Repression of the **microRNA**, miR156, by miR159 was found to regulate the timing of the juvenile-to-adult transition in *Arabidopsis* (Guo et al. 2017). The juvenile to adult transition is a key developmental event in plant life cycle, and it is regulated by a decrease in the expression of a conserved microRNA miR156/157 (Xu et al. 2018). As microRNAs have been identified as master regulators of developmental timing in both plants and animals, they might also be involved in the transition from juvenile to adult wood production in trees.

Summary

- The transition from juvenile wood (core wood) to adult wood (outer wood), which occurs gradually during tree development, is regulated by hormonal signals and their gradients along the plant axis, with opposing mechanisms in conifers *versus* angiosperms.
- Gibberellin is a key controlling signal in juvenile-adult transition. GA promotes the transition from the juvenile to the adult stage in young conifer trees by inducing the reproductive phase and cone production, while GA can delay the reproductive phase in angiosperms and may promote rejuvenation of adult woody angiosperms.

- Root signals arriving to the shoot apical meristem (SAM) delay the transition from the juvenile to the adult phase. Grafting adult branches on young juvenile rootstocks can promote rejuvenation of both buds and cambium that start to produce typical juvenile features.
- Since the cambium is influenced and produces wood in response to the hormonal signals it receives, the vague term "cambial age" (i.e., young cambia produce juvenile wood and old cambia produce mature wood) should be avoided. The nature of the produced wood does not depend on the chronological age of the vascular cambium but on the hormonal signals and carbohydrates supply that it receives; with the appropriate hormonal stimulation, an adult cambium can start to produce juvenile wood.

References and Recommended Readings[1]

Aloni R (2001) Foliar and axial aspects of vascular differentiation - hypotheses and evidence. J Plant Growth Regul 20: 22–34.

Aloni R (2007) Phytohormonal mechanisms that control wood quality formation in young and mature trees. In: *The Compromised Wood Workshop 2007*. K Entwistle, P Harris, J Walker (eds). The Wood Technology Research Centre, University of Canterbury, Christchurch, New Zealand, pp 1–22.

Aloni R (2013) The role of hormones in controlling vascular differentiation. In:, *Cellular Aspects of Wood Formation,* J Fromm (ed). Springer-Verlag, Berlin, pp 99–139.

Aloni R (2015) Ecophysiological implications of vascular differentiation and plant evolution. Trees 29: 1–16.

Aloni R, Aloni E, Langhans M, Ullrich CI (2006) Role of cytokinin and auxin in shaping root architecture: regulating vascular differentiation, lateral root initiation, root apical dominance and root gravitropism. Ann Bot 97: 883–893.

Aloni R, Langhans M, Aloni E, Dreieicher E, Ullrich CI (2005) Root-synthesized cytokinin in *Arabidopsis* is distributed in the shoot by the transpiration stream. J Exp Bot 56: 1535–1544.

*Aloni R, Tollier T, Monties B (1990) The role of auxin and gibberellin in controlling lignin formation in primary phloem fibers and in xylem of *Coleus blumei* stems. Plant Physiol 94: 1743–1747.

Aloni R, Zimmermann MH (1983) The control of vessel size and density along the plant axis - a new hypothesis. Differentiation 24: 203–208.

*Anfodillo T, Deslauriers A, Menardi R, Tedoldi L, Petit G, Rossi S (2012) Widening of xylem conduits in a conifer tree depends on the longer time of cell expansion downwards along the stem. J Exp Bot 63: 837–845.

Binenbaum J, Weinstain R, Shani E (2018) Gibberellin localization and transport in plants. Trends Plant Sci 23: 410–421.

Bockstette SW, Thomas BR (2019) Impact of genotype and parent origin on the efficacy and optimal timing of $GA_{4/7}$ stem injections in a lodgepole pine seed orchard. New Forest 51: 421–434.

Campbell L, Turner S (2017) Regulation of vascular cell division. J Exp Bot 68: 27–43.

Dayan J, Voronin N, Gong F, Sun T-p, Hedden P, Fromm H, Aloni R (2012) Leaf-induced gibberellin signaling is essential for internode elongation, cambial activity, and fiber differentiation in tobacco stems. Plant Cell 24: 66–79.

---

[1] Papers of particular interest for suggested reading have been highlighted (with *)

*Fischer U, Kucukoglu M, Helariutta Y, Bhalerao RP (2019) The dynamics of cambial stem cell activity. Annu Rev Plant Biol 70: 293–319.

Frydman VM, Wareing PF (1973) Phase change in *Hedera helix* L. II. The possible role of roots as a source of shoot gibberellin-like substances. J Exp Bot 24: 1139–1148.

Frydman VM, Wareing PF (1974) Phase change in *Hedera helix* L. III. The effects of gibberellins, abscisic acid and growth retardants on juvenile and adult ivy. J Exp Bot 25: 420–429.

Goldschmidt EE, Samach A (2004) Aspects of flowering in fruit trees. *Proc 9th IS on Plant Bioregulators*, SM Kang et al. (eds). Acta Hort 653, pp 23–27.

*Guo C, Xu Y, Shi M, Lai Y, Wu X, Wang H, Zhu Z, Poethig RS, Wu G (2017) Repression of miR156 by miR159 regulates the timing of the juvenile-to-adult transition in *Arabidopsis*. Plant Cell 29: 1293–1304.

Hess T, Sachs T (1972) The influence of a mature leaf on xylem differentiation. New Phytol 71: 903–914.

Miyashima S, Sebastian J, Lee JY, Helariutta Y (2013) Stem cell function during plant vascular development. EMBO J 32: 178–93.

*Moore JR, Cown DJ (2017) Corewood (juvenile wood) and its impact on wood utilisation. Curr For Rep 3: 107–118.

Mutz R, Guilley E, Sauter UH, Nepveu G (2004) Modelling juvenile-mature wood transition in Scots pine (*Pinus sylvestris* L.) using nonlinear mixed-effects models. Ann For Sci 61: 831–841.

Oles V, Panchenko A, Smertenko A (2017) Modeling hormonal control of cambium proliferation. PLoS One 12: e0171927.

Pharis RP, Ross SD, McMullan E (1980) Promotion of flowering in the Pinaceae by gibberellins III. Seedlings of Douglas fir. Physiol Plant 50: 119–126.

Pharis RP, Webber JE, Ross SD (1987) The promotion of flowering in forest trees by gibberellin A_{47} and cultural treatments: A review of the possible mechanisms. Forest Ecol Manag 19: 65–84.

Ragni L, Nieminen K, Pacheco-Villalobos D, Sibout R, Schwechheimer C, Hardtke CS (2011) Mobile gibberellin directly stimulates *Arabidopsis* hypocotyl xylem expansion. Plant Cell 23: 1322–1336.

Regnault T, Davière JM, Wild M, Sakvarelidze-Achard L, Heintz D, Carrera Bergua E, Lopez Diaz I, Gong F, Hedden P, Achard P (2015) The gibberellin precursor GA12 acts as a long-distance growth signal in *Arabidopsis*. Nat Plants 1: 15073.

Rodriguez-Zaccaro FD, Valdovinos-Ayala J, Percolla MI, Venturas MD, Pratt RB, Jacobsen AL (2019) Wood structure and function change with maturity: Age of the vascular cambium is associated with xylem changes in current-year growth. Plant Cell Environ 42: 1816–1831.

Sorce C, Giovannelli A, Sebastiani L, Anfodillo T (2013) Hormonal signals involved in the regulation of cambial activity, xylogenesis and vessel patterning in trees. Plant Cell Rep 32: 885–898.

Taiz L, Zeiger E, Møller IM, Murphy A (2018) *Fundamentals of Plant Physiology*. Sinauer, Sunderland, MA.

Tal I, Zhang Y, Jørgensen ME, Pisanty O, Barbosa IC, Zourelidou M, Regnault T, Crocoll C, Olsen CE, Weinstain R, Schwechheimer C, Halkier BA, Nour-Eldin HH, Estelle M, Shani E (2016) The *Arabidopsis* NPF3 protein is a GA transporter. Nat Commun 7: 11486.

*Wareing P F, Frydman VM (1976) General aspects of phase change with special reference to Hedera helix L. Acta Hort 56: 57–68.

Xu H, Cao D, Feng J, Wu H, Lin J, Wang Y (2016) Transcriptional regulation of vascular cambium activity during the transition from juvenile to mature stages in *Cunninghamia lanceolata*. J Plant Physiol 200: 7–17.

*Xu Y, Zhang L, Wu G (2018) Epigenetic regulation of juvenile-to-adult transition in plants. Front Plant Sci 9: 1048.

Zobel BJ, van Buijtenen JP (1989) *Wood Variation, its Causes and Control*. Springer-Verlag, Berlin.

Zobel BJ, Sprague JR. **1998**. General concepts of juvenile wood. In: *Juvenile Wood in Forest Trees*. Springer, Berlin, pp 1–20.

The Control of Tracheid Size, Vessel Widening and Density Along the Plant Axis

13

The hydraulic performance of plants is crucially affected by tracheid and vessel diameter (Tyree and Zimmermann 2002; Lucas et al. 2013; Rosell et al. 2017; Williams et al. 2019), which also affects xylem pathology and wood adaptation (Aloni 1987, 2015; Aloni and Ullrich 2008). Therefore, it is important to understand the mechanisms that control the diameter and density of these vascular conduits in plants.

A well-documented phenomenon is the downward gradual and continuous increase in tracheid and vessel size from leaves to roots. This widening in vessel diameter was found along leaves from the tip to the base of the leaf (Colbert and Evert 1982; Russell and Evert 1982; Lechthaler et al. 2019). A continuous gradual increase in vessel diameter and vessel length was demonstrated from twigs to branches, downward along the stem and into the roots of *Acer rubrum* trees (Zimmermann and Potter 1982). This basic pattern was found in dicotyledons as well as monocotyledons (Tomlinson and Zimmermann 1967; Carlquist 1976; Zimmermann 1983). The tracheids and vessels are narrow at the leaves, and their diameter increases gradually downward and continuously along the stem (Bailey 1958; Carlquist 1975; Zimmermann and Potter 1982; Aloni and Zimmermann 1983; Sorce et al. 2013; Olson et al. 2014; Lazzarin et al. 2016; Rosell et al. 2017; Williams et al. 2019) and the root (Riedl 1937; Fahn 1964), with a decrease in vessel density from leaves to roots, which was reported for many species (Fegel 1941; Carlquist 1976; Aloni and Zimmermann 1983; Leitch 2001; Sorce et al. 2013; Zhao 2015). The general increase in conduit size can also be detected in the wood (secondary xylem) along the radial direction, from pit to bark, as was descried by the pioneering observations of Sanio (1872).

Xylem conduits expand gradually (Fig. 13.1) until they deposit their thick and rigid secondary wall. Similarly, along the stem axis, the phloem's sieve elements increase in diameter from leaves to root (Fig. 13.2) at rates comparable to those of xylem conduits (Petit and Crivellaro 2014). The sieve elements likely have a similar mechanism to that of tracheary elements, but instead of secondary cell wall deposition; the sieve elements lose their nucleus, which stops their expansion.

© Springer Nature Switzerland AG 2021
R. Aloni, *Vascular Differentiation and Plant Hormones*,
https://doi.org/10.1007/978-3-030-53202-4_13

Fig. 13.1 Transverse section in a vascular bundle of *Coleus blumei*, showing a differentiating vessel with a nucleus and its nucleolus (*arrow*) in the vessel widening phase that lasts until the secondary wall is deposited, which stops cell expansion. Cambium, c. Bar = 50 μm

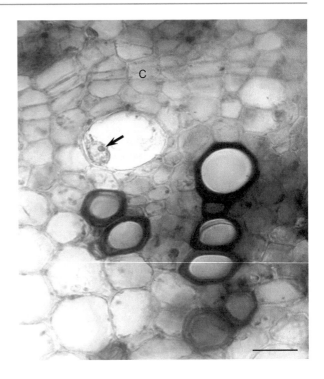

Knowing that vessels and sieve tubes are directly induced by auxin (as discussed in Chaps. 3 and 5), the documented polar gradient pattern of conduit width and density raises the questions: what is the long-distance auxin signaling mechanism that controls conduit width and density along the plant axis? If this mechanism is based on decreasing auxin concentrations from leaves to roots, it raises additional basic question: what is the evidence for a decreasing gradient of auxin concentration along the plant axis from leaves to roots? or does auxin act as a **morphogen** (Turing 1952; Wolpert 1969), which stimulates its own degradation (by conjugation and catabolism) along its downward polar transport to induce **gradients**, or is it a self-enhancing signal that promotes its transport (Sachs 1981) and thus forms gradients, which provides positional information to the cells through which it flows? The known nature of auxin polar transport makes it a perfect candidate to promote and control the polar cellular patterns described above in the vascular system, which will be discussed in this chapter.

There are contradictory hypotheses regarding the control of vascular element size. The first called the *hypothesis of tracheid diameter regulation* (Larson 1969) suggests that the gradients of tracheid diameter throughout the stem are "positively" regulated by parallel stimulating gradients of auxin. This hypothesis, which was based on photoperiodic studies of growth ring formation in *Pinus*, ascribes the formation of wide tracheids (like those in the earlywood) to high levels of auxin associated with leaf development and shoot elongation and the formation of narrow tracheids (as in latewood) to low levels of auxin associated with cessation of shoot

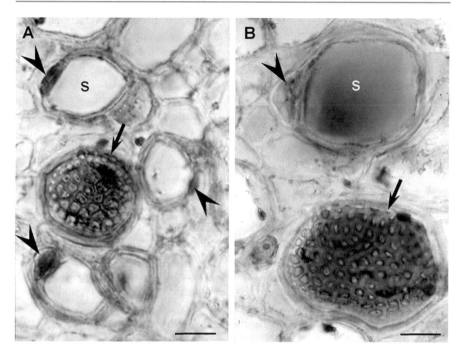

Fig. 13.2 Transverse sections of the phloem along the same continuous vascular bundle passing through young (**A**) and mature (**B**) internodes of an intact stem of *Luffa cylindrica*, stained with lacmoid (Aloni 1979), showing the substantial increase of sieve elements in the older internode (**B**), with the increasing distance from young leaves. Sieve element (s), sieve plate (*arrow*), companion cell (*arrowhead*). Bars = 50 μm

growth (Larson 1964, 1969). However, as tracheids are controlled by the combined stimulation of both auxin and gibberellin (Kalev and Aloni 1998), which are altered during the growth season, as auxin gradually decreases and gibberellin (which reduces cell width) gradually increases. Therefore, the possible role of gibberellin from maturing leaves (Dayan et al. 2012) in regulating tracheid size along the season should be considered.

Experimental results obtained with the ring-porous species *Robinia pseudoacacia* by Digby and Wareing (1966) indeed show a linear increase in vessel diameter with increasing auxin concentration from 1 mg/l to 1000 mg/l (with gibberellic acid at 100 mg/l in all the treatments; the GA was added to stimulate cambial activity, following Wareing (1958)). Although this result seems to support the idea about a "positive" correlation between auxin level and vessel width, however, this increase in vessel diameter could perhaps be ascribed to the disparity in rates between gibberellin and auxin. At low auxin (1 mg/l), the gibberellin (100 mg/l) was the dominant signal, while under high auxin (1000 mg/l), the effect of GA which decreases vessel diameter was minimized.

The hypothesis that proposes a positive correlation between auxin level and tracheid diameter in *Pinus* (Larson 1969) and vessel diameter in *Robinia* (Digby and

Wareing 1966) is contrary to what one would expect from the overall pattern of vascular element size along the plant. The smallest vascular elements differentiate near the young auxin-producing leaves, where the highest auxin concentrations are expected, while the largest elements are formed in the roots, at the greatest distance from the auxin sources. This polar pattern in conduit size along plants is well known. The pioneering observation of Grew (1682) showed that vessels in the roots are generally wider than those in the trunk. Much later, Sanio (1872) subsequently discovered the general increase in tracheid size along the secondary xylem of *Pinus sylvestris* in branches and stem – an increase in tracheid size proceeding outward from the inner growth ring through a number of annual growth rings until a constant size is attained. He also found that tracheids are smaller in branches and bigger in the trunk (Sanio 1872). These Sanio findings were later confirmed in *Sequoia sempervirens* by Bailey (1958), who clearly demonstrated a continuous increase in the length and width of secondary tracheids proceeding from branches to trunk and downward into the roots. Interestingly, the increase in tracheid diameter from leaves to roots is positively correlated with an increase in the duration of tracheid expansion in the same direction (Denne 1972; Saks and Aloni 1985; Anfodillo et al. 2012).

To resolve the apparent contradiction between auxin concentration and the pattern of conduit size, and to better explain the mechanism controlling the general increase in conduit size and decrease in vessel density from leaves to roots, Aloni and Zimmermann (1983) proposed the ***auxin gradient hypothesis*** (that was first called the ***six-point hypothesis***) suggesting that the auxin hormone descending from young leaves to root tips acts as a morphogenetic signal which forms a long-distance decreasing concentration gradient mechanism that controls conduit width and density along the plant axis.

Morphogen gradients were suggested as mechanisms for assigning positional information within a field. Accordingly, models explaining how positional information is set up and how it is interpreted by developing cells were proposed for different patterns and biological systems in both animals and plants (Turing 1952; Wolpert 1969, 2016; Meinhardt 2009; Bhalerao and Fischer 2014).

In vascular differentiation, the polar transport of the bioactive auxin is the morphogenetic signal that is suggested to create such a gradual polar gradient in the vascular cambium providing directional and location information to the differentiating cells (i.e., tracheids, vessels, fibers, and sieve tubes) along the morphogenetic field. To explain the mechanism that controls the gradual patterns of increasing conduit size and decreasing vessel density from leaves to roots, Aloni and Zimmermann (1983) proposed the following six points of the ***auxin gradient hypothesis***:

1. Downward polar flow of auxin from leaves to roots establishes a **gradient of decreasing auxin concentration** in this direction.
2. Local structural or physiological **obstruction** to auxin flow results in a local increase in auxin concentration.

3. The **distance from the source of auxin** to the differentiating cells controls the amount of auxin flowing through the differentiating cells at a given time, thus determining the cell's position in the gradient.
4. The **rate of conduit differentiation** is positively correlated with the amount of auxin that the differentiating cells receive; consequently, the duration of the differentiation process increases along the decreasing auxin gradient from leaves to roots.
5. The **final size** of a conduit is determined by the rate of cell differentiation. Since cell expansion ceases after the secondary wall is deposited, high-auxin concentrations near the young leaves induce narrow vessels because of their rapid differentiation, allowing only limited time for cell widening. Conversely, slow differentiation further down permits more cell expansion before secondary wall deposition and therefore results in wide tracheary elements at the base of the stem. Hence, decreasing auxin concentration from leaves to roots leads to an increase in conduit size in this direction.
6. **Vessel density** is controlled by and positively correlates with the auxin concentration; consequently, vessel density decreases from leaves to roots.

The *auxin gradient hypothesis* was experimentally confirmed by showing that various auxin concentrations applied to decapitated stems induce substantial gradients of increasing vessel diameter and decreasing vessel density from the auxin source toward the roots (Figs. 13.3, 13.4 and 13.5). High-auxin concentration yielded numerous vessels that remained narrow because of their rapid differentiation; low-auxin concentration resulted in slow differentiation and therefore in fewer and wider vessels (Aloni and Zimmermann 1983). Similarly, in intact untreated *Pinus pinea* seedlings, the number of primary and secondary tracheids decreased, and their radial diameter increased from cotyledons to roots. Rapid differentiation immediately beneath the cotyledons resulted in narrow tracheids, while slow development at the root yielded wide tracheids (Saks and Aloni 1985).

Studies on transgenic plants with altered levels of IAA confirmed the general relations between IAA concentration and vessel size and density. Thus, auxin over-producing plants (i.e., over-expressing the *iaaM* gene) contained many more vessel elements than did control plants, and their vessels were narrow (Klee et al. 1987); conversely, plants with lowered IAA levels (i.e., expressing the *iaaL* gene as an anti-auxin gene) contained fewer vessels of generally larger size (Romano et al. 1991).

There are three evidences that support the role of auxin as a morphogen that establishes a decreasing concentration gradient, which provides positional information to the cells along the vascular cambium and the differentiating phloem and xylem.

1. Quantitative measurements of auxin concentrations along the cambium of the pine trees *Pinus sylvestris and P. contorta* revealed that the downward polar movement of auxin along the stem occurs in a **wave-like pattern**. To quantify the native IAA concentrations along the cambial zone, the auxin from the cambial region of the pine stems was collected into agar strips from axial series of

Fig. 13.3 Transverse sections along the same internode, the second internode above the cotyledons of *Phaseolus vulgaris* after 3 weeks of treatment with auxin (0.1% naphthaleneacetic acid (NAA) applied in a lanoline paste) renewed every 3 days, on the top of the internode after the shoot above it was excised. The sections were stained with a mixture of safranin and alcian green which gave red color to the lignified xylem cells, observed and photographed under a light microscope. (**A**) 5 mm below the site of NAA application, showing massive xylem formation with 7 layers of vessels induced by 7 renewed auxin applications, characterized by many narrow vessels in high density. (**B**) 40 mm below the site of auxin application, showing a substantial decrease in xylem formation, characterized by wide vessels in low density, organized in bundles. Note the change of vessel patterns from layers (**A**) to bundles (**B**). Bars = 100 μm. (From Aloni and Zimmermann 1983)

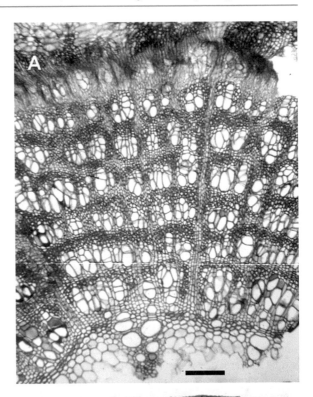

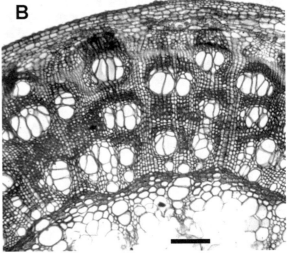

6-millimeter-long sections, and the IAA was carefully quantified by the highest-quality quantification analyzes (Wodzicki et al. 1987). The quantitative results from *P. sylvestris* and *P. contorta* showed a wave-like pattern of oscillating IAA concentrations along the vascular cambium, evidently demonstrating that the

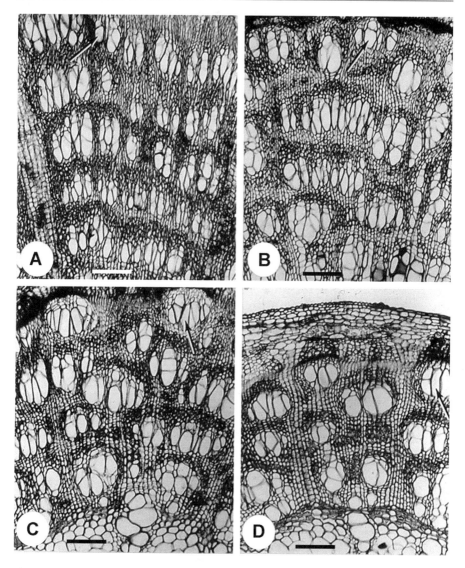

Fig. 13.4 Transverse sections along the same internode of *Phaseolus vulgaris*, showing the gradual changes in patterns of the induced vessels with increasing distance from the auxin source, after 3 weeks of treatment with 0.1% NAA applied on the top of the internode after the shoot above it was excised. The auxin was replaced and renewed every 3 days. (**A**) 5 mm below the site of NAA application, (**B**) 10 mm, (**C**) 20 mm, and (**D**) 40 mm below the site of auxin application. The photographs show the gradual increase in vessel diameter and decrease in vessel density with increasing distance from the auxin source, as well as gradual changes of vessel pattern from layers (**A** and **B**) to bundles (**C** and **D**). *Arrow* marks a late-formed secondary vessel. Bars = 100 μm. (From Aloni and Zimmermann 1983)

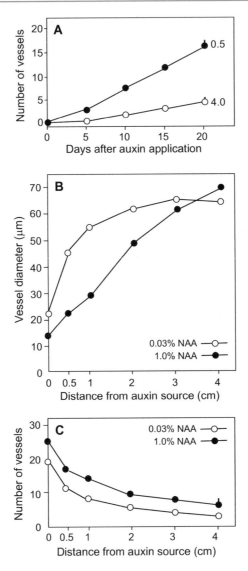

Fig. 13.5 Effects of applied auxin concentration (0.03% NAA, 0.1% NAA, or 1.0% NAA w/w in lanolin, renewed every 3 days) on secondary vessel differentiation in the second internode above the cotyledons of *Phaseolus vulgaris*, observed after 3 weeks of hormonal applications, on the top of the internode after the shoot above it was excised. (**A**) Effect of distance (0.5 and 4.0 cm) from 0.1% NAA application site on the rate of secondary vessel formation, showing intensive vessel differentiation near (0.5 cm) the site of auxin application. (**B**) Effect of 0.03% and 1.0% NAA on the radial diameter of the late-formed secondary vessels, along the studied internode, showing the substantial increase in vessel diameter with increasing distance from the applied auxin. (**C**) Effect of 0.03% and 1.0% NAA on the number of secondary vessels induced along a xylem radius, as affected by distance from auxin source. *Vertical bars* indicate standard errors which are comparable at all points. (From Aloni and Zimmermann 1983)

polar auxin moved downward in a wave-like pattern along the cambium region of the pine stems (Wodzicki et al. 1984, 1987). This downward wavy pattern of polar IAA transport can provide morphogenetic information along the plant axis, which can inform the cambium initials and their differentiating phloem and xylem derivatives about their location along a decreasing auxin gradient from the auxin-producing young leaves downward to the roots (Wodzicki et al. 1984, 1987; Zajaczkowski et al. 1984).

2. There is a limited molecular knowledge how auxin gradients are regulated, and there is a need for more studies in this direction for better understanding of both IAA conjugation (Ludwig-Müller 2011) and IAA catabolism (Peer et al. 2013; Pěnčík et al. 2013) in establishing auxin gradients along plants. (i) Auxin can be conjugated to amino acids and sugars (Ludwig-Müller 2011) to form either (a) irreversible degradation products that may promote a gradual decreasing gradient along the plant axis or (b) storage products that can possibly build auxin reservoir in the foliage. (ii) Auxin gradients can also be promoted by auxin catabolism, through the oxidation pathway of auxin to the IAA catabolite **2-oxindole-3-acetic acid (oxIAA)** (Peer et al. 2013; Pěnčík et al. 2013; Zhao 2018) leading to gradual IAA degradation along the plant. It was shown that oxIAA is inactive in auxin signaling and in auxin transport across cell membranes, thus demonstrating that oxidation of IAA to oxIAA is capable of modulating developmentally important auxin gradients and auxin maxima/minima along plants (Pěnčík et al. 2013).

3. Along trees, there is a steady **downward increase in cambium circumference** that gradually dilutes IAA concentration, which also contributes to the gradual decrease in the auxin signal along the plant axis from leaves to roots.

The duration of cell differentiation along the stem of a *Picea abies* tree demonstrated that the extent of the expansion phase is positively correlated with the lumen area of the tracheids and that the lumen area of the conduits from the top of the tree to its base was linearly dependent on the time during which the differentiating tracheids remained in the expansion phase (Anfodillo et al. 2012). This study shows that at the top of the tree's trunk (9 m from the ground) the tracheid expansion time was 7 days, at 6 m above ground the cells expanded for 14 days, and at 3 m for 19 days (Anfodillo et al. 2012). Therefore, the tracheids at the base of the tree, which have the longest period of cell expansion before secondary wall deposition, become the widest conduits (Anfodillo et al. 2012; Sorce et al. 2013) as predicted by the *auxin gradient hypothesis* (Aloni and Zimmermann 1983). Measurements in *Picea mariana* trees also demonstrated that the duration of cell expansion phase determined the final width of tracheids (Buttò et al. 2019). Analysis of tracheid developmental kinetics in conifer trees showed that the duration of cell enlargement contributes 75% of the increase in tracheid diameter (Cuny et al. 2014).

A recent study of conduit diameter in the earth's tallest tree species, *Sequoia sempervirens*, *Sequoiadendron giganteum*, and *Eucalyptus regnans*, that were 86–105 m tall and exceeded 85% of the maximum height for each species showed the typical gradual increase in conduit width along the upper parts of their shoots.

However, at the base of their trunks, below about 60 m from the tree tops, conduit diameters approached their maximum size, as they do not continue to expand, demonstrating that at the base of these giant trees, there is a limitation to conduit widening (Williams et al. 2019).

Analysis of vessel density in various plant species showed a gradual decrease in vessel density from leaves to roots (Aloni and Zimmermann 1983; Leitch 2001; Olson et al. 2014; Zhao 2015). Confirming the idea that vessel density is controlled by auxin concentration, high-auxin concentrations (near the young leaves) induce greater density, while low concentrations (toward the roots) diminish density (Aloni and Zimmermann 1983).

The polar auxin gradient mechanism regulates the gradual downward vessel widening along plants which continuously increases vessel diameter with increasing stem length (Aloni and Zimmermann 1983) which indicates a correlation between vessel diameter and stem length. This expected correlation was quantitatively confirmed by Olson et al. (2014). Their data shows that vessel widening, as predicted by Aloni and Zimmermann (1983), with distance from the stem apex across angiosperm order, habits, and habitats, is usually explained by stem length. Compared with climate and biomes, stem length is by far the main driver of global variation in mean vessel diameter. The universal vessel diameter-stem length correlation (Olson et al. 2014) seems to indicate that the regulation of the vessel widening mechanism by the polar auxin gradient (Aloni and Zimmermann 1983) is likely a universal plant mechanism. Nevertheless, the sizes of vessel diameter and of stem length by themselves are not causative factors, and their nice statistical correlations are based on the polar auxin gradient; therefore the picture presented by Olson et al. (2014) is incomplete because they do not consider the biologically causing factor, namely, the auxin gradient mechanism (Aloni and Zimmermann 1983), that induces and regulates the vessel diameter-stem length correlation.

Summary

- Tracheid and vessel diameter determine the hydraulic performance of plants. Conduit width gradually increases, while vessel density decreases downward, from the young leaves toward the roots. Sieve elements and fibers show the same downward gradual increase in their width.
- The *auxin gradient hypothesis* proposes that the auxin hormone flowing polarly from young leaves to root tips acts as a morphogenetic signal which forms a long-distance decreasing IAA concentration gradient that controls conduit width and density along the plant axis.
- Along the gradient, the distance from the auxin-producing young leaves to the differentiating cells positively correlates the amount of auxin flowing through the differentiating cells; consequently, the duration of the differentiation process increases along the decreasing auxin gradient from leaves to roots.
- The final size of a conduit is determined by the rate of cell differentiation. Since cell expansion ceases after the secondary wall is deposited, high-auxin concentrations near the young leaves induce narrow tracheids and vessels because of

their rapid differentiation, allowing only limited time for cell widening. Conversely, slow differentiation further down permits more cell expansion before secondary wall deposition and therefore results in wide tracheary elements at the base of the stem.

- Vessel density is controlled by and positively correlates with the auxin concentration; consequently, vessel density decreases from leaves to roots.
- Evidence for a downward auxin decreasing morphogenetic gradient comes from quantitative measurements of auxin concentrations along the cambium of pine trees, showing that the polar auxin movement along the cambium occurs in a wave-like pattern; limited molecular studies on downward IAA conjugation and IAA catabolism; and the contribution of the gradual and steady downward increase in cambium circumference that gradually dilutes IAA concentration.
- The *auxin gradient hypothesis* was confirmed, while stimulating research yielding interesting results; e.g., in the earth's tallest trees, the typical gradual increase in conduit width occurred along the upper parts of their shoots. However, at the base of their trunks, below about 60 m from the tree tops, conduit diameters approached their maximum size, and they do not continue to expand, demonstrating that at the base of these giant trees, there is a limitation to conduit widening.

References and Recommended Reading[1]

Aloni R (1979) Role of auxin and gibberellin in differentiation of primary phloem fibers. Plant Physiol 63: 609–614.

Aloni R (1987) Differentiation of vascular tissues. Annu Rev Plant Physiol 38: 179–204.

Aloni R (2015) Ecophysiological implications of vascular differentiation and plant evolution. Trees 29: 1–16.

Aloni R, Ullrich CI (2008) Biology of crown gall tumors. In: *Agrobacterium*, T Tzfira, V Citovsky (eds). Springer, New York, pp. 565–591.

*Aloni R, Zimmermann MH (1983) The control of vessel size and density along the plant axis - a new hypothesis. Differentiation 24: 203–208.

*Anfodillo T, Deslauriers A, Menardi R, Tedoldi L, Petit G, Rossi S (2012) Widening of xylem conduits in a conifer tree depends on the longer time of cell expansion downwards along the stem. J Exp Bot 63: 837–845.

Bailey IW (1958) The structure of tracheids in relation to the movement of liquids, suspensions and undissolved gases. In: *The Physiology of Forest Trees*, KV Thimann (ed), Ronald, New York, pp. 71–82.

Bhalerao RP, Fischer U (2014) Auxin gradients across wood-instructive or incidental? Physiol Plant 151: 43–51.

Buttò V, Rossi S, Deslauriers A, Morin H (2019) Is size an issue of time? Relationship between the duration of xylem development and cell traits. Ann Bot 123: 1257–1265.

Carlquist S (1975) *Ecological strategies of xylem evolution*. University of California Press, Berkeley.

[1] Papers of particular interest for suggested reading have been highlighted (with *).

Carlquist S (1976) Wood anatomy of Roridulaceae: ecological and phylogenetic implications. Am J Bot 63: 1003–1008.

Colbert JT, Evert RF (1982) Leaf vasculature in sugar cane (*Saccharum officinarum* L.). Planta 156: 136–151.

Cuny HE, Rathgeber CB, Frank D, Fonti P, Fournier M (2014) Kinetics of tracheid development explain conifer tree-ring structure. New Phytol 203: 1231–1241.

Dayan J, Voronin N, Gong F, Sun T-p, Hedden P, Fromm H, Aloni R (2012) Leaf-induced gibberellin signaling is essential for internode elongation, cambial activity, and fiber differentiation in tobacco stems. Plant Cell 24: 66–79.

Denne MP (1972) A comparison of root and shoot-wood development in conifer seedlings. Ann Bot 36: 579–587.

Digby J, Wareing PF (1966) The effect of applied growth hormones on cambial division and the differentiation of cambial derivatives. Ann Bot 30: 539–548.

Fahn A (1964) Some anatomical adaptations of desert plants. Phytomorphology 14: 93–102.

Fegel CA (1941) Comparative anatomy and varying physical properties of trunk, branch and root wood in certain northeastern trees. Bull NY State College of Forestry at Syracuse Univ, Vol. 14, No 2b, Tech Publ 55: 1–20.

Grew N (1682) *The Anatomy of Plants with an Idea of a Philosophical History of Plants*. Rawlins, London.

Kalev N, Aloni R (1998) Role of auxin and gibberellin in regenerative differentiation of tracheids in *Pinus pinea* seedlings. New Phytol 138: 461–468.

Klee HJ, Horsch RB, Hinchee MA, Hein MB, Hoffmann MB (1987) The effects of overproduction of two *Agrobacterium tumefaciens* T-DNA auxin biosynthetic gene products in transgenic petunia plants. Genes Dev 1: 86–96.

Larson PR (1964) Some indirect effects of environment on wood formation. In: *Formation of Wood in Forest Trees*, MH Zimmermann (ed). Academic, New York, pp. 345–365.

Larson PR (1969) *Wood formation and the concept of wood quality*. Sch For Bull 74, Yale univ, New Haven.

Lazzarin M, Crivellaro A, Mozzi G, Williams C, Dawson T, Anfodillo T (2016) Tracheid and pit anatomy vary in tandem in a tall giant sequoia. IAWA J 37: 172–185.

Lechthaler S, Colangeli P, Gazzabin M, Anfodillo T (2019) Axial anatomy of the leaf midrib provides new insights into the hydraulic architecture and cavitation patterns of *Acer pseudoplatanus* leaves. J Exp Bot 70: 6195–6201.

Leitch MA (2001) Vessel-element dimensions and frequency within the most current growth increment along the length of *Eucalyptus globules* stems. Trees 15: 353–357.

Lucas WJ, Groover A, Lichtenberger R, Furuta K, Yadav SR, Helariutta Y, He XQ, Fukuda H, Kang J, Brady SM, Patrick JW, Sperry J, Yoshida A, López-Millán AF, Grusak MA, Kachroo P (2013) The plant vascular system: evolution, development and functions. J Integr Plant Biol 55: 294–388.

Ludwig-Müller J (2011) Auxin conjugates: their role for plant development and in the evolution of land plants. J Exp Bot 62: 1757–1773.

Meinhardt H (2009) Models for the generation and interpretation of gradients. Cold Spring Harb Perspect Biol 1: a001362.

*Olson ME, Anfodillo T, Rosell JA, Petit G, Crivellaro A, Isnard S, León-Gómez C, Alvarado-Cárdenas LO, Castorena M (2014) Universal hydraulics of the flowering plants: vessel diameter scales with stem length across angiosperm lineages, habits and climates. Ecol Lett 17: 988–997.

Peer WA, Cheng Y, Murphy AS (2013) Evidence of oxidative attenuation of auxin signalling. J Exp Bot 64: 2629–2639.

*Pěnčík A, Simonovik B, Petersson SV, Henyková E, Simon S, Greenham K, Zhang Y, Kowalczyk M, Estelle M, Zažímalová E, Novák O, Sandberg G, Ljung K (2013) Regulation of auxin homeostasis and gradients in *Arabidopsis* roots through the formation of the indole-3-acetic acid catabolite 2-oxindole-3-acetic acid. Plant Cell 25: 3858–3870.

Petit G, Crivellaro A (2014) Comparative axial widening of phloem and xylem conduits in small woody plants. Trees 28: 915–921.

Riedl H (1937) Bau und Leistungen des Wurzelholzes. Jb Wiss Bot 85: 1–75 (English translation available from: National Translations Center, 35 West 33rd St, Chicago, IL 60616).

Romano CP, Hein MB, Klee HJ (1991) Inactivation of auxin in tobacco transformed with the indoleacetic acid-lysine synthetase gene of *Pseudomonas savastanoi*. Genes Dev 5: 438–446.

Rosell JA, Olson ME, Anfodillo T (2017) Scaling of xylem vessel diameter with plant size: causes, predictions, and outstanding questions. Curr For Rep 3: 46–59.

Russell SH Evert RF (1982) Leaf vasculature in maize (*Zea mays* L.) Bot Soc Am Misc Publ 162: 23.

Sachs T (1981) The control of patterned differentiation of vascular tissues. Adv Bot Res 9: 151–262.

Saks Y, Aloni R (1985) Polar gradients of tracheid number and diameter during primary and secondary xylem development in young seedlings of *Pinus pinea* L. Ann Bot 56: 771–778.

Sanio K (1872) Über die Grösse der Holzzellen bei der gemeinen Kiefer (*Pinus sylvestris*). Jahrb Wiss Bot 8: 401–420.

Sorce C, Giovannelli A, Sebastiani L, Anfodillo T (2013) Hormonal signals involved in the regulation of cambial activity, xylogenesis and vessel patterning in trees. Plant Cell Rep 32: 885–898.

Tomlinson PB, Zimmermann MH (1967) The 'wood' of monocotyledons. IAWA Bull 2: 4–24.

Turing AM (1952) The chemical basis of morphogenesis. Phil Trans R Soc B London 237: 37–72.

*Tyree MT, Zimmermann MH (2002) *Xylem Structure and the Ascent of Sap*, 2nd edn. Springer, Berlin.

Wareing PF (1958) Interaction between indole-acetic acid and gibberellic acid in cambial activity. Nature 181: 1744–1745.

*Williams CB, Anfodillo T, Crivellaro A, Lazzarin M, Dawson TE, Koch GW (2019) Axial variation of xylem conduits in the Earth's tallest trees. Trees 33: 1299–1311.

*Wodzicki TJ, Abe H, Wodzicki AB, Pharis RP, Cohen JD (1987) Investigations on the nature of the auxin-wave in the cambial region of pine stems: validation of IAA as the auxin component by the Avena coleoptile curvature assay and by gas chromatography-mass spectrometry-selected ion monitoring. Plant Physiol 84: 135–143.

Wodzicki TJ, Knegt E, Wodzicki AB, Bruinsma J (1984) Is indolyl-3-acetic acid involved in the wave-like pattern of auxin efflux from *Pinus sylvestris* stem segments? Physiol Plant 44: 122–126.

Wolpert L (1969) Positional information and the spatial pattern of cellular differentiation. J Theoret Biol 25: 1–47.

Wolpert L (2016) Chapter thirty-five - positional information and pattern formation. Curr Top Devel Biol 117: 597–608.

Zajaczkowski S, Wodzicki TJ, Romberger JA (1984) Auxin waves and plant morphogenesis. In: *Encyclopedia of Plant Physiology (New Series). Hormonal Regulation of Development, II, Vol 10*, TK Scott (ed). Springer-Verlag, Berlin, pp. 244–262.

Zhao X (2015) Effects of cambial age and flow path-length on vessel characteristics in birch. J For Res 20: 175–185.

Zhao Y (2018) Essential roles of local auxin biosynthesis in plant development and in adaptation to environmental changes. Ann Rev Plant Biol 69: 417–435.

Zimmermann MH (1983) *Xylem Structure and the Ascent of Sap*. Springer, Berlin, Heidelberg.

Zimmermann MH, Potter D (1982) Vessel-length distribution in branches, stem and roots of *Acer rubrum* L. IAWA Bull 3: 103–109.

Circular Vascular Tissues, Vessel Endings and Tracheids in Organ Junctions

14

The differentiation of vessels in the form of a closed ring (Fig. 14.1) is a result of the movement of auxin in circular routes. Such circular vessels demonstrate that the induced vessel elements in these rings responded to the circular flux, rather than to a gradient or concentration of the auxin signal (Sachs and Cohen 1982). **Circular vessels**, and circular sieve tubes, differentiate in sites where the polar auxin movement occurs in circular patterns, possibly in organ junctions, grafts, above wounds, above interruptions to auxin flow, and where cells are in between opposite polarities. They develop in parenchyma tissues with low polarity like of tissue cultures and galls (see Chaps. 20 and 21, Figs. 20.2C, 21.3A, 21.10A, B) (Aloni et al. 1995, 1998; Dorchin et al. 2002; Ullrich et al. 2019), root junctions (Sachs and Cohen 1982), base of suppressed buds (Aloni and Wolf 1984), and branch junctions (Lev-Yadun and Aloni 1990; Kurczyńska and Hejnowicz 1991). Circular vessels are often produced in the upper side of branch junctions (Fig. 14.2) and their size and frequency increase continuously with age and branch width (Lev-Yadun and Aloni 1990). The circular vessels do not function in water transport and can actually interrupt water flow pathways and reduce water transport in the upper side of branch junctions. Therefore, the long-distance water transport into a branch may occur preferably through the branch sides and it is the lower side. This was evident in the ring-porous oak *Quercus velutina*, in which most (up to 92%) of water flow occurred in the lower region of the branch, where most of the wide earlywood vessel differentiated, but not in the diffuse-porous maple *Acer saccharum* tree. In maple, most of the conductive tissue developed in the upper side of the branches (due to tension wood formation), which was equally or more conductive than the lower side (Aloni et al. 1997).

Larson and Isebrands (1978) found that at the base of the leaf petiole in *Populus deltoides* trees, there is a constriction zone where vessel diameters are narrower. This constriction zone is located at the short cell that will become the abscission zone. This decrease in vessel diameter at the leaf base causes a hydraulic segmentation at the leaf/stem junction. A comparable decrease in vessel element size occurs in the leaf-stem junction of *Arabidopsis thaliana* (see Chap. 2, Fig. 2.27). Likewise,

© Springer Nature Switzerland AG 2021
R. Aloni, *Vascular Differentiation and Plant Hormones*,
https://doi.org/10.1007/978-3-030-53202-4_14

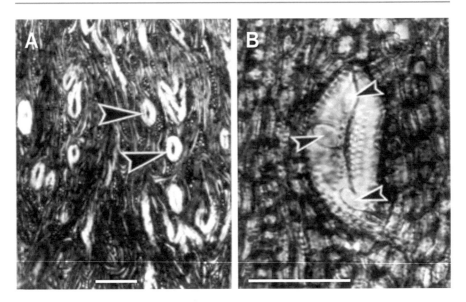

Fig. 14.1 Experimentally induced circular vessels that differentiated above a "butterfly" wound (above an upside down open triangle wound forming a V shape) made in the bark and cambium of the maple *Acer saccharum* (**A**) and the oak *Quercus velutina* (**B**). (**A**) Showing a few circular vessels that differentiated in maple, due to its active cambial activity. Two of the circular vessels are marked by *arrowheads*. (**B**) Close-up view on a circular vessel in the oak, the perforations of the three vessel elements are marked by *arrowheads*. Bars = 100 μm (**B**); 200 μm (**A**). (From Aloni et al. 1997)

but more radical is the hydraulic constriction zone that occurs at the leaf-stem junction in palms, where the vessels of the stem must function and survive for many years, but the leaves are disposable. Palms cannot replace their large metaxylem vessels in the trunk because they do not produce cambium (Zimmermann and Tomlinson 1965; Zimmermann and Sperry 1983). In the palm *Rhapis excels*, narrow protoxylem **tracheids** differentiate between the vessels of the stem and those of the leaf, thus protecting the primary large metaxylem vessels of the trunk from embolism that might originate in drying leaves (Zimmermann and Sperry 1983; Zimmermann 1983; Sperry 1985). The differentiation of tracheids between the leaf vessels and the stem vessels at the leaf-stem junction enable trunk survival, by creating a **hydraulic "safety zone"** that prevents embolism in the vessels of the trunk.

Zimmermann (1978, 1983) found a sharp drop in vessel diameter just above the branch junction in dicotyledonous trees. He also observed that many branch vessels just proximal of the branch attachment to stem were non-conducting, often occluded by gums, likely caused by branches swaying in the wind. In addition, by analyzing vessel-length distribution, in many species he found that there are more **vessel endings** at a branch junction than along clear length of the axes. These junction obstacles to water flow, which form the hydraulic segmentation of branches from the main stem, led Zimmermann (1978, 1983) to propose his *segmentation hypothesis* suggesting that the branch junction is a *"bottleneck"* for water transport from stem

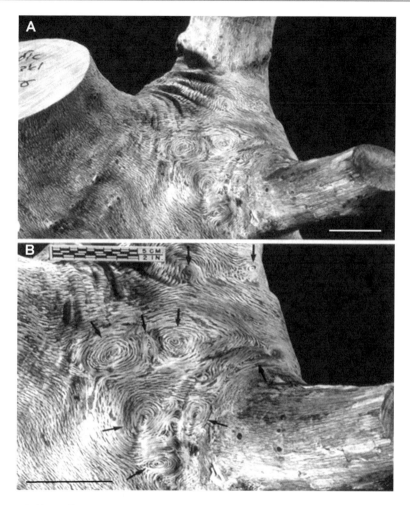

Fig. 14.2 Naturally occurring circular xylem patterns at the junctions of large branch, exposed by bark removal in the oak *Quercus ithaburensis*. (**A**) Late-formed wood patterns showing a few large circular vessels. (**B**) Close-up view of the same circular vessels (*arrows*). Bars = 50 mm. (From Lev-Yadun and Aloni 1990)

to branch, which gives priority of water supply to the leading stem over the branches (Zimmermann 1983).

Zimmermann's (1983) concept of *hydraulic segmentation* has been supported by structural and hydraulic studies on shoots of many plant species (Zimmermann and Sperry 1983; Sperry 1985; Salleo and LoGullo 1986; Luxová 1986; Tyree and Sperry 1988; Tyree 1988; Aloni and Griffith 1991; Aloni et al. 1997; Chen et al. 2018; do Amaral et al. 2019). However, although hydraulic constriction due to vessel narrowing in branch junctions was frequently recorded, hydraulic segmentation is not common in every species (Tyree and Alexander 1993), and trees with dichotomous branching patterns have no junction constriction (Tyree and Ewers 1991).

Auxin induces and regulates the differentiation of the "bottlenecks" in junctions of both the leaf-stem and the branch-stem. Stem nodes and branch junctions are important physiological sites, where hormonal signals are concentrated due to the nodal anatomy structure. In these sites, the cells are short and therefore have more cell membranes per distance along the hormonal pathway. The cell membranes in these short cells slow the flow of the hormonal signals originating in leaves, resulting in high hormonal concentration at the nodes and junctions (see in Chap. 2, Fig. 2.27). The *six-point hypothesis* (Aloni and Zimmermann 1983) (see Chap. 13) explains the mechanism how the size of a conduit is controlled. The second point of the hypothesis proposes that "local structural or physiological obstruction to auxin flow results in a local increase in auxin concentration." At the base of the leaf petiole (Larson and Isebrands 1978) and the base of branch junctions (Zimmermann 1978, 1983), the local increase in auxin concentration, due to their short cells, induces high rate of cell differentiation locally. The rapid cell maturation at the nodes promotes early deposition of secondary cell walls that limits cell expansion, allowing only limited time for cell widening. Therefore, the vessel elements at the base of the petiole and at the branch junction are narrow and form "bottlenecks" which limit and retard water flow into leaves and branches. Thus, the *auxin gradient hypothesis* (orininally named the *six-point hypothesis* Aloni and Zimmermann 1983) explains the mechanism why and how leaf/stem and branch/stem junctions produce narrow vessel elements that form the "bottlenecks," while the **segmentation hypothesis** (Zimmermann 1978, 1983) explains the hydraulic segmentation along the shoot, caused by the narrow conduits at both the leaf/stem and branch/stem junctions, which slow water flow into leaves and branches, thus giving priority of water supply to the main stem and its apical bud.

In addition to the "bottleneck" caused by narrow vessels in branch junction regulated by the local high-auxin concentration mechanism, explained above by the *auxin gradient hypothesis* (Aloni and Zimmermann 1983), the circular auxin flows (Sachs and Cohen 1982) in the upper side of branch junctions induce spiral vascular tissues and circular vessels (Fig. 14.2) (Lev-Yadun and Aloni 1990; Kurczyńska and Hejnowicz 1991), which can further reduce water transport through branch junctions and contribute to the hydraulic segmentation of branches from stem.

The young-leaf biomass, which produces auxin, regulates the development of a branch and its success in competition with other branches. Greater IAA production induces more longitudinal vessels, or tracheids, parallel to the wood grain which gives the branch improved access to the water resources of the tree. Thus IAA export from a branch regulates branch vigor (Kramer and Borkowski 2004; Kramer 2006).

Pruning lateral branches of young trees improves the growth of their main stem and their wood quality. Pruning branch/stem junctions stops their circular pattern formation, resulting in the production of uniform trunk wood without knots in the post-pruning annual rings and optimization of mechanical wood properties at the original junction sites, which is important for the wood industry.

Cereals adapted to stress of drought and extremely cold weather produce a *hydraulic "safety zone"* in their root-shoot junctions, which protect the vessels of their roots from embolism originating in the shoot (Luxová 1986; Aloni and Griffith 1991). A study of root-shoot junctions in six cereal species (Aloni and Griffith 1991) revealed that there are two types of hydraulic architectures in cereal roots: (**i**) a completely unsafe system typical to mesic conditions, where the vessels of the roots are continuous with the vessels in the shoots (Fig. 14.3A, B), *versus* (**ii**) a very safe root vessels adapted to stress, as in winter rye, in which the vessels of the roots are separated from those of the shoots by non-perforated tracheary elements (Fig. 14.3C). The xylem anatomy of the seminal roots is generally correlated with the species-specific overall root morphology. Rye, wheat, and barley, which develop four to six seminal roots, show a high degree of vascular segmentation resulting in the formation of safe root vessels, while maize, sorghum and oats, which typically develop a primary seminal root, contain unsafe root vessels that are continuous through the mesocotyl and through the first node (Aloni and Griffith 1991). These finding show that Zimmermann's (1983) *segmentation concept*, which was developed following studies on woody perennial trees, can be broadened to include herbaceous plants, demonstrating the formation of vessel endings and tracheids

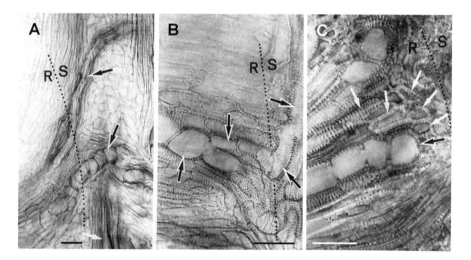

Fig. 14.3 Photomicrographs of the root-shoot junction, showing longitudinal thick sections of the root-shoot junction. The tissues are all unstained and cleared in lactic acid. All photographs are longitudinal views of the junction with the border between the root (R) and shoot (S) tissues delineated by a dotted line. (**A**) Adventitious root junction in corngrass, which is a corngrass mutation of maize (*Zea Mays, Cg* mutant) showing the entry of two metaxylem vessels from the root (*black arrows*) into the shoot. The lower part of a longitudinal vessel of the shoot is marked by a *white arrow*. (**B**) Close-up view of a continuous metaxylem vessel (*arrows*) in the junction between an adventitious root and the shoot in corngrass. (**C**) Root-shoot junction of a seminal root in winter rye (*Secale cereal* cv. Musketeer) showing the upper portion of the central metaxylem vessel, which ends at an imperforate wall (*black arrow*), and the small tracheids with simple pitting (*white arrows*), which connect the vessel of the root to vessels in the shoot. Bars = 50 μm (**C**), 100 μm (**A**, **B**). (From Aloni and Griffith 1991)

(Fig. 14.3C) in junction between the shoot and the seminal roots of spring barley (Luxová 1986; Aloni and Griffith 1991), winter rye, and wheat (Aloni and Griffith 1991) which are adapted to stress. This primary xylem structure in cereals adapted to stress indicates that a "safety zone" exists at their root/shoot junctions, through which gaseous emboli and fungal spores cannot pass. In conclusion, these findings show that Zimmermann's (1983) concept of hydraulic segmentation of lateral organs from the main stem as well as the "safety zone" of leaf/stem junction in palms (Zimmermann and Tomlinson 1965; Zimmermann and Sperry 1983; Zimmermann 1983; Sperry 1985) can also be applied to the hydraulic architecture and "safety zone" occurring between the root/stem junctions in cereals adapted to stress.

Summary

- Throughout the plant body, the junctions of organs are important physiologically sites, especially under stress conditions. Junctions are sites of high hormonal concentrations where hormone flows merge. At the junctions, auxin may move in circular routes that induce vessels, tracheids, sieve tubes, and fibers in circular patterns. Circular vessels do not function in water transport and may interrupt longitudinal water flow.
- The local increase in auxin concentration at the junction induces rapid vascular cell differentiation resulting in narrow conduits, vessel endings, and tracheids between vessels, which form hydraulic constriction zones at the stem/organ junctions, creating hydraulic "safety zones" that prevent embolism in the primary vessels of palm stems and in cereals roots.
- The junctions are "bottleneck" sites which can restrict water flow to branches forming hydraulic segmentation of the plant body, giving priority in water and nutrient supply to the shoot apical bud of the leader, on the expense of lateral branches. Therefore, under serious water shortages, branches can dry and die while the leader survives.

References and Recommended Readings[1]

*Aloni R, Alexander JD, Tyree MT (1997) Natural and experimentally altered hydraulic architecture of branch junctions in *Acer saccharum* Marsh. and *Quercus velutina* Lam. trees. Trees 11: 255–264.

*Aloni R, Griffith M (1991) Functional xylem anatomy in root-shoot junctions of six cereal species. Planta 184: 123–129.

Aloni R, Pradel KS, Ullrich CI (1995) The three-dimensional structure of vascular tissues in *Agrobacterium tumefaciens*-induced crown galls and in the host stems of *Ricinus communis* L. Planta 196: 597–605.

---

[1] Papers of particular interest for suggested reading have been highlighted (with *)

Aloni R, Wolf A (1984) Suppressed buds embedded in the bark across the bole and the occurrence of their circular vessels in *Ficus religiosa*. Am J Bot 71: 1060–1066.

Aloni R, Wolf A, Feigenbaum P, Avni A, Klee HJ (1998) The *Never ripe* mutant provides evidence that tumor-induced ethylene controls the morphogenesis of *Agrobacterium tumefaciens*-induced crown galls on tomato stems. Plant Physiol 117: 841–847.

Aloni R, Zimmermann MH (1983) The control of vessel size and density along the plant axis - a new hypothesis. Differentiation 24: 203–208.

Chen T, Feng B, Fu W, Zhang C, Tao L, Fu G (2018) Nodes protect against drought stress in rice (*Oryza sativa*) by mediating hydraulic conductance. Environ Exp Bot 155: 411–419.

do Amaral FM, Joffily A, Vieira RC (2019) Three-dimensional structure of stem vascular connections with leaf and adventitious root in *Asplundia brachypus* (Drude) Harling (Cyclanthaceae). Flora 258: 151442.

Dorchin N, Freidberg, A, Aloni R (2002) Morphogenesis of stem gall tissues induced by larvae of two cecidomyiid species (Diptera: Cecidomyiidae) on *Suaeda monoica* (Chenopodiaceae). Can J Bot 80: 1141–1150.

Kramer EM (2006) Wood grain pattern formation: a brief review. J Plant Growth Regul 25: 290–301.

Kramer EM, Borkowski MH (2004) Wood grain patterns at branch junctions: modeling and implication. Trees 18: 493–500.

Kurczyńska EU, Hejnowicz Z (1991) Differentiation of circular vessels in isolated segments of *Fraxinus excelsior*. Physiol Plant 83: 275–280.

Larson PR, Isebrands JG (1978) Functional significance of the nodal constricted zone in *Populus deltoides*. Can J Bot 56: 801–804.

*Lev-Yadun S, Aloni R (1990) Vascular differentiation in branch junction: circular patterns and functional significance. Trees 4: 49–54.

Luxová M (1986) The hydraulic safety zone at the base of barley roots. Planta 169: 465–470.

*Sachs T, Cohen D (1982) Circular vessels and the control of vascular differentiation in plants. Differentiation 21: 22–26.

Salleo S, LoGullo MA (1986) Xylem cavitation in nodes and internodes of whole *Chorisia inignis* HB et K plants subjected to water stress: relations between xylem conduit size and cavitation. Ann Bot 58: 431–441.

Sperry JS (1985) Xylem embolism in the palm *Rhapis excelsa*. IAWA Bull ns 6: 283–292.

Tyree, MT (1988) A dynamic model for water flow in a single tree. Tree Physiol 4: 195–217.

Tyree MT, Alexander JD (1993) Hydraulic conductivity of branch junctions in three temperate tree species. Trees 7: 156–159.

*Tyree MT, Ewers FW (1991) The hydraulic architecture of trees and other woody plants. New Phytol 119: 345–360.

Tyree MT, Sperry JS (1988) Do woody plants operate near the point of catastrophic xylem dysfunction caused by dynamic water stress? Answers from a model. Plant Physiol 88: 574–580.

Ullrich CI, Aloni R, Saeed MEM, Ullrich W, Efferth T (2019) Comparison between tumors in plants and human beings: Mechanisms of tumor development and therapy with secondary plant metabolites. Phytomedicine 64: 153081.

Zimmermann MH (1978) Hydraulic architecture of some diffuseporous trees. Can J Bot 56: 2286–2295.

*Zimmermann MH (1983) *Xylem Sructure and the Ascent of Sap*. Springer-Verlag, Berlin.

Zimmermann MH, Sperry JS (1983) Anatomy of the palm *Rhapis excelsa*. IX. Xylem structure of the leaf insertion. J Arnold Arbor 64: 599–609.

Zimmermann MH, Tomlinson PB (1965) Anatomy of the palm *Rhapis excelsa*. I. Mature vegetative axis. J Arnold Arbor 46: 160–180.

Ray Differentiation: The Radial Pathways 15

The radial component of the secondary vascular tissues is the vascular ray system. The rays serve as radial transport pathways between the xylem and the phloem and vice versa. Usually, the appearance of rays indicates the transition from procambium to cambium (Larson 1994) and as the plant's stem and root expand and become wider by cambial activity, the ray system is enlarged. Parallel to the well-documented gradual increase in tracheids, vessel, and sieve-tube width (see Chap. 13), there is a gradual increase in the size of the vascular rays which become larger with increasing distance from the young leaves or from the pith (Fig. 15.1A, B). In conifers, the ray initials occupy about 10% of the cambium surface, whereas in woody angiosperms, the ratios are more variable ranging from 0% rays (in rayless wood) to about 25% ray volume. A substantial increase in ray dimensions occurs in response to wounding. This wound effect on ray size can be experimentally induced by ethylene stimulation (Figs. 15.1C and 15.2B, C) (Lev-Yadun and Aloni 1995; Aloni et al. 2000).

Vascular rays are an important component of the secondary body. The rays are aeration pathways that enable radial movement of gases, e.g., the centrifugal flow of **ethylene**, synthesized in the sapwood (the active water transporting secondary xylem) of trees (Eklund 1990; Ingemarsson et al. 1991) outside toward the bark. Rays are essential for regeneration after injury, as their parenchyma cells start to proliferate and form callus on the cut surface and cover the exposed wound. As discussed earlier, in hardwood trees, parenchyma cells that also originate from the rays penetrate through the pits into damaged air-field vessels to form tyloses (De Micco et al. 2016) which plug these non-functional vessels and prevent movement of fungi and bacteria along them (see Chap. 2, Fig. 2.14B; and Chap. 19, Fig. 19.7B). Toward winter, the rays become loaded with starch and other reserve materials (Fig. 15.2) that are kept in the vascular rays during winter dormancy (see in Chap. 2, Fig. 2.31) and are released in spring.

The continuous radial pattern of the ray system indicates that a continuous inducing signal regulates their development. Bünning (1952, 1965) proposed that rays are distributed in regular patterns in the secondary body of plants due to inhibitory

© Springer Nature Switzerland AG 2021
R. Aloni, *Vascular Differentiation and Plant Hormones*,
https://doi.org/10.1007/978-3-030-53202-4_15

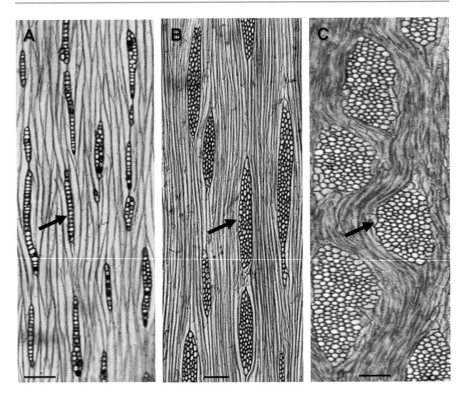

Fig. 15.1 Photomicrographs of longitudinal tangential sections in the secondary xylem of *Melia azedarach*, showing the naturally occurring increase in the size of vascular rays (*arrows*) with increasing distance from the pith (**A**, **B**) and a substantial increase in ray dimensions in response to wounding (**C**). (**A**) Narrow and long vascular rays differentiated near the pith. (**B**) Wider and larger rays developed away from pith. (**C**) Much larger rays differentiated following wounding, likely in response to wound-induced ethylene. Bars = 100 μm

influence of each ray on the neighboring rays. However, Bünning's hypothesis that rays act as meristemoids that have mutual inhibition on each other contradicts a basic feature of the ray system, namely, that rays tend to fuse (Lev-Yadun and Aloni 1995) (e.g., Fig. 15.2C). Detailed histological investigations of conifers wood revealed that fusiform initials that were in contact with rays survived in the cambium for longer periods than those that were away from rays, which tend to become shorter (Bannan 1951, 1953). These observations led Ziegler (1964) to suggest that the rays, which are radial transporting pathways, are induced and regulated by a radial moving signal. Ziegler's (1964) idea that the rays are induced by a radial signal flow is supported by the development of radial "rays" inside a vascular ray (Lev-Yadun and Aloni 1991) (Fig. 15.3). Lev-Yadun and Aloni (1995) suggested that the mechanism of ray initiation is regulated by the gaseous hormone **ethylene**. Existing rays drain xylem-synthesized ethylene toward the phloem. When fusiform cambial initials are not associated with rays, they are exposed to higher levels of ethylene among the rays, become shorter, divide, and become new ray initials.

Fig. 15.2 Photomicrographs of cross sections in the secondary xylem of *Hibiscus cannabinus* young stems, in an intact control (**A**) and following the effect of ethylene, originating from 1% ethrel (2-chloroethylphosphonic acid) in lanolin (**B, C**). The cytoplasm in the ray cells (marked by *arrows*) is dark due to accumulation of starch and other reserve materials (toward winter). All photographs are at the same orientation (cambium in the upper side), from the same experiment and internode age. Showing (**A**) Naturally occurring narrow rays in an intact stem. (**B**) Substantial increase in ray width induced by ethylene (released from 1% ethrel). (**C**) Fusion of rays promoted by the ethylene stimulation. Bars = 50 μm. (From Aloni et al. 2000)

Application of the ethylene releasing agent, ethrel (2-chloroethylphosphonic acid), increases ray size and ray-cell number (Savidge 1988; Aloni et al. 2000) (Fig. 15.2B, C). The ethylene is naturally synthesized in the xylem (Eklund 1990; Ingemarsson et al. 1991), specifically in maturing tracheary elements (Pesquet and Tuominen 2011). From the differentiating vessels, the ethylene moves centrifugally through the cambium (from the maturing tracheary elements towards the phloem) and regulates the differentiation of the ray system. Therefore, it was suggested that ethylene flow is the major radial signal that promotes the initiation of rays and their regulation (size and spacing) in the cambium (Lev-Yadun and Aloni 1995; Aloni et al. 2000).

Lev-Yadun and Aloni (1991) suggested that ethylene influences ray initiation and ray size by its known negative effect on the polar auxin transport (Mattoo and Aharoni 1988 and references therein). The disturbance to the longitudinal polar auxin flow enables the radial signal flow of ethylene to effectively induce the initiation of new rays and promote the enlargement of existing rays.

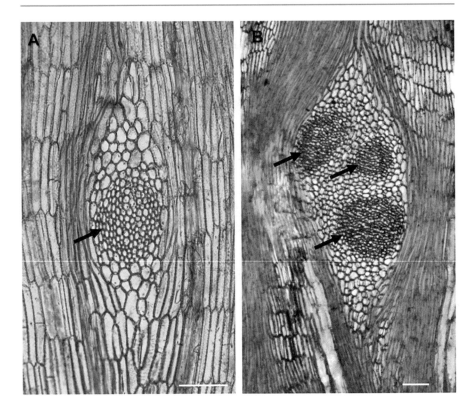

Fig. 15.3 Photomicrographs of longitudinal tangential sections in the secondary xylem of *Suaeda monoica* showing stages in the development of a polycentric vascular ray, with one (**A**) and three (**B**) radially elongated "centers" that form a kind of radial "rays" inside a ray. Each "center" of small parenchyma cells is marked with an *arrow*. (**A**) Section located away from the cambium showing an early stage of ray development with one "center." (**B**) More developed and larger polycentric ray in a section located close to cambium. The radially continuous narrow radial "rays" inside the large polycentric ray demonstrate that the three "centers" are the result of individual radial preferable pathways of a radial signal flow occurring inside the large polycentric vascular ray. Bars = 75 μm. (From Lev-Yadun and Aloni 1991)

Usually, the vascular rays are built of parenchyma cells forming longitudinal ray shapes (Fig. 15.1A, B). Their elongated structure is likely shaped by axial moving signals (mainly auxin) (Lev-Yadun and Aloni 1995). However, radially oriented tracheids, vessels, and sieve tubes can also differentiate inside the rays (Esau 1965; Fahn 1979, 1990; Lev-Yadun and Aloni 1991), indicating that the hormonal signals that induce these radially oriented vascular elements, namely, auxin and gibberellin, can flow radially through the rays. The reason why the rays are usually built of parenchyma cells only is likely due to the flow of ethylene through the rays, which antagonizes or reduces the effects of the other flowing hormonal signals. However, ethylene is positively involved in the induction of radial resin ducts (see Chap. 17).

The increase in tracheids and vessel size from young leaves to roots was explained by the decreasing auxin concentration gradient in this direction (Aloni and

Zimmermann 1983). The increase in ray size with increasing distance from young leaves, or from the pith, is likely due to the changes in the relations between the axial and radial signal flows. Due to the decreasing auxin concentration gradient from leaves to roots, the effect of polar auxin flow on rays is gradually reduced with increasing distance from young leaves. Therefore, the effect of the centrifugal ethylene flow, originating in the sapwood, gradually becomes more dominant resulting in the gradual increase in ray size with increasing distance from the young leaves or the pith (Lev-Yadun and Aloni 1995).

Summary

- Vascular rays serve as radial transport pathways between the xylem and the phloem and vice versa. They are induced by a radial moving signal, which also induces the development of radial "rays" inside a vascular ray.
- The gaseous hormone ethylene, which is naturally synthesized in the living wood where sap flows (sapwood), specifically in maturing tracheary elements, is the primary hormonal signal that induces the rays, and regulates their initiation, enlargement, and pattern.
- Ethylene influences ray initiation and ray size by its known negative effect on the polar auxin transport. The disturbance to the longitudinal polar auxin flow enables the centrifugal signal flow of ethylene through the cambium to effectively induce the initiation of new rays and promote the enlargement of existing rays.
- The typical increase in ray size with increasing distance from young leaves, or from the pith, is regulated by gradual changes in the relations between the longitudinal polar auxin flow and the radial ethylene flow. The downward gradient of decreasing auxin concentration from leaves to roots enables the centrifugal ethylene flow, originating in the sapwood, to progressively become more dominant resulting in the gradual increase in ray size with increasing distance from the young leaves, or the pith.
- Wounding that substantially boost ethylene production significantly increases the dimensions of rays. This increase in ray size by wounding can be experimentally induced in intact stems by applying ethylene or ethrel that release ethylene.

References and Recommended Readings[1]

*Aloni R, Feigenbaum P, Kalev N, Rozovsky S (2000) Hormonal control of vascular differentiation in plants: the physiological basis of cambium ontogeny and xylem evolution. In: *Cell and Molecular Biology of Wood Formation*, RA Savidge, JR Barnett, R Napier (eds). BIOS Scientific Publishers, Oxford, pp. 223–236.

Aloni R, Zimmermann MH (1983) The control of vessel size and density along the plant axis - a new hypothesis. Differentiation 24: 203–208.

---

[1] Papers of particular interest for suggested reading have been highlighted (with *)

Bannan MW (1951) The annual cycle of size changes in the fusiform cambial cells of *Chamaecyparis* and *Thuja*. Can J Bot 29: 421–437.

Bannan MW (1953) Further observations on the reduction of fusiform cambial cells in *Thuja occidentalis*. Can J Bot 31: 63–74.

Bünning E (1952) Morphogenesis in plants. Surv Biol Progr 2: 105–140.

Bünning E (1965) Die Entstehung von Mustern in der Entwicklung von Pflanzen. Handb Pflanzenphysiol 15: 383–408.

De Micco, Balzano A, Wheeler E, Baas P (2016) Tyloses and gums: a review of structure, function and occurrence of vessel occlusions. IAWA J 37: 186–205.

Eklund L (1990) Endogenous levels of oxygen, carbon dioxide and ethylene in stems of Norway spruce trees during one growing season. Trees 4: 150–154.

Esau, K. (1965) *Plant Anatomy*. 2nd edn, John Wiley, New York.

Fahn A (1979) *Secretory Tissues in Plants*. Academic Press, London.

Fahn A (1990) *Plant Anatomy*. 4th edn, Pergamon Press, Oxford.

*Ingemarsson BSM, Lundqvist E, Eliasson L (1991) Seasonal variation in ethylene concentration in wood of *Pinus sylvestris* L. Tree Physiol 8: 273–279.

Larson PR (1994) *The vascular cambium: Development and Structure*, Springer Verlag, Berlin, Heidelberg.

*Lev-Yadun S, Aloni R (1991) Polycentric vascular rays in *Suaeda monoica* and the control of ray initiation and spacing. Trees 5: 22–29.

*Lev-Yadun S, Aloni R (1995) Differentiation of the ray system in woody plants. Bot Rev 61: 45–84.

Mattoo AK, Aharoni N (1988) Ethylene and plant senescence. In: *Senescence and Aging in Plants*, LD Nooden, AC Leopold (eds). Academic Press, San Diago, pp. 241–280.

*Pesquet E, Tuominen H (2011) Ethylene stimulates tracheary element differentiation in *Zinnia elegans* cell cultures. New Phytol 190: 138–149.

Savidge RA (1988) Auxin and ethylene regulation of diameter growth in trees. Tree Physiology 4: 401–414.

Ziegler H (1964) Storage, mobilization and distribution of reserve material in trees. In: *The Formation of Wood in Forest Trees*, MH Zimmermann (ed). Academic Press, New York, pp. 303–320.

Environmental Adaptation of Vascular Tissues

16

Vascular plants grow in different environments, ranging from deserts to rain forests and from arctic regions to the tropics. Comparative anatomical studies (e.g., Baas et al. 1983; Baas and Carlquist 1985; Wheeler et al. 2007; De Micco et al. 2008; Wheeler and Baas 2019) reveal similarities in structure of the vascular system in plants grown in extreme habitats versus ones grown in favorable environments. Desert (Carlquist and Hoekman 1985; Fahn et al. 1986), arctic, and alpine shrubs (Carlquist 1975) show high density of very narrow vessels. Such vascular systems are typical for extreme habitats and are deemed adaptive safety mechanisms against drought and freezing (Baas et al. 2004; Lucas et al. 2013). Conversely, forest trees and lianas, which characterize the tropics and rain forests, have low density vessels of very wide diameter (Carlquist 1975; Zimmermann 1983; Ewers 1985; Tyree and Sperry 1989) at the base of their trunk, which affords maximal efficiency of water conduction (Ellmore and Ewers 1985; Tyree and Ewers 1991; Tyree and Zimmermann 2002; Williams et al. 2019) and is considered to be an adaptation to mesic conditions. Environmental regulation of vascular differentiation at the molecular level following drought, extreme temperature, and salt and mechanical stresses should also be considered when studying the role of the environment in vascular adaptation (Lucas et al. 2013; Agustí and Blázquez 2020).

The influence of various environmental factors on the anatomy of the xylem was extensively reviewed by Creber and Chaloner (1984), but the mechanisms whereby the vascular system of plants respond or adapt to their environment remained unknown. Therefore, in order to explain the adaptation of plants' vascular systems to the environment, Aloni (1987) suggested that the environment controls the plant's vascular system through its control of plant's development, height, and shape. To explain how the ecological conditions control the size and frequency of vessels and fibers in plants, Aloni (1987) proposed the following tripartite *vascular adaptation hypothesis*:

1. Reduction of a growth factor in the plant's immediate environment limits the final size of the plant resulting in small and suppressed shoot, whereas favorable

© Springer Nature Switzerland AG 2021
R. Aloni, *Vascular Differentiation and Plant Hormones*,
https://doi.org/10.1007/978-3-030-53202-4_16

conditions that do not restrict plant development allow the plant to attain its appropriate shape and maximal height.

2. The duration of the growth period affects the rate of plant development. In extreme and limiting habitats the active growth period is relatively short and results in small plants, whereas stable and moderately comfortable conditions like those found in the humid tropics allow growth activity throughout the year, thereby enabling more growth and consequently large and well developed plants.

3. The height of the plant and the degree of its branching determine gradients of auxin along the plant's axis. An increase in plant's height and a diminution of its branching enhances the length of the decreasing IAA gradients from the young auxin-producing leaves to the lower parts of the stem. In small shrubs, which are typical to extremely cold, dry, and saline habitats, as well as in grazing areas, or locations with insufficient soil for the roots, the distances from the young leaves to the roots are very short and no substantial decreasing gradient of auxin can be formed. Therefore, the concentrations of IAA along these small plants are relatively high and result in the differentiation of numerous very small vessels in the greatest densities, as predicted by the *auxin-gradient hypothesis* (Aloni and Zimmermann 1983) accompanied by the production of small fibers with thick secondary walls, as stipulated by Aloni (1979). Conversely, in the large trees and in long lianas, the very great distances from the young auxin-producing leaves to the roots enable a substantial decrease in auxin concentrations in the lower parts of the stem and in the roots, where it leads to slow conduit differentiation that allows more cell expansion resulting in very wide vessels at low density, along with large fibers.

To experimentally analyze the hypothesis, *Hibiscus cannabinus* plants were exposed to insufficient available soil by decreasing their pot size for limiting root space and consequently root development. As expected, reduced pot size decreased shoot development and limited plant height (Fig. 16.1A) which shortened the decreasing auxin gradient. The anatomical results (Fig. 16.1B–D) demonstrate that decreasing root space resulted in decreasing secondary vessel diameter. The unstressed *Hibiscus* plants, in the largest pot, produced the widest secondary vessels and largest secondary xylem fibers (Fig. 16.1B); the intermediate pot size resulted in intermediate vessel diameter (Fig. 16.1C); and the plants in the smallest pots, which were the shortest, produced the narrowest vessels and narrowest fibers with thickest cell walls (Fig. 16.1D). These predicted results support the *vascular adaptation hypothesis* (Aloni 1987).

As a dicot plant, the vessels in *Hibiscus* were secondary vessels that originated from cambium. This raises the basic question, whether (**i**) the *vascular adaptation hypothesis* is relevant only to conifers and dicotyledons that continuously produce secondary tracheids and secondary vessels from their cambium, or (**ii**) the hypothesis is applicable also to primary vessels in monocotyledons? To find experimental answer to this question, corn (*Zea mays*) seeds were germinated and grown in large *versus* small pots, creating similar limiting conditions as described for *Hibiscus* (Fig. 16.1). The results (Fig. 16.2) demonstrate that the corn stems of plants grown in the small pots did respond to the environmental stress. While the basal internodes

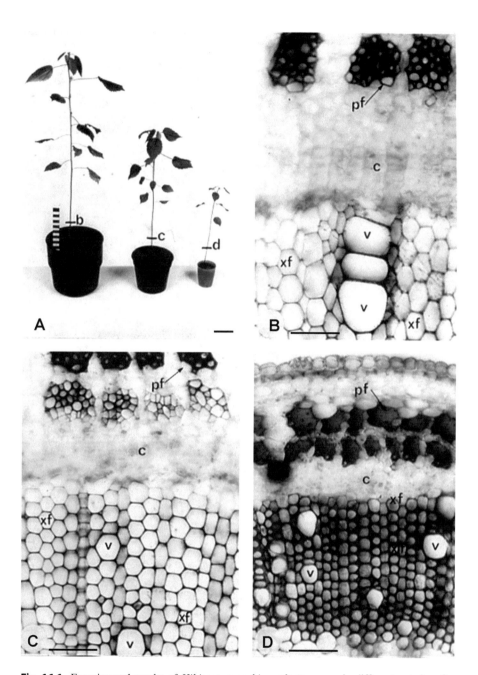

Fig. 16.1 Experimental results of *Hibiscus cannabinus* plants grown in different pot sizes for limiting root development (**A**), and cross sections (**B–D**) done in their stems (from (**B**) to (**D**) in Fig. 16.1A) at the same distance from soil surface. (**A**) Decrease in pot size limited plant development and maximal stem height; the smallest pot enable limited plant development and resulted in the shortest plant. (**B**) The secondary vessels (v) and secondary xylem fibers (xf) were the widest in the tallest stem. (**C**) Intermediate stem length produced intermediate diameters in the secondary vessel and xylem fiber. (**D**) The shortest stem was characterized by the narrowest vessels and fibers. The latter produce thick secondary walls, as is known to be induced in fibers by high-auxin concentration (Aloni 1979). In (**B**)–(**D**): c, cambium; pf, phloem fibers; xf, xylem fibers, v, secondary vessel. Bars = 100 μm (**B–D**), 50 mm (**A**) (From Aloni 1988)

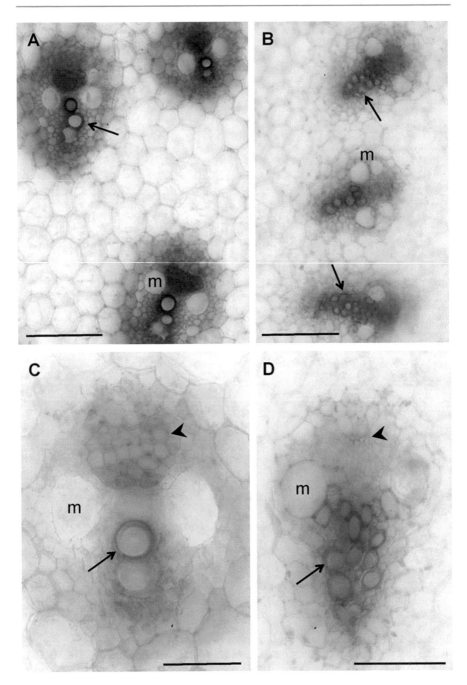

Fig. 16.2 Cross sections from the middle of the basal internode of corn (*Zea mays*) grown under: (**A**, **C**) favorable conditions in a large pot, *versus* (**B**, **D**) stress conditions, due to limited root space in a small pot, showing the effect of stress on the differentiation of primary xylem vessels and phloem sieve tubes (which might be influenced also by abscisic acid (ABA) from the root meri

of the unstressed *Zea* plants (grown in large pots) produced the typical few (2–3) normal size diameter protoxylem vessels per bundle (Fig. 16.2A, C), the identical internodes of the stressed *Zea* plants (grown in small pots) produced many (up to about 15) very narrow protoxylem vessels (Fig. 16.2B, D) which can have an adaptive safety advantage for water transport under drought and freezing conditions. These striking results demonstrate that in short plants, there is no enough stem length to establish a decreasing gradient of auxin; consequently, the concentration of auxin at the stem base is high (because it is close to the auxin-producing leaves), which therefore induces many vessels that differentiate rapidly and consequently have only a short widening phase (see Chap. 13) resulting in narrow vessels (Fig. 16.2B, D). The high-auxin concentration at the basal internodes of the stressed corn also caused the differentiation of very narrow primary sieve elements (*arrowhead* in Fig. 16.2D) compared with the normal ones in the plants grown in the large pot (*arrowhead* in Fig. 16.2C).

Another important feature of vascular adaptation in plants is their vascular segmentation occurring between leaf/stem, branch/stem, and stem/root junction, which is explained by the *segmentation hypothesis* proposed by Zimmermann (1983) (See Chap. 14).

Olson and Rosell (2013) measurements confirmed the *vascular adaptation hypothesis* (Aloni 1987) on a large scale (of 237 species of over 40 angiosperm orders across a wide range of habits and habitats), demonstrating that vessel diameter is proportional to both stem length and stem diameter. Their results show that plant size is related to climate, leading to the vessel-climate relationship; vessels are narrower in drier communities because dry land plants are on average smaller than the wider vessels found in large forest trees and long lianas grown under favorable conditions (Olson and Rosell 2013). However, their statistical correlation between vessel diameter and plant size does not explain the biological causative factors regulating this relationship. Therefore, the picture presented by Olson and Rosell (2013) is incomplete, since they ignore the well-known adapting auxin gradient mechanism causing the correlation between plant size and vessel diameter that is shaped by the environment, which is already explained and predicted by the *vascular adaption hypothesis* (Aloni 1987). Likewise, the review of Rosell et al. (2017) clarifies that the main driver of vessel diameter is plant size, specifically the length of the stem, which are influenced by temperature and water availability that determine maximum plant height. Their measurements and evidence strongly confirm the *vascular*

Fig. 16.2 (continued) stems, see Sect. 3.5). (**A**) Typical primary vascular bundles with two or three protoxylem vessels (*arrow*), and expanding metaxylem vessels (m) that their living cytoplasm was stained light-blue by lacmoid. (**B**) The stressed plant developed vascular bundles with many very narrow protoxylem vessels (*arrows*) and two or four metaxylem (m) vessels, (**C**) Magnified view of a typical vascular bundle with two functional protoxylem vessels (*arrow*), two expanding metaxylem vessels (before secondary wall deposition), and normal size sieve elements in the phloem (*arrowhead*). (**D**) Magnified bundle affected by stress with many very narrow protoxylem vessels (*arrow*), expanding metaxylem (m) vessels and very narrow sieve tubes (*arrowhead*). Bars = 100 μm (**C, D**), 200 μm (**A, B**). (By R. Aloni and J. Mattsson, unpublished)

adaptation hypothesis (Aloni 1987) although they ignore the biological role of the auxin gradient mechanism in controlling vessel diameter along the stem (see also Chap. 13).

Similarly, the environment regulates the adaptation of *Eucalyptus* trees in native forests. Growth measurements of *Eucalyptus* tree across temperate and sub- tropical mesic Australia (along a gradient in mean annual precipitation from 558 to 2105 mm and mean annual temperature of 6.4–22.4 °C) show that the climate determines the potential tree's trunk diameter (measured at breast height) which is positively correlated with tree height. The most productive forests in the study region are located in cool, moist areas of southeastern Australia and are among those with the highest biomass on earth, where the world's tallest angiosperm, *Eucalyptus regnans*, grows (attains heights over 100 m). In this cool moist forest, the dense forest causes intense competition for light among mature trees on the more productive sites, leading to intense competition for light, which causes self-thinning and drives rapid height growth. In the drier forests, tree size and stand density appear to be constrained by water availability. In these dry sclerophyll forests, adult eucalypts suppress juvenile eucalypts through competition for water (Prior and Bowman 2014). Also, in eucalypts, vessel diameter gradually increases and vessel density decreases from leaves to roots (Leitch 2001), indicating that the tallest trees likely have the widest vessels at their base, allowing efficient water supply to these largest trees. However, in the tallest *E. regnans* trees, vessel diameter approach their maximum width, at about 60 m from tree tops and do not continue additional vessel widening at the base of these giant trees (Williams et al. 2019), possibly due to secondary wall strength limitation. Furthermore, dominant trees are known to have more intense cambial activity than suppressed trees, promoting the largest trees to produce more wood for improved water supply and crown support (Rathgeber et al. 2011).

Summary

- Plants grown in extreme suppressing habitats are characterized by very narrow vessels in high density. Such vascular systems are considered adaptive safety mechanisms against drought and freezing.
- Conversely, forest trees and lianas, which characterize the tropics and rain forests, have low density vessels of very wide diameter at the base of their stems, which affords maximal efficiency of water conduction.
- The *vascular adaptation hypothesis* proposes that the environment controls the plant's vascular system through its control of plant's development, height, and shape. Limiting conditions suppress plant growth and shorten the active growth period, which restrict plant development resulting in small plants. Conversely, favorable conditions allow growth activity throughout the year, enabling more growth and consequently well-developed plants and maximal height.
- The height of the plant and the degree of its branching determine gradients of auxin along the plant's axis. In small shrubs, which are typical to extreme stressful environmental conditions, the distances from the young leaves to the roots are very short and no substantial decreasing gradient of auxin can be formed.

Therefore, the concentrations of IAA along these small plants are relatively high and result in the differentiation of numerous very narrow vessels in the greatest densities. Conversely, in the large trees and in long lianas, the very great distances from the young auxin-producing leaves to the roots enable a substantial decrease in auxin concentrations in their lower parts, which leads to slow conduit differentiation that allows more cell expansion, resulting in very wide vessels in low density at their base.

- Experimentally stressed small *Hibiscus cannabinus* plants produced many narrow secondary vessels as predicted by the hypothesis. Likewise, suppressed short *Zea mays* plants produced many narrow protoxylem vessels instead of the few typical wider vessels in unstressed plants.
- The *vascular adaptation hypothesis* was confirmed experimentally and by analyzing the correlation between plant size and vessel diameter on a large scale of collected species from a wide range of growth conditions.

References and Recommended Readings[1]

*Agustí J, Blázquez MA (2020) Plant vascular development: mechanisms and environmental regulation. Cell Mol Life Sci 77: 3711–3728.

Aloni R (1979) Role of auxin and gibberellin in differentiation of primary phloem fibers. Plant Physiol 63: 609–614.

*Aloni R (1987) Differentiation of vascular tissues. Annu Rev Plant Physiol 38: 179–204.

Aloni R (1988) Vascular differentiation within the plant. In: *Vascular Differentiation and Plant Growth Regulators*, LW Roberts, PB Gahan, R Aloni, Springer-Verlag, Berlin, Heidelberg, pp. 39–62.

*Aloni R, Zimmermann MH (1983) The control of vessel size and density along the plant axis - a new hypothesis. Differentiation 24: 203–208.

Baas P, Carlquist S (1985) A comparison of the ecological wood anatomy of the floras of southern California and Israel. IAWA Bull ns 6: 349–353.

Baas P, Ewers FW, Davis SD, Wheeler EA (2004) Evolution of xylem physiology. In: *The Evolution of Plant Physiology*. AE Hemsley, I Poole (eds) Linnean Soc Sympm Series, no. 21, Academic Press, Amsterdam, pp. 273–295.

Baas P, Werker E, Fahn A (1983) Some ecological trends in vessel characters. IAWA Bull ns 4: 141–159.

Carlquist S, (1975) *Ecological Strategies of Xylem Evolution*. Univ Calif Press, Berkeley.

Carlquist S, Hoekman DA (1985) Ecological wood anatomy of the woody southern Californian flora. IAWA Bull 6: 319–346.

Creber GT, Chaloner WG (1984) Influence of environmental factors on the wood structure of living and fossil trees. Bot Rev 50: 357–448.

De Micco V, Aronne G, Baas P (2008) Wood anatomy and hydraulic architecture of stems and twigs of some Mediterranean trees and shrubs along a mesic-xeric gradient. Trees 22: 643–655.

Ellmore GS, Ewers FW (1985) Hydraulic conductivity in trunk xylem of elm, *Ulmus americana*. IAWA Bull ns 6: 303–307.

Ewers FW (1985) Xylem structure and water conduction in conifer trees, dicot trees and liana. IAWA Bull ns 6: 309–317.

[1] Papers of particular interest for suggested reading have been highlighted (with *)

Fahn A, Werker E, Baas P (1986) *Wood Anatomy and Identification of Trees and Shrubs from Israel and Adjacent Regions*. Israel Acad Sci, Jerusalem.

Leitch MA (2001) Vessel-element dimensions and frequency within the most current growth increment along the length of *Eucalyptus globules* stems. Trees 15: 353–357.

Lucas WJ, Groover A, Lichtenberger R, Furuta K, Yadav SR, Helariutta Y, He XQ, Fukuda H, Kang J, Brady SM, Patrick JW, Sperry J, Yoshida A, López-Millán AF, Grusak MA, Kachroo P (2013) The plant vascular system: evolution, development and functions. J Integr Plant Biol 55: 294–388.

Olson ME, Rosell JA (2013) Vessel diameter-stem diameter scaling across woody angiosperms and the ecological causes of xylem vessel diameter variation. New Phytol 197: 1204–1213.

Prior LD, Bowman DM (2014) Across a macro-ecological gradient forest competition is strongest at the most productive sites. Front Plant Sci 5: 260

Rathgeber CB, Rossi S, Bontemps JD (2011) Cambial activity related to tree size in a mature silver-fir plantation. Ann Bot 108: 429–438.

Rosell JA, Olson ME, Anfodillo T (2017) Scaling of xylem vessel diameter with plant size: causes, predictions, and outstanding questions. Curr For Rep 3: 46–59.

Tyree MT, Ewers FW (1991) The hydraulic architecture of trees and other woody plants. New Phytol 119: 345–360.

Tyree MT, Sperry JS (1989) Vulnerability of xylem to cavitation and embolism. Annu Rev Plant Physiol 40: 19–36.

Tyree MT, Zimmermann MH (2002) *Xylem Structure and the Ascent of Sap*, 2nd edn. Springer, Berlin.

Wheeler E, Baas P (2019) Wood evolution: Baileyan trends and functional traits in the fossil record. IAWA J 40: 488–529.

Wheeler EA, Baas P, Rodgers S (2007) Variations in dicot wood anatomy: a global analysis based on the insidewood database. IAWA J 28: 229–258.

*Williams CB, Anfodillo T, Crivellaro A, Lazzarin M, Dawson TE, Koch GW (2019) Axial variation of xylem conduits in the Earth's tallest trees. Trees 33: 1299–1311

Zimmermann MH (1983) *Xylem Structure and the Ascent of Sap*. Springer-Verlag, Berlin.

Resin Glands and Traumatic Duct Formation in Conifers

17

Resin storage structures lined with resin-secreting epithelial cells are a common feature in conifers (Fahn 1990; Evert 2006). The resin ducts and glands protect the tree from insects and their associated pathogens (Keeling and Bohlmann 2006; Ralph et al. 2007; Clark et al. 2012; Erbilgin et al. 2017) as well as from extensive ungulate browsing, due to high foliar content of monoterpenoids that deter browsing, providing an avenue for resistance selection in young trees (Foster et al. 2016). Resin duct and glands might contain different protecting terpenoids, even in the same plant. For example, in *Thuja plicata*, the juvenile needles contain a single longitudinal terpenoid duct with (+)-sabinene and (−)-α-pinene as prevalent monoterpenoids. In contrast, adult-scale mature leaves contain resin glands (Fig. 17.1A) that have markedly different monoterpenoid profile from needles, with α-thujone as the most prevalent monoterpenoid. Both ducts and glands are close to the epidermis and the vascular tissues. Glands produced in the shoot are directly connected to the central vascular tissues by the transfusion short tracheids (see Chap. 2, Figs. 2.23 and 2.24) originating from parenchyma cells (Aloni et al. 2013; Foster et al. 2016), which were experimentally shown to be induced by auxin (Aloni et al. 2013), likely indicating auxin production in the resin glands during their differentiation. The involvement of auxin in the differentiation of the resin duct system is likely true for other conifers, as their resin ducts are induced in connection to the vascular tissues, or inside their vascular system (Fig. 17.1B). However, auxin is not the specific signal that induces resin duct differentiation, as will be discussed below.

In the vascular tissues of Pinaceae, there are genera (e.g., *Pinus*, *Picea*, *Larix*, and *Pseudotsuga*) which produce resin ducts as a natural feature (Fig. 17.1B) and in response to injury will also produce traumatic resin ducts; other genera (e.g., *Abies*, *Tsuga*, *Cederus*, and *Pseudolarix*) produce only traumatic resin ducts in response to wounding; and there are genera (e.g., *Cupressus* and *Juniperus*) which never produce any resin ducts (Fahn 1990; Evert 2006). In pine species, the natural-occurring resin duct system is built of longitudinal and radial (inside large vascular rays) ducts. The largest number of resin ducts is produced when the cambium of an injured branch is intensively active (Fahn and Zamski 1970; Fahn 1990). In *Pinus*

© Springer Nature Switzerland AG 2021
R. Aloni, *Vascular Differentiation and Plant Hormones*,
https://doi.org/10.1007/978-3-030-53202-4_17

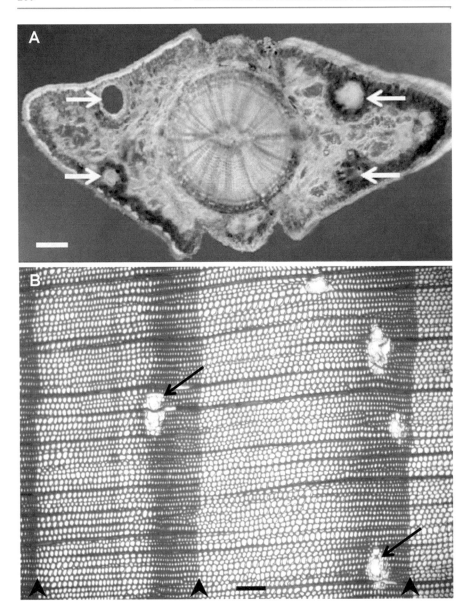

Fig. 17.1 Transverse sections showing resin storage structures (*arrows*) in intact mature leaves (**A**) and stems (**B**). (**A**) Showing four rounded resin glands (*white arrows*) from which only the gland in the upper left side was cut in the middle, in lateral scale leaves in unstained secondary shoot of *Thuja plicata*. (**B**) Longitudinal resin ducts (*black arrows*) occurring naturally at the late-wood in the trunk of *Pinus halepensis* stained with safranin, which stained the tracheids and rays in red. The photomicrograph shows two complete annual rings. The borders between the years are marked by *arrowheads*. Note that the wider annual wood ring (on the *right side*), which likely differentiated in a rainy year, produced a few resin ducts, clearly more than in the narrow annual wood ring (*left*) produced in a dry year. Bars = 100 μm (**B**); 200 μm (**A**). (**A** is adapted from Foster et al. 2016)

halepensis (Fig. 17.1B), auxin which enhances radial growth of wood also promoted resin-duct formation (Fahn and Zamski 1970); however this resin duct formation is not a direct auxin effect because the resin ducts developed only about 1 month after the auxin application.

Application of the ethylene-releasing agent, ethrel (2-chloroethylphosphonic acid), to *P. halepensis* seedlings promoted the production of longitudinal resin ducts in their wood (Yamamoto and Kozlowski 1987). As **ethylene** is the specific signal which induces the vascular rays (see Chap. 15), and promotes traumatic-resin-duct formation (Hudgins and Franceschi 2004; Hudgins et al. 2006), it likely promotes resin duct differentiation in the large vascular rays of conifers, like in pines.

Jasmonate, more specifically, methyl jasmonate (MeJA), which activates defense-related genes in plants (Okada et al. 2015; Wang et al. 2019), seems to be the primary and specific signal which induces traumatic resin-duct formation in conifers (Martin et al. 2002; Fäldt et al. 2003; Hudgins et al. 2003; Hudgins and Franceschi 2004; Huber et al. 2005; Miller et al. 2005; Erbilgin et al. 2006; Zeneli et al. 2006; Krokene et al. 2008; Zulak et al. 2010; Schmidt et al. 2011), and this jasmonate-induced defense response can be mediated by ethylene (Hudgins and Franceschi 2004). This was evident in studies on *Pseudotsuga menziesii* which showed that the MeJA-induced ethylene production earlier and 77-fold higher than wounding. Pre-treatment of *P. menziesii* stems with an ethylene response inhibitor (1-methylcyclopropene) inhibited the MeJA and wound responses (Hudgins and Franceschi 2004).

Severe stress and wounding might induce large resin cavities which damage the wood for technological and industrial use. This damage could likely be prevented in trees with low ethylene sensitivity. Similarly, we demonstrated that crown-gall tumor development which is also regulated by ethylene was inhibited on the ethylene-insensitive tomato, the *Never ripe* mutant (see Chap. 21; Fig. 21.8) (Aloni et al. 1998). In vitro production of pine trees from shoots is well established (Hargreaves et al. 2005). Regularly, to increase survival of adventitious originated plants, the culture jars are aerated to reduce the build-up of ethylene concentrations during the culture process. Therefore, I propose that conifer trees with lowered sensitivity to ethylene would be selected during the tissue culture process by keeping the jars closed, or almost closed. Thus, only tissues of ethylene insensitive trees will survive. Various lines with different sensitivities to ethylene can be selected and will be analyzed for traumatic resin-duct formation and afterward for their resistance under field conditions. It is expected that selected tree lines with lowered ethylene sensitivity will show decreased response to wounding and consequently limited traumatic resin-duct formation. Lowering ethylene sensitivity in conifers may reduce their insect resistance. This issue might be solved genetically by introducing into the selected trees toxin genes against insects (Gordon et al. 2007; Gurevitz et al. 2007). For example, transgenic *Pinus radiata* trees containing a *Bacillus thuringiensis* toxic gene displayed variable levels of resistance to insect damage, with one transgenic line being highly resistant to feeding damage (Grace et al. 2005). Anti-attractant treatments against insects may also be used (Erbilgin et al. 2007).

Summary

- Resin ducts and glands that protect the tree from insects and their associated pathogens are common in conifers. They are associated with the vascular tissues and therefore auxin is likely involved in their development. Their differentiation is promoted by stress, wounding and ethylene.
- Jasmonate, specifically, methyl jasmonate (MeJA), which activates defense-related genes in plants, is the primary and specific signal that induces traumatic resin-duct formation in conifers. This jasmonate-induced defense response can be mediated by ethylene.

References and Recommended Readings[1]

Aloni R, Foster A, Mattsson J (2013) Transfusion tracheids in the conifer leaves of *Thuja plicata* (Cupressaceae) are derived from parenchyma and their differentiation is induced by auxin. Am J Bot 100: 1949–1956.

Aloni R, Wolf A, Feigenbaum P, Avni A, Klee HJ (1998) The *Never ripe* mutant provides evidence that tumor-induced ethylene controls the morphogenesis of *Agrobacterium tumefaciens*-induced crown galls on tomato stems. Plant Physiol 117: 841–847.

Clark EL, Huber DP, Carroll AL (2012) The legacy of attack: implications of high phloem resin monoterpene levels in lodgepole pines following mass attack by mountain pine beetle, *Dendroctonus ponderosae* Hopkins. Environ Entomol 41: 392–398.

Erbilgin N, Cale JA, Lusebrink I, Najar A, Klutsch JG, Sherwood P, Enrico Bonello P, Evenden ML (2017) Water-deficit and fungal infection can differentially affect the production of different classes of defense compounds in two host pines of mountain pine beetle. Tree Physiol 37: 338–350.

Erbilgin N, Gillette NE, Mori SR, Stein JD, Owen DR, Wood DL (2007) Acetophenone as an anti-attractant for the western pine beetle, *Dendroctonus brevicomis* LeConte (Coleoptera: Scolytidae). J Chem Ecol 33: 817–823.

Erbilgin N, Krokene P, Christiansen E, Zeneli G, Gershenzon J (2006) Exogenous application of methyl jasmonate elicits defenses in Norway spruce (*Picea abies*) and reduces host colonization by the bark beetle *Ips typographus*. Oecologia 148: 426–436.

Evert RF (2006) *Esau's Plant Anatomy, Meristems, Cells, and Tissues of the Plant Body - their Structure, Function, and Development*. Wiley & Sons, Hoboken, NJ.

Fahn A (1990) *Plant Anatomy*, 4th edn. Pergamon press, Oxford.

Fahn A, Zamzki E (1970) The influence of pressure, wind, wounding and growth substances on the rate of resin duct formation in *Pinus halepensis* wood. Isr J Bot 19: 429–446.

Fäldt J, Martin D, Miller B, Rawat S, Bohlmann J (2003) Traumatic resin defense in Norway spruce (*Picea abies*): methyl jasmonate-induced terpene synthase gene expression, and cDNA cloning and functional characterization of (+)-3-carene synthase. Plant Mol Biol 51: 119–133.

*Foster AJ, Aloni R, Fidanza M, Gries R, Gries G, Mattsson J (2016) Foliar phase changes are coupled with changes in storage and biochemistry of monoterpenoids in western redcedar (*Thuja plicata*). Trees 30: 1361–1375.

Gordon D, Karbat I, Ilan N, Cohen L, Kahn R, Gilles N, Dong K, Stuhmer W, Tytgat J, Gurevitz M (2007) The differential preference of scorpion alpha-toxins for insect or mammalian sodium channels: Implications for improved insect control. Toxicon 49: 452–472.

---

[1] Papers of particular interest for suggested reading have been highlighted (with *)

Grace LJ, Charity JA, Gresham B, Kay N, Walter C (2005) Insect-resistant transgenic *Pinus radiate*. Plant Cell Rep 24: 103–111.

Gurevitz M, Karbat I, Cohen L, Ilan N, Kahn R, Turkov M, Stankiewicz M, Stuhmer W, Dong K, Gordon D. (2007) The insecticidal potential of scorpion beta-toxins. Toxicon 49: 473–489.

Hargreaves CL, Grace LJ, van der Maas SA, Menzies MI, Kumar S, Holden DG, Foggo MN, Low CB, Dibley MJ (2005) Comparative in vitro and early nursery performance of adventitious and axillary shoots from epicotyls of same zygotic embryo of control-pollinated *Pinus radiata*. Can J For Res 35: 2629–2641.

*Huber DP, Philippe RN, Madilao LL, Sturrock RN, Bohlmann J (2005) Changes in anatomy and terpene chemistry in roots of Douglas-fir seedlings following treatment with methyl jasmonate. Tree Physiol 25: 1075–1083.

Hudgins JW, Christiansen E, Franceschi VR (2003) Methyl jasmonate induces changes mimicking anatomical defenses in diverse members of the Pinaceae. Tree Physiol 23: 361–371.

*Hudgins JW, Franceschi VR (2004) Methyl jasmonate-induced ethylene production is responsible for conifer phloem defense responses and reprogramming of stem cambial zone for traumatic resin duct formation. Plant Physiol 135: 2134–2149.

Hudgins JW, Ralph SG, Franceschi VR, Bohlmann J (2006) Ethylene in induced conifer defense: cDNA cloning, protein expression, and cellular and subcellular localization of 1-aminocyc lopropane-1-carboxylate oxidase in resin duct and phenolic parenchyma cells. Planta 224: 865–877.

Keeling CI, Bohlmann J. (2006) Genes, enzymes and chemicals of terpenoid diversity in the constitutive and induced defence of conifers against insects and pathogens. New Phytol 170: 657–75.

Krokene P, Nagy NE, Solheim H (2008) Methyl jasmonate and oxalic acid treatment of Norway spruce: anatomically based defense responses and increased resistance against fungal infection. Tree Physiol 28: 29–35.

Martin D, Tholl D, Gershenzon J, Bohlmann J (2002) Methyl jasmonate induces traumatic resin ducts, terpenoid resin biosynthesis, and terpenoid accumulation in developing xylem of Norway spruce stems. Plant Physiol 129: 1003–1018.

Miller B, Madilao LL, Ralph S, Bohlmann J (2005) Insect-induced conifer defense. White pine weevil and methyl jasmonate induce traumatic resinosis, de novo formed volatile emissions, and accumulation of terpenoid synthase and putative octadecanoid pathway transcripts in Sitka spruce. Plant Physiol 137: 369–382.

Okada K, Abe H, Arimura G (2015) Jasmonates induce both defense responses and communication in monocotyledonous and dicotyledonous plants. Plant Cell Physiol 56: 16–27

Ralph SG, Hudgins JW, Jancsik S, Franceschi VR, Bohlmann J (2007) Aminocyclopropane carboxylic acid synthase is a regulated step in ethylene-dependent induced conifer defense. Full-length cDNA cloning of a multigene family, differential constitutive, and wound- and insect-induced expression, and cellular and subcellular localization in spruce and Douglas fir. Plant Physiol 143: 410–424.

*Schmidt A, Nagel R, Krekling T, Christiansen E, Gershenzon J, Krokene P (2011) Induction of isoprenyl diphosphate synthases, plant hormones and defense signalling genes correlates with traumatic resin duct formation in Norway spruce (*Picea abies*). Plant Mol Biol 77: 577–590.

Wang J, Wu D, Wang Y, Xie D (2019) Jasmonate action in plant defense against insects. J Exp Bot 70: 3391–3400.

Yamamoto F, Kozlowski TT (1987) Effect of ethrel on growth and stem anatomy of *Pinus halepensis* seedlings. IAWA Bull ns 8: 11–19.

Zeneli G, Krokene P, Christiansen E, Krekling T, Gershenzon J (2006) Methyl jasmonate treatment of mature Norway spruce (Picea abies) trees increases the accumulation of terpenoid resin components and protects against infection by Ceratocystis polonica, a bark beetle-associated fungus. Tree Physiol 26: 977–988.

Zulak KG, Dullat HK, Keeling CI, Lippert D, Bohlmann J (2010) Immunofluorescence localization of levopimaradiene/abietadiene synthase in methyl jasmonate treated stems of Sitka spruce (*Picea sitchensis*) shows activation of diterpenoid biosynthesis in cortical and developing traumatic resin ducts. Phytochemistry 71: 1695–1699.

Hormonal Control of Reaction Wood Formation

Gymnosperm and angiosperm trees differ in the nature of their **reaction wood** (see Sect. 2.8) produced following shoot bending for uplifting of the stem to its straight vertical position (Timell 1986; Sheng and Fukuju 2007; Donaldson and Singh 2016b; Felten and Sundberg 2013; Fournier et al. 2014; Fischer et al. 2019). In branches, reaction wood formation enables to reorient their growth at the tree's crown for optimal exposure of each branch to sun light, as well as to replace a damaged apical shoot by reorientation of an adjacent branch, which becomes the new leader. The formation of reaction wood is often accompanied by promoting cambial cell division in the reaction wood side, whereas cell division at the opposite side, called **opposite wood**, is more or less inhibited (see Sect. 2.8; Fig. 2.28).

In gymnosperm trees, the reaction wood is called **compression wood** that develops in the lower side of curved stem. Compression wood is characterized by short, rounded tracheids that have thick walls with increased lignin content and increased microfibril angles (Chap. 2, Fig. 2.29B) that push the stem to upright position.

In angiosperm trees, reaction wood is called **tension wood**, which is produced on the upper side of bent stems (Chap. 2, Fig. 2.30). The tension wood is characterized by few narrow vessels and many gelatinous fibers (G-fibers) with an inner gelatinous cell wall layer (G-layer) that consists of almost pure cellulose with microfibrils that are oriented parallel to the long cell axis. The G-layer induces longitudinal shrinkage which acts as a "muscle" that pulls the stem to vertical position. The swelling of the crystalline cellulose microfibrils of the G-layer produces the shrinkage that brings the stem to upright position (Jourez et al. 2001; Clair et al. 2006, 2011; Donaldson and Singh 2016a, b).

The study of reaction wood formation raises two basic questions: (**i**) how does the stem **sense gravity**? and (**ii**) what is the **response mechanism** for stimulating the cambium to start producing reaction wood?

Gravity sensing in plants occurs in specialized cells called statocytes, which are loaded with dense amyloplasts rich in starch that serve as statoliths. Under the influence of gravity the amyloplasts sink to the bottom of the statocyte. Statocytes that are located in the root cup direct root growth downward (positive geotropism),

© Springer Nature Switzerland AG 2021
R. Aloni, *Vascular Differentiation and Plant Hormones*,
https://doi.org/10.1007/978-3-030-53202-4_18

while in a young stem the statocytes that direct shoot growth upward (negative geotropism) build a starch layer of cells at the periphery of the vascular tissues, which is a continuous layer to the endodermis of the root and enable the stem to sense gravity. Statocytes are also found in the secondary phloem of old trunks. The sedimentation of the statoliths in response to gravity enables the stem the perception of its orientation and response (Toyota et al. 2013). A change in root or stem orientation is followed by a relatively rapid modification in the location of the auxin efflux carrier protein PIN-FORMED3 (PIN3) on the plasma membrane due to the actin-dependent cycling mechanism (Friml et al. 2002) (see Sect. 3.1.3). The PIN3 is a component of the lateral auxin efflux transport system that regulates tropic growth by modifying auxin flow direction promoting gravibending.

Auxin is a major regulator of tension wood formation. In an elegant experimental study, Gerttula et al. (2015) demonstrated in young *Populus* stems that following 4 days of gravistimulation, the auxin efflux carrier protein PIN3 in the plasma membrane of the statocytes loaded with starch-statoliths, was preferentially localized toward the ground in both the upper side (where tension wood will be produced) and in the lower side (at the opposite wood location). This means that because the PIN3-expressing statocytes are peripheral to the cambium, auxin (detected by DR5::GUS staining) was directed by PIN3 to the cambial zone in the upper stem side (to promote tension wood production), but at the lower stem side, the PIN3 directed the auxin toward the cortex (away from the cambium), thus preventing IAA from the cambium of the opposite wood. Consequently, the lateral auxin transport toward the ground by PIN3-expressing cells elicited differential growth responses of the cambium to gravistimulation, with IAA flowing into the cambium and promoting activity only in the upper (tension wood) side, *versus* absence of detectable IAA at the lower (opposite wood) side. Gibberellin was found to potentiate auxin responsiveness and had a synergistic promoting effect with auxin on G-fiber differentiation and tension wood formation (Gerttula et al. 2015). These interesting results are a promising starting point to continue and uncover the entire mechanism that regulates tension wood formation in angiosperms.

Conflicting results were reported in the early studies on the role of auxin in reaction wood formation. Changes in exogenous auxin stimulation promoted or inhibited reaction wood formation. Thus, a complete ring of tension wood was locally developed in the stem of *Acer rubrum* below the application of the auxin transport inhibitor 2,3,5-tri-iodobenzoic acid (TIBA). However, the induced tension wood was restricted to a narrow region of only 5–6 mm below the treatment site, but was never formed above the TIBA application site. The TIBA-induced local tension wood was characterized by narrow vessels and development of G-fibers with thick G-layer (Cronshaw and Morey 1965; Morey and Cronshaw 1968a). Application of gibberellic acid (GA_3) to the *Acer* stems promoted cambial activity but with no clear effect on secondary xylem formation; a treatment of TIBA with gibberellic acid (TIBA + GA) gave similar results to the effect of TIBA alone (Morey and Cronshaw 1968b).

In horizontally placed *Acer rubrum* seedlings, the development of tension wood was inhibited by auxin, especially when applied along the upper side of the axis, indicating that the development of tension wood appears to be correlated with a reduced auxin level on the upper side of the stem (Cronshaw and Morey 1968).

A general concept was then developed, based on the auxin and auxin inhibitor application experiments, that tension wood in angiosperms is likely induced by a decrease in auxin concentration at the upper side of bent stems, whereas compression wood of gymnosperm is likely a result of an increased auxin concentration at their lower side (Timell 1986; Little and Savidge 1987). However, measurements of endogenous IAA concentrations did not support this suggested general concept (Wilson et al. 1989; Funada et al. 1990; Hellgren et al. 2004).

Application of the auxin transport inhibitors N-1-naphthylphthalamic acid (NPA) or methyl-2-chloro-9-hydroxyfluorene-9-carboxylic acid (CF) (in a lanolin paste) as a ring around young stem of *Pinus sylvestris* induced the formation of compression wood formation above the ring, consistent with the idea that high auxin concentrations triggering compression wood formation (Sundberg et al. 1994). The level of free IAA was dramatically decreased below the ring, indicating that, as expected, the polar transport of endogenous IAA was inhibited by the treatment. Surprisingly, the free IAA level above the ring, where compression wood was formed, was also slightly lower than in control shoots. The author suggested that acropetally transported NPA and CF induce compression wood formation by interacting with the NPA receptor in differentiating tracheids, thereby locally increasing IAA in these cells (Sundberg et al. 1994).

Wilson et al. (1989) did not detect significant changes in endogenous IAA concentrations between the upper and lower sides of Douglas-Fir branches that were intact, or reoriented down experimentally. There was no correlation, or negative correlation in segments reoriented down between auxin concentration and compression wood formation or the rate of new tracheid production. Likewise, Funada et al. (1990) detected higher IAA contents in cambial region tissues forming compression wood compared to tissues produced in normal wood after 1 week, but not after 4 weeks, in gravistimulated *Cryptomeria* trees. Furthermore, Hellgren et al. (2004) studied endogenous IAA distribution across the cambial region tissues in both *Populus tremula* and *Pinus sylvestris* bent trees producing reaction wood, by tangential cryosectioning combined with sensitive gas chromatography-mass spectrometry analysis. Their study demonstrates that the formation of tension wood in poplar, or compression wood in pine, occurred without any obvious alterations in IAA concentration in the cambial region.

However, the recent promising evidence from tension wood research with *Populus* stems shows that auxin transport is modified by PIN3 at the early stage of tension wood formation, directing IAA to the cambium of the upper side where tension wood is produced, but preventing IAA from the cambium of the opposite wood (Gerttula et al. 2015; Groover 2016) promise new insights in tension wood research. Nevertheless, the entire mechanisms of compression wood *versus* tension wood formation remain unclear and hopefully will be uncovered by new molecular findings.

Gibberellin is a major regulator of tension wood formation. GA fed through the transpiration stream induced tension wood formation in weeping cherry, resulting in branches changing from the weeping to the upright form (Nakamura et al. 1994).

GA application to horizontal *Fraxinus mandshurica* seedlings increased G-fiber differentiation and tension wood formation and promoted stem's upright growth, while the GA inhibitor, uniconazole-P, which inhibited upright position, decreased xylem formation but did not inhibit G-fiber formation (Jiang et al. 1998). Exogenously applied GA was found to induce tension wood in a variety of angiosperm trees (Funada et al. 2008). GA, which is the specific hormonal signal that induces fiber differentiation (Aloni 1979; Dayan et al. 2012) (see Sects. 5.2.3 and 19.2), promotes G-fiber differentiation, their elongation, and tension wood formation in bent stems of *Acacia mangium seedling* (Nugroho et al. 2012, 2013). GA promotes upward stem growth (negative shoot gravitropism) by the formation of tension wood in the upper side of tilted *A. mangium* seedlings. While inhibitors of GA synthesis (both paclobutrazol and uniconazole-P applied to the soil in which the seedlings were growing) suppress the formation of tension wood and strongly inhibited the return to vertical growth (Nugroho et al. 2012), application of paclobutrazol, or uniconazole-P, inhibited the increase in the thickness of gelatinous layers and prevented the elongation of G-fibers in the tension wood of inclined stems (Nugroho et al. 2013).

Application of gibberellin with low auxin (1% GA_3 + 0.1% NAA in lanolin) to hop (*Humulus lupulus*) (which is primarily used as a bittering, flavoring, and stability agent in the beer industry) climbing stems from which the leaves were removed caused a substantial increase in gelatinous fiber quantity in both the secondary phloem and secondary xylem. The produced G-fibers did not differ in their structure or pattern formation from the G-fibers produced in intact plants (see Chap. 2, Fig. 2.31) (K. Sims, R. Aloni, J. Mattsson, unpublished results).

Ethylene is produced in differentiating wood tissues, stimulates and promotes cambial activity to produce more wood, initiates new vascular rays and enlarges existing rays (see Chap. 15), decreases vessel diameter, and interrupts fiber differentiation in normal and tumorous tissues (see Chap. 21, Figs. 21.6B, C, 21.10A) during secondary xylem formation (Savidge 1988; Lev-Yadun and Aloni 1995; Aloni et al. 1998; Andersson-Gunnerås et al. 2003; Love et al. 2009; Aloni 2013; Seyfferth et al. 2019; Ullrich et al. 2019).

Nelson and Hillis (1978) measured in horizontally grown *Eucalyptus gomphocephala* seedlings higher ethylene levels at the base of their stems than in vertically grown seedlings. The upper half of the horizontal stems had greater amounts of internal and emanated ethylene than their lower half and half stems of vertical seedling. The upper half of horizontal stem had greater radial growth than the lower half, forming tension wood in the upper half. While radial growth in vertical seedling was symmetrical and they contained negligible tension wood. Nelson and Hillis (1978) results suggest that the horizontal orientation of the *Eucalyptus* stems induced tensile stress, which promoted ethylene production in the upper half, resulting in increased cambial activity forming more wood, mainly tension wood, in the upper side. Jiang et al. (2009) detected increased ethylene evolution from buds of tilted *Fraxinus mandshurica* seedlings than in upright ones. Application of ethylene inhibitors, aminoethoxyvinylglycine (AVG), or $AgNO_3$, to horizontally placed 1-year-old stems did not affect gelatinous layer formation. The results show that ethylene increases the quantity of xylem production, but does not affect G-layer formation in fibers of *F. mandshurica* seedlings.

On the other hand, exogenously applied ethylene, or its precursor 1-aminocyclo propane-1-carboxylic acid (ACC), to *Populus* trees induced G-layers rich in cellulose in their G-fibers, typical to tension wood. The ACC or ethylene application did not induce G-fibers in the ethylene-insensitive *Populus* trees, demonstrating a direct signaling role of ethylene on G-layer formation in the G-fibers (Felten et al. 2018). A genome-wide transcriptome profiling revealed ethylene-dependent genes in tension-wood differentiation, showing that ethylene regulates transcriptional responses related to the amount of G-fiber formation and their properties during tension-wood formation (Seyfferth et al. 2019).

Brassinosteroids play a foundational role in the regulation of secondary growth and wood formation in *Populus*, through the regulation of cell differentiation and secondary cell wall biosynthesis. BRs were recently found to also regulate tension wood formation. Inhibition of BR synthesis resulted in decreased growth and secondary vascular differentiation (Du et al. 2020).

Summary

- Gymnosperm and angiosperm trees differ in the nature of their reaction wood produced following shoot bending for uplifting of the stem to its straight vertical position; gymnosperms produce compression wood in the lower side of the trunk, while angiosperms produce tension wood in the upper side.
- In young *Populus* stems, sedimentation of the statoliths (in the starch layer located at the periphery of the vascular tissues) in response to gravity changes enables the stem the perception of its orientation, which is followed by a rapid modification in the location of the auxin efflux carrier protein PIN3 that is localized toward the ground. Therefore, in inclined stems, PIN3 directs the auxin to the cambial zone in the upper stem side, which activates the cambium and promotes tension wood formation, while in the lower side, PIN3 directs the auxin to the cortex, thus preventing auxin from the cambium of the opposite wood.
- Gibberellin, which is the specific inducing signal for fiber differentiation, promotes tension wood formation and induces gelatinous fibers during tension wood differentiation in inclined stems and in climbing stems.
- Ethylene promotes cambial activity. Inclined stems of angiosperm trees produce more ethylene at their upper side where tension wood differentiates; the ethylene promotes gelatinous layer formation in the tension wood fibers.
- The three major hormonal signals of vascular differentiation: auxin, gibberellin, and ethylene are involved in reaction wood formation. It is possible that the hormonal signals act in combinations, have cross talks, and likely act synergistically. Cytokinin promotes cambial cell division activity along the stem side where the cambium produces reaction wood.
- Despite serious attempts to understand the mechanisms that regulate reaction wood formation in both conifers and hardwoods, and recently also at the molecular level, the fundamental question was not answered: why gymnosperms *versus*

angiosperms choose such different mechanisms to solve their wood responses to gravity. More efforts are needed to uncover the basic mechanisms that induce and regulate both types of reaction wood.

References and Recommended Readings[1]

Aloni R (1979) Role of auxin and gibberellin in differentiation of primary phloem fibers. Plant Physiol 63: 609–614.

Aloni R (2013) The role of hormones in controlling vascular differentiation. In: *Cellular Aspects of Wood Formation*, J Fromm (ed). Springer, Berlin, pp. 99–139.

Aloni R, Wolf A, Feigenbaum P, Avni A, Klee HJ (1998) The *Never ripe* mutant provides evidence that tumor-induced ethylene controls the morphogenesis of *Agrobacterium tumefaciens*-induced crown galls on tomato stems. Plant Physiol 117: 841–847.

Andersson-Gunnerås S, Hellgren JM, Björklund S, Regan S, Moritz T, Sundberg B (2003) Asymmetric expression of a poplar ACC oxidase controls ethylene production during gravitational induction of tension wood. Plant J 34: 339–349.

Clair B, Alméras T, Pilate G, Jullien D, Sugiyama J, Riekel C (2011) Maturation stress generation in poplar tension wood studied by synchrotron radiation microdiffraction. Plant Physiol 155: 562–570.

Clair B, Alméras T, Yamamoto H, Okuyama T, Sugiyama J (2006) Mechanical behavior of cellulose microfibrils in tension wood, in relation with maturation stress generation. Biophys J 91: 1128–1135.

Cronshaw J, Morey PR (1965) Induction of tension wood by 2,3,5-tri-iodobenzoic acid. Nature 205: 816–818.

Cronshaw J, Morey PR (1968) The effect of plant growth substances on development of tension wood in horizontal inclined stems of *Acer rubrum* seedlings. Protoplasma 65: 379–391.

Dayan J, Voronin N, Gong F, Sun TP, Hedden P, Fromm H, Aloni R (2012) Leaf-induced gibberellin signaling is essential for internode elongation, cambial activity, and fiber differentiation in tobacco stems. Plant Cell 24: 66–79.

Donaldson LA, Singh AP (2016a) Formation and structure of compression wood. In: *Cellular Aspects of Wood Formation*, J Fromm (ed). Springer, Berlin, pp. 225–256.

*Donaldson LA, Singh AP (2016b) Reaction wood. In: *Secondary Xylem Biology, Origins, Functions, and Applications*, YS Kim, R Funada, AP Singh (eds). Elsevier/Academic press, Amsterdam, Boston, pp. 100–109.

Du J, Gerttula S, Li Z, Zhao ST, Liu YL, Liu Y, Lu MZ, Groover AT (2020) Brassinosteroid regulation of wood formation in poplar. New Phytol 225: 1516–1530.

Felten J, Sundberg B (2013) Biology, chemistry and structure of tension wood. In: *Cellular Aspects of Wood Formation*, J Fromm (ed). Springer, Berlin, pp. 203–224.

Felten J, Vahala J, Love J, Gorzsás A, Rüggeberg M, Delhomme N, Leśniewska J, Kangasjärvi J, Hvidsten TR, Mellerowicz EJ, Sundberg B (2018) Ethylene signaling induces gelatinous layers with typical features of tension wood in hybrid aspen. New Phytol 218: 999–1014.

*Fischer U, Kucukoglu M, Helariutta Y, Bhalerao RP (2019) The dynamics of cambial stem cell activity. Annu Rev Plant Biol 70: 293–319.

*Fournier M, Almeras T, Clair B, Gril J. 2014. Biomechanical action and biological functions. In: *The biology of reaction wood*. B Gardiner, J Barnett, P Saranpaa, J Grill (eds). Springer, Heidelberg, New York, pp. 139–170.

Friml J, Wisniewska J, Benkova E, Mendgen K, Palme K. 2002. Lateral relocation of auxin efflux regulator PIN3 mediates tropism in *Arabidopsis*. Nature 415: 806–809.

---

[1] Papers of particular interest for suggested reading have been highlighted (with *)

*Funada R, Miura T, Shimizu Y, Kinase T, Nakaba S, Kubo T, Sano Y (2008) Gibberellin-induced formation of tension wood in angiosperm trees. Planta 227: 1409–1414.

Funada R, Mizukami E, Kubo T, Fushitani M, Sugiyama T (1990) Distribution of indole-3-acetic acid and compression wood formation in the stems of inclined *Cryptomerica japonica*. Holzforschung 44: 331–334.

*Gerttula S, Zinkgraf M, Muday GK, Lewis DR, Ibatullin FM, Brumer H, Hart F, Mansfield SD, Filkov V, Groover A (2015) Transcriptional and hormonal regulation of gravitropism of woody stems in *Populus*. Plant Cell 27: 2800–2813.

*Groover A (2016) Gravitropisms and reaction woods of forest trees - evolution, functions and mechanisms. New Phytol 211: 790–802.

Hellgren JM, Olofsson K, Sundberg B (2004) Patterns of auxin distribution during gravitational induction of reaction wood in poplar and pine. Plant Physiol 135: 212–220.

Jiang S, Furukawa I, Honma T, Mori M, Nakamura T, Yamamoto F (1998) Effect of applied gibberellin and uniconazole-P on gravitropism and xylem formation in horizontally positioned *Fraxinus mandshurica* seedlings. J Wood Sci 44: 385–391.

Jiang S, Xu K, Zhao N, Zheng SX, Ren YP, Gao YB, Gu S (2009) Ethylene evolution changes in tilted *Fraxinus mandshurica* Rupr. var. *japonica* Maxim. seedlings in relation to tension wood formation. J Integr Plant Biol 51: 707–713.

Jourez B, Riboux A, Leclercq A (2001) Anatomical characteristics of tension wood and opposite wood in young inclined stems of poplar (*Populus euramericana* CV "Ghoy"). IAWA J 22: 133–157.

Lev-Yadun S, Aloni R (1995) Differentiation of the ray system in woody plants. Bot Rev 61: 45–84.

Little CHA, Savidge RA (1987) The role of plant growth regulators in forest tree cambial growth. Plant Growth Regul 6: 137–169.

Love J, Björklund S, Vahala J, Hertzberg M, Kangasjärvi J, Sundberg B. (2009) Ethylene is an endogenous stimulator of cell division in the cambial meristem of *Populus*. Proc Natl Acad Sci USA 106: 5984–5989.

Morey PR, Cronshaw J (1968a) Developmental changes in the secondary xylem of *Acer rubrum* induced by various auxins and 2,3,5-tri-iodobenzoic acid. Protoplasma 65: 287–313.

Morey PR, Cronshaw J (1968b) Developmental changes in the secondary xylem of *Acer rubrum* induced by gibberellic acid, various auxins and 2,3,5-tri-iodobenzoic acid. Protoplasma 65: 315–326.

Nakamura T, Saotome M, Ishiguro Y, Itoh R, Higurashi S, Hosono M, Ishii Y (1994) The effects of GA3 on weeping of growing shoots of the Japanese cherry, *Prunus spachiana*. Plant Cell Physiol 35: 523–527.

Nelson ND, Hillis WE (1978) Ethylene and tension wood formation in *Eucalyptus gomphocephala*. Wood Sci Technol 12: 309–315.

*Nugroho WD, Nakaba S, Yamagishi Y, Begum S, Marsoem SN, Ko JH, Jin HO, Funada R (2013) Gibberellin mediates the development of gelatinous fibres in the tension wood of inclined *Acacia mangium* seedlings. Ann Bot 112: 1321–1329.

Nugroho WD, Yamagishi Y, Nakaba S, Fukuhara S, Begum S, Marsoem SN, Ko JH, Jin HO, Funada R (2012) Gibberellin is required for the formation of tension wood and stem gravitropism in *Acacia mangium* seedlings. Ann Bot 110: 887–895.

Savidge RA (1988) Auxin and ethylene regulation of diameter growth in trees. *Tree Physiology* 4: 401–414.

*Seyfferth C, Wessels BA, Gorzsás A, Love JW, Rüggeberg M, Delhomme N, Vain T, Antos K, Tuominen H, Sundberg B, Felten J (2019) Ethylene signaling is required for fully functional tension wood in hybrid Aspen. Front Plant Sci 10: 1101.

Sheng D, Fukuju Y (2007) An overview of the biology of reaction wood formation. J Integr Plant Biol 49: 131–143.

Sundberg B, Tuominen H, Little C. 1994. Effects of the indole-3-acetic acid (IAA) transport inhibitors N-1-naphthylphthalamic acid and morphactin on endogenous IAA dynamics in relation to compression wood formation in 1-year-old *Pinus sylvestris* (L.) shoots. Plant Physiol 106: 469–476.

Timell TE (1986) *Compression Wood in Gymnosperms*, Vol 2. Springer-Verlag, Heidelberg, pp. 983–1262.

Toyota M, Ikeda N, Sawai-Toyota S, Kato T, Gilroy S, Tasaka M, Morita MT. 2013. Amyloplast displacement is necessary for gravisensing in *Arabidopsis* shoots as revealed by a centrifuge microscope. Plant J 76: 648–660.

Ullrich CI, Aloni R, Saeed MEM, Ullrich W, Efferth T (2019) Comparison between tumors in plants and human beings: Mechanisms of tumor development and therapy with secondary plant metabolites. Phytomedicine 64: 153081.

Wilson BF, Chien CT, Zaerr JB (1989) Distribution of endogenous indole-3-acetic acid and compression wood formation in reoriented branches of Douglas-fir. Plant Physiol 91: 338–344.

Hormonal Control of Wood Evolution

19

Epigenetics shows us that environmental signals can regulate how plant cells develop. The environment induces protein changes in plant cells which regulate gene expression by different epigenetic means such as histone modification, DNA methylation, silencing, and anti-silencing molecular mechanisms (Moissiard and Voinnet 2004; Bologna and Voinnet 2014; Ríos et al. 2014). Evidence shows that epigenetic modifications can be transmitted across generations (Feng et al. 2010). The environment control over gene expression and inheritance indicates that environmental changes, mediated by hormonal signals, regulate plant adaptation and evolution. Three major evolutionary mechanisms controlling vascular cell types and xylem tissue patterns induced by modifications in the hormonal stimulation and the environment will be clarified in this chapter.

19.1 The Development of Perforations

Vessels are more efficient conduits of water than tracheids, since water flows through vessel elements occur via openings, namely, perforations, rather than diffusion through the primary cell walls, through the bordered pits of tracheids (Tyree and Zimmermann 2002). Tracheids appeared in ancient land plants about 470 million years ago, while vessel elements were recorded much later, about 140 million years ago (Gerrienne et al. 2011; Evert and Eichhorn 2013; Harrison and Morris 2017), or even 250 million years ago (Li et al. 1996), and became dominant in angiosperms. Naturally occurring perforated tracheids were very rarely found in conifer trees (Bannan 1958). Vessel elements have evolved independently from tracheids in several diverse groups of plants, making them an excellent example of parallel evolution (Bailey 1944). Understanding this major evolutionary progress of perforation formation (Bailey and Tupper 1918; Bailey 1944) poses a major challenge to uncover the mechanism that has shaped the gradual evolutionary progress from tracheids to vessels.

© Springer Nature Switzerland AG 2021
R. Aloni, *Vascular Differentiation and Plant Hormones*,
https://doi.org/10.1007/978-3-030-53202-4_19

In response to the challenge, a working hypothesis was suggested for studying the mechanism that induces and regulates perforation formation in tracheary elements, proposing that the polar auxin flow, which induces the conduits, is also the signal which induces and controls perforation formation. Consequently, if there is a need for auxin flow to induce perforations, it is likely that preventing the flow might promote vessel ending formation. Accordingly, in experiments in which auxin flow was substantially reduced by stem decapitation (Indig and Aloni 1989), **vessel endings** were induced (instead of perforation) in the stem internodes a few millimeters below the cut (Chap. 2, Fig. 2.14A). On the other end, application of high-auxin concentration to hypocotyls of young *Pinus pinea* seedlings (Fig. 19.1) induced the development of perforations in the pine's tracheids (Aloni et al. 2000; Aloni 2013a). Therefore, these **tracheids with perforations** (Fig. 19.2) support the idea that auxin has controlled the evolutionary development of vessels from tracheids.

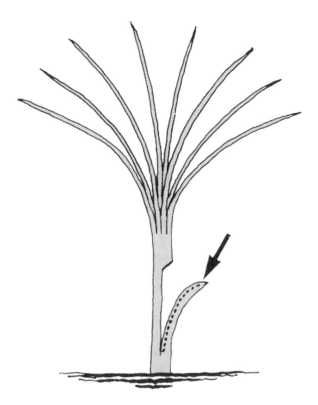

Fig. 19.1 Diagram showing the experimental method for inducing regenerative tracheids from young parenchyma cells in the hypocotyl of *Pinus pinea* seedlings. Half of the hypocotyl is separated from the leaves by a cut. The *arrow* marks the site where a lanolin paste with auxin, gibberellin, or a mixture of auxin and gibberellin is applied to the partly separated half, which remains attached to the root at its base. The *dotted line* inside the isolated half of the hypocotyl marks the sites where regenerative tracheids differentiate. The exposed tissues are separated and protected with parafilm. (Following Kalev and Aloni 1998)

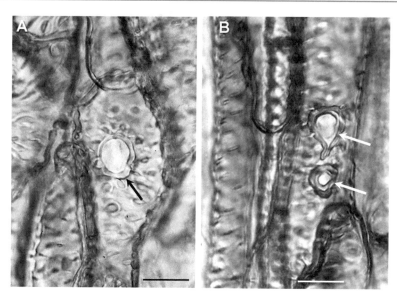

Fig. 19.2 Longitudinal views of tracheids with perforations (*arrows*) induced experimentally by auxin (0.5% 1-naphthaleneacetic acid (NAA) in lanolin w/w) in the hypocotyl of young *Pinus pinea* seedlings (as described in Fig. 19.1). (**A**) A single perforation developed in the tracheid's upper cell wall. (**B**) Two openings developed on the side wall toward the lower side of the perforated tracheid, with a relatively thick rim around the openings. Bar = 50 μm. (From Aloni 2013a)

The early fossil evidence which shows that an increase in auxin stimulation promoted perforation formation is the fossil report of Permian vessel elements in stems of the Late Permian plant from China (about 250 million years ago) that developed large leaves and contained vessel elements with perforations in its stem (Li et al. 1996). It is likely that this early plant that developed large leaves produced more auxin that induced the perforations in its tracheary elements. However, I would like to suggest that it is more likely that during xylem evolution, the transporting tissues in the highly developed plants have become more sensitive to the auxin stimulation (Barbez et al. 2012) rather than an increase in auxin stimulation occurred in the diverse groups of plants. Such an increase in vascular sensitivity to the polar auxin stimulation resulted in the development of vessel perforations (openings between vessel elements), starting with various patterns of **scalariform perforation plates** found in primitive plants to the most developed one, namely, the **simple perforation plate**, which is the most efficient water transporting wide opening of the highly developed plants.

19.2   Evolution of Vessels and Fibers from Tracheids

Xylem fibers, like vessels, have originated from tracheids of more primitive plants (Bailey and Tupper 1918; Evert and Eichhorn 2013). IAA movement through the cambium of conifer trees promotes the differentiation of tracheids from cambium initials (Savidge 1996; Uggla et al. 1996). In isolated half hypocotyls of young pine seedlings (Fig. 19.1), replacing the cotyledons by auxin application induced very short regenerative tracheids (Fig. 19.3), which originated from very young paren-chyma cells, while there was need for both auxin and gibberellin for inducing the differentiation of long tracheids (Fig.19.4) (Kalev and Aloni 1998; Aloni et al. 2000; Aloni 2013a). Furthermore, in young stems of *Ephedra campylopoda*, which regu-larly produces tracheids, vessels, and fibers (Fig. 19.5A), a gibberellin application promoted fiber formation (Fig. 19.5B), while auxin application induced vessel dif-ferentiation (Fig. 19.5C) with no fibers (by Pua Feigenbaum and R. Aloni; pub-lished in Aloni 2013a).

Collectively, all these experimental results indicate that during vascular evolu-tion, the original hormonal mechanism that induced the differentiation of tracheids in primitive plants, a combination of both auxin and gibberellin, has become more specific in higher plants. Thus, from the ancient inducing mechanism for typically elongated tracheids (of auxin with gibberellin), each vascular element in higher plants is mainly induced and regulated by one specific hormone. **Auxin** by itself

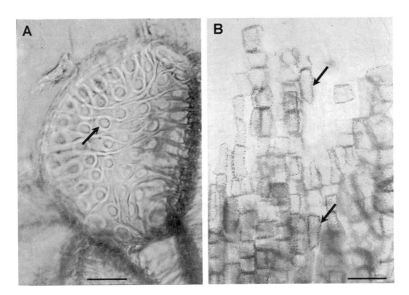

Fig. 19.3 Longitudinal views of experimentally induced regenerative tracheids in the hypocotyl of young *Pinus pinea* seedlings induced by auxin (0.1% NAA in lanolin) observed in thick sections cleared in lactic acid. (**A**) Close view on a short tracheid with typical pattern of lignified secondary wall thickenings (*arrow*). (**B**) Under low magnification, showing the pattern of numerous short tracheids below the site of auxin application, produced across the hypocotyl. The auxin induced only short tracheids (*arrows*). Bars = 25 µm (**A**), 100 µm (**B**)

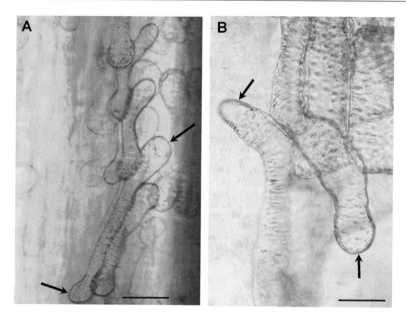

Fig. 19.4 Longitudinal views of regenerated tracheids in thick sections cleared with lactic acid in hypocotyls of *Pinus pinea*, demonstrating the effect of gibberellic acid (1.0% GA$_3$ + 0.1% NAA) with low auxin (*gibberellin by itself did not induce tracheids*) on tracheid elongation. (**A**) Substantial elongation of tracheids by intrusive growth of their upper and lower ends (*arrows*). Note the swelling of the tracheids' growing points. (**B**) Continuous intrusive growth of two regenerative tracheid tips (*arrows*) that moved away from each other. Bars = 50 μm (**B**), 100 μm (**A**). (From Kalev and Aloni 1998 (**A**); From Aloni 2013a (**B**))

induces short vessel elements (Jacobs 1952; Sachs 1981; Aloni 2010; Scarpella and Helariutta 2010), whereas **gibberellin**, in the presence of auxin, has become the specific signal which induces elongated fibers (Aloni 1979, 1987; Dayan et al. 2012). This means that the well-known evolutionary transition from tracheids to fibers and vessel elements reflects the hormonal specialization that has occurred during plant evolution (Fig. 19.6) (Aloni 2013a, b).

During plant evolution there is a general **increase in tissue sensitivity** to hormonal signals (Aloni R, unpublished observations). For example, the increased sensitivity to auxin will be discussed below regarding the development of ring porosity (Aloni 1991, 2001, 2013a).

19.3 Evolution of Ring-Porous Wood from Diffuse-Porous Xylem in Limiting Environments

An extreme adaptation of the tree's vegetative body to selective pressures in limiting environments is the development of ring-porous wood in temperate deciduous hardwood trees. Fossil records indicate that the ring-porous wood structure has

Fig. 19.5 Cross sections in *Ephedra campylopoda* stems, characterized by a relatively primitive vascular system built of tracheids (*white arrowheads*), vessels (*red arrows*), and fibers (*red arrowheads*), in the intact stem (**A**), demonstrating the role of gibberellin (**B**), and auxin (**C**), applied along 1 month (renewed every 3 days) on the differentiation of fibers and vessels, respectively. (**A**) The intact stem with vascular system built of tracheids, vessels, and fibers. (**B**) Gibberellin (1% GA$_3$ in lanolin, with no auxin) induced tracheids in the xylem (with no vessels) and many fibers in the phloem. (**C**) Auxin (0.2% NAA) induced continuous layers of mainly vessels (with no fibers). Bars = 100 μm. (From Aloni 2013a)

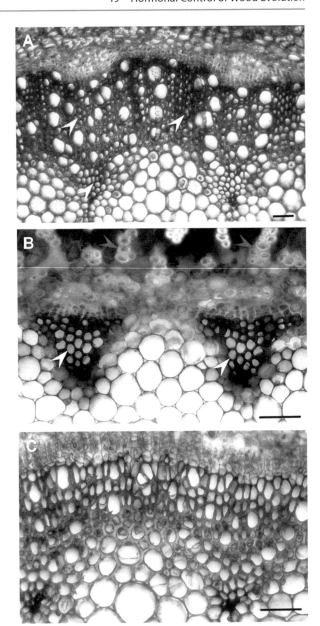

developed under various environmental stresses especially during the past 50 million years, when the global climates have been undergoing active changes (Evert and Eichhorn 2013; Wheeler and Baas 2019). Deciduous trees lose their leaves during periods of extreme environmental conditions, which is promoted by abscisic acid (ABA) (Hou et al. 2006), and then their cambium becomes dormant. The perception of seasonal changes is absorbed in leaves that sense environmental cues,

Fig. 19.6 Schematic diagram illustrating the role of auxin (IAA) and gibberellin (GA) in shaping the evolution of vessel elements and fibers from the long tracheids of primitive plants. The tracheids characterized by bordered pits are induced by a mixture of both IAA and GA (**a**, **b**). During plant evolution, GA has become the specific signal for fibers with simple pits (**c**, **d**), and IAA the inducing signal for short vessel elements with perforation plates (**e–g**)

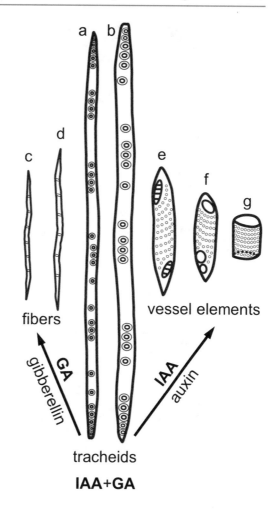

like the gradual day-length changes, which prepares the plant for winter. Toward winter, the ABA promotes the blockage of the plasmodesmata (Tylewicz et al. 2018) which initiates dormancy of deciduous trees by communication shutdown, when symplastic intercellular movement through plasmodesmata is blocked. The blockage of the plasmodesmata makes the dormant cambium nonresponsive to growth signals during an occasional sunny and warm day during winter (Tylewicz et al. 2018). The release from dormancy to hormonal responsiveness is stimulated by auxin from the buds (Aloni 1991; Aloni et al. 1991; Aloni and Peterson 1997) and likely promoted by cytokinin from the roots.

In temperate deciduous hardwood trees, the size differences of vessels in the earlywood and latewood are quite marked, and two main xylem categories can be distinguished: **diffuse-porous wood** and **ring-porous wood**. In diffuse-porous wood, the vessels are more or less uniform in size (Figs. 19.7A and 19.8A), whereas in ring-porous wood, the vessels produced at the beginning of the growth season are

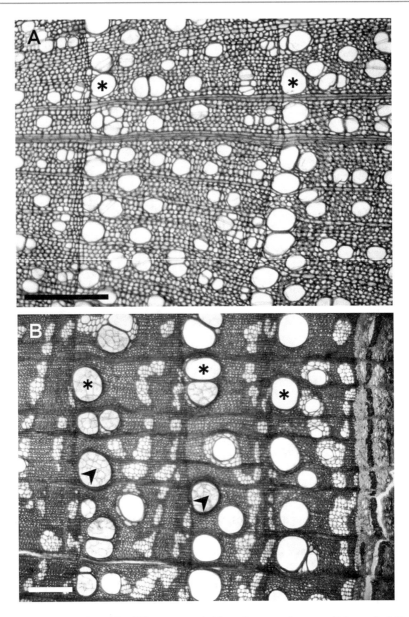

Fig. 19.7 Cross sections of a diffuse-porous (with tendency to ring porosity) wood of *Styrax officinalis* (**A**), in comparison with a ring-porous wood of *Robinia pseudoacacia* (**B**). *Asterisks* mark earlywood vessels in both photographs. (**A**) In the diffuse-porous pattern, there is a gradual decrease in vessel diameter during the annual growth season, and the vessels remain functional (open) for a few years. (**B**) in the ring-porous wood, all the very wide earlywood vessels are functional in the recent season (near the green-stained phloem, at the right side), but tend to become embolized and therefore blocked by the penetration of parenchyma cells to form tyloses (*arrowheads*) in the wood of previous years. The recent year produced wide earlywood vessels followed by a few narrow vessels in the latewood, while the previous years' latewood contained mainly fibers and large groups of parenchyma cells, with almost no vessels. Bar = 500 μm

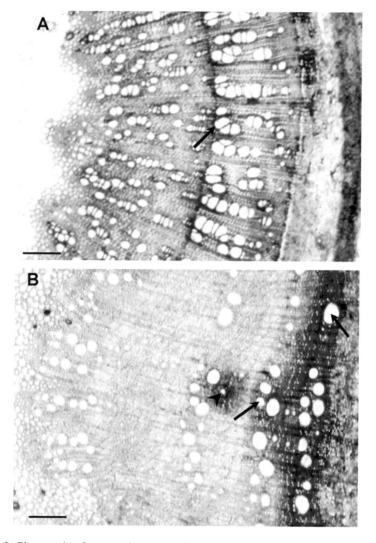

Fig. 19.8 Photographs of cross sections comparing typical secondary xylem production and demonstrating xylem conductivity at the base of intact branches of the diffuse-porous maple tree *Acer saccharum* (**A**) and the ring-porous oak tree *Quercus velutina* (**B**) grown at the same site. The red regions near the vessels were stain by safranin, which moved upward with the red-colored water through an intact branch junction and stained the conductive vessels and their surrounding cells. The *arrows* mark actively transporting earlywood vessels. (**A**) Shows that almost all the vessels in the 2-year-old diffuse-porous branch transported water. (**B**) In a 4-year-old ring-porous branch, all the earlywood vessels of the fourth year were active (*right arrow*); some earlywood vessels of the third year remained active (*left arrow*). The *arrowhead* marks the only water transporting latewood vessel of the second season. Both photographs are at the same magnification, Bar = 100 μm. (From Aloni et al. 1997)

significantly wider (Figs. 19.7B and 19.8B) than those produced at the end of the season (Evert 2006). Earlywood vessels in ring-porous trees can be huge (width of up to 500 μm and length of the entire tree) and therefore are very efficient in water conductance, but their size makes them vulnerable. The wide earlywood vessels usually function during one season, and then they become occluded by tyloses (see Chap. 2, Fig. 2.14B) or gum to plug the air-filled vessel and prevent possible penetration damage (Aloni et al. 1997; Tyree and Zimmermann 2002; Evert 2006) (Fig. 19.7B). Tyloses formation in earlywood vessels occurs earlier under drier conditions (Pérez-de-Lis et al. 2018). When tree species that have already developed ring porosity during evolution grow under favorable conditions, they can reach large sizes, although they usually show a slow growth pattern in comparison with faster-growing diffuse-porous trees.

The challenge to understand the mechanisms that have shaped these vessel patterns during the evolution of temperate deciduous hardwood trees requires elucidation of the roles of **tissue sensitivity to auxin** (Trewavas 1983; Bradford and Trewavas 1994; Barbez et al. 2012) and the specific hormonal signaling in these trees (Aloni 1991, 2001, 2013a). It has been suggested that ring-porous trees have originated from diffuse-porous species (Aloni 1991; Wheeler and Baas 1991). The development of ring porosity has probably arisen independently multiple times during the diversification of angiosperms, and different lineages might therefore have modified mechanisms in different families.

To explain how ring-porous wood has developed during the evolution of temperate deciduous hardwood trees, Aloni (1991) proposed the *limited-growth hypothesis*, suggesting that during the evolution of temperate deciduous hardwood trees, the ring-porous species have developed from diffuse-porous species under selective pressures in limiting environments which resulted in limited vegetative growth. It was further postulated that under extreme environmental conditions, the natural selection for ring-porous wood has led to a decrease in the intensity of vegetative growth, accompanied by reduced auxin levels. The latter was followed by an increase in the sensitivity of the cambium to relatively low-auxin stimulation; these changes created the conditions which enable wide earlywood vessel differentiation during a short duration in the beginning of the growth season (Aloni 1991), as will be clarify below.

Evidence that supports the hypothesis comes from observations that a diffuse-porous tree (*Populus euphratica*) and a ring-porous tree (*Quercus ithaburensis*) can change their porosity under opposite environmental conditions (Liphschitz 1995). Thus, under stress conditions when extension growth is suppressed, both tree species produced narrow annual rings characterized by ring-porous wood (as predicted by the *limited-growth hypothesis*), whereas under favorable conditions when extensive growth is intensive, both species produce wide annual rings with diffuse-porous wood (Liphschitz 1995).

Cytokinins from the root tips (Aloni et al. 2005, 2006; Matsumoto-Kitano et al. 2008; Nieminen et al. 2008) increase the sensitivity of the cambium to the auxin signal originating in young leaves (Baum et al. 1991; Aloni 1993, 1995; Aloni et al. 2003). Cytokinin prevents the usually rapid occurring IAA conjugation (Coenen and Lomax 1997); therefore, elevated CK concentration enables the transport of extremely low IAA concentrations via the cambium, which may explain the increased sensitivity of the cambium to the auxin hormone. Experimental evidence from transformed plants (Zhang et al. 1995; Eklöf et al. 1997) supports the idea that reduced auxin concentrations can elevate cytokinin concentration, which would enhance tissue sensitivity to the auxin signal (Trewavas 1983; Aloni 1991; Bradford and Trewavas 1994; Barbez et al. 2012). The experiments demonstrate that auxin or cytokinin modifies the content of the other hormone by affecting its rate of synthesis. Reduced IAA concentration increases free cytokinin level (Palni et al. 1988; Zhang et al. 1995; Eklöf et al. 1997). This, in turn, enhances cambium sensitivity to extremely low-concentration IAA streams originating in swelling buds and creates the special physiological conditions that enable the differentiation of very wide earlywood vessels during a limited period of time in early spring (Aloni 1991, 2001).

The increased cambium sensitivity to IAA in ring-porous trees enables early cambium reactivation at the beginning of the growth season before bud break. This was evident in stem diameter measurements of the ring-porous *Zelkova serrata* saplings in their leafless state, showing stem swelling 2–6 weeks before bud opening. During this developmental stage, actively dividing cambial cells, and immature earlywood vessels derived from them, are very soft, as they have not yet deposited their hard secondary cell walls (Yoda et al. 2003).

The substantial increase in cambial sensitivity to auxin in ring-porous trees created the special internal conditions that enable them to respond to initial flows of **extremely low IAA concentrations** originating in dormant-looking (before swelling) buds a few weeks before bud break (Aloni 1991, Aloni and Peterson 1997, Aloni et al. 1997), stimulating slow vessel differentiation, which permits more cell expansion, promoting widening of differentiating vessels before their secondary wall deposition (as expected by the *auxin gradient hypothesis* of Aloni and Zimmermann 1983; see Chap. 13), resulting in the formation of very wide earlywood vessels. Therefore, the first wide earlywood vessels of ring-porous trees are initiated 6 to 2 **weeks before** the onset of leaf expansion (Suzuki et al. 1996; Sass-Klaassen et al. 2011; Takahashi et al. 2013; Lavrič et al. 2017; Puchałka et al. 2017) and cause stem swelling before bud opening (Yoda et al. 2003). The pattern of earlywood vessel maturation in the ring-porous hardwoods, *Quercus serrata* and *Robinia pseudoacacia*, progressed downward. The first mature earlywood vessel elements appeared at bud break, first at the top of the stem, and continue downward to the lower parts of the stem (Kudo et al. 2015).

Conversely, in diffuse-porous species, the first earlywood vessels are initiated 2 to 7 **weeks after** the onset of leaf expansion (Suzuki et al. 1996; Takahashi et al. 2013), and because of the low cambium sensitivity in diffuse-porous trees, their cambium requires high-auxin concentrations (from fast-growing young leaves) for reactivation. These results explain the old report of Priestley and Scott (1936) who found that in a deciduous ring-porous tree, the cambium undergoes extremely fast reactivation before bud break, which occurs almost simultaneously in the branches and along the trunk. This is why the bark of deciduous ring-porous trees may be peeled a few days before any bud swelling can be observed in spring. (The bark can be removed along the newly formed cell layer of reactivated cambium because it possesses new thin radial cell walls following early cambial cell divisions.) Conversely, a deciduous diffuse-porous species requires several weeks for a "wave" of cambial reactivation to extend from the twigs of a large tree downward to the base of its trunk (Priestley and Scott 1936).

An opposite explanation for the differentiation of wide earlywood vessel in ring-porous trees was suggested by Wareing (1951) who studied cambial reactivation and wood formation in ring-porous *versus* diffuse-porous trees. Wareing suggested that the characteristic pattern of early rapid spread of cambium reactivation and the development of wide earlywood vessels in ring-porous species "is due to the presence in the cambium of a reserve of auxin-precursor, which makes possible the rapid spread of wide-vessel formation throughout the tree, at an early stage of development of the buds" (although the widening of the wide earlywood vessels in ring-porous trees is a slow expansion process extending along a few weeks before the onset of leaf expansion, see in Suzuki et al. 1996). Wareing also mentioned the possibility that the dormant-looking buds of ring-porous trees that initiated cambial reactivation "were no longer 'physiologically' dormant" (Wareing 1951), in other words, that the dormant-looking buds possibly started to produce the auxin hormone.

Evidently, an extremely low-auxin concentration (0.003% 1-naphthaleneacetic acid (NAA) in lanolin w/w) applied to disbudded shoots of *Melia azedarach* trees induced wide earlywood vessels (Fig. 19.9B) in the deciduous ring-porous trees (Aloni 1991, 2001), but this low-auxin concentration was not strong enough to stimulate any earlywood vessel differentiation in deciduous diffuse-porous trees (Aloni 2001). This was true also with 0.01% NAA (in lanolin) that induced more medium-size earlywood vessels (Fig. 19.9C) but was not strong enough to induce earlywood vessels in diffuse-porous trees. On the other hand, a high-auxin concentration (0.1% NAA in lanolin) induced rapid differentiation of narrow earlywood vessel in the ring-porous trees, because of fast secondary wall deposition that prevented vessel expansion, and therefore remained narrow vessels (Fig. 19.9D). The high-auxin concentration induced earlywood vessel differentiation in diffuse-porous trees (Aloni 2001). These results clearly demonstrate that the wide earlywood vessels of a ring-porous tree are induced by extremely low-auxin stimulation before bud swelling, while in diffuse-porous trees, there is a need for high-auxin concentration for inducing earlywood vessel differentiation from fast-growing leaves (Aloni 1991, 2001, 2013a).

The auxin produced by the buds and young leaves induces early vessel differentiation first immediately below the buds. Complete early differentiation of the earlywood vessels occurs first in the upper stem region than progresses to the middle and lower regions during bud swelling in the ring-porous *Quercus serrata* seedlings (Kudo et al. 2018). During this downward earlywood vessel differentiation process, the developing buds and young shoot organs are supplied by the functional networks of previous year's narrow latewood vessels, while the wide earlywood vessels of the current year differentiate and mature (Kudo et al. 2018).

Cambium reactivation and earlywood vessel formation is influenced by temperature. Localized heating by wrapping an electric heating ribbon around stems of the

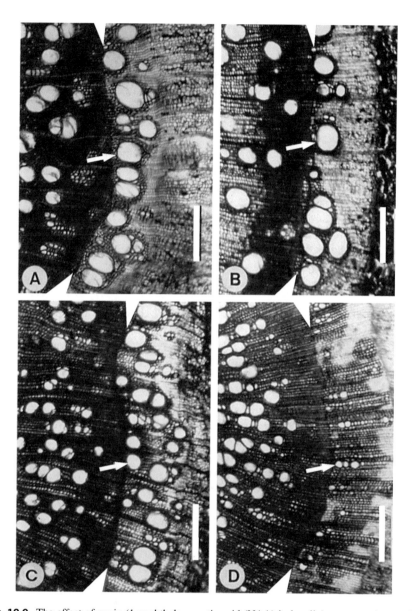

Fig. 19.9 The effect of auxin (1-naphthaleneacetic acid (NAA) in lanolin) concentration on the width of earlywood vessel differentiation is shown in transverse sections in stems of the ring-porous tree *Melia azedarach*. All photomicrographs were taken from the same experiment; run in Tel Aviv from February 15 to March 15, 1986; and are presented in the same orientation and magnification (Bars = 250 μm). All the sections were taken 50 mm below the apical bud, which was left intact (**A**) or was replaced by a range of auxin concentrations: low, 0.003% NAA (**B**); medium, 0,01% NAA (**C**); or high, 0.1% NAA (**D**). The auxin was applied in the form of a lanolin paste, which was renewed every 3 days. The photomicrographs show a substantial decrease in the diameter of the earlywood vessels (*white arrows*) with increasing auxin concentration (**B–D**). The low auxin concentration induced wide vessels (**B**). The two higher concentrations induced many more xylem cells (along a radius) with narrower vessels (**C, D**). The highest auxin concentration tested (0.1% NAA) resulted in very narrow earlywood vessels (**D**). The borderline between the latewood of 1985 (*left*) and the new earlywood of 1986 (*right*) is marked with *white triangles*. The experiment was repeated three times (in 1984, 1985, and 1986) with 5–10 stems per treatment, yielding the same results. (From Aloni 1991)

deciduous ring-porous seedlings of *Quercus serrata* induced early cambial reactivation and differentiation of narrow vessel elements sooner than in non-heated seedlings and before bud break. The results of this study show that the elevated temperature was a direct trigger for cambial reactivation and differentiation of first vessel elements. Bud growth was not essential for cambial reactivation and differentiation of first vessel elements in the ring-porous seedlings (Kudo et al. 2014). In the ring-porous chestnut trees grown on the Swiss Alps, the earlywood vessels were positively affected by warm temperatures of early spring (April), at the time of resumption of shoot growth, likely affecting cambial sensitivity to auxin (Fonti et al. 2007). Rising temperature before bud break increased the expression of genes involved in polar auxin transport (Schrader et al. 2003), providing evidence that sensitivity to auxin can be modulated by temperature. When triggered later in the season, this higher cell sensitivity to the IAA signal would result in smaller vessels as a consequence of a faster process of differentiation (Aloni and Zimmermann 1983).

Diffuse-porous species start the growth season a few weeks earlier than ring-porous trees and have a longer growth season which is characterized by continuous production of young leaves during a few months (Aloni et al. 1997). Conversely, ring-porous trees which are late leafing trees (Lechowicz 1984) produce young leaves for only a short period of a few weeks, and later they have mainly mature leaves (Aloni et al. 1997). Because young diffuse-porous trees possess greater growth intensity, they might produce more xylem per year than young ring-porous trees (Aloni et al. 1997). In diffuse-porous trees, the continuous development of new young-IAA-producing leaves along the growth season stimulates continuous production of new vessels along the entire growth season with relatively thin-wall fibers, whereas in ring-porous trees, the dominating mature leaves, which produce the gibberellin signal (Dayan et al. 2012), induce the development of numerous well-developed hard lignified fibers during most of the growth season with only a few narrow vessels. These diverse earlywood and latewood properties in ring-porous wood, namely, the soft wide earlywood vessels *versus* the numerous hard latewood fibers, affect lumber stability and can have major effects on wood and fiber utilization.

It should be clarified that for inducing the wide earlywood vessels in ring-porous trees, there is a requirement of early very low-auxin flow from dormant-looking buds early in the growth season along the very sensitive cambium as explained above by the *limited-growth hypothesis* (Aloni 1991). These unique conditions allow the early slow widening process typical for earlywood vessel formation in the branches and trunk produced at the start of the new growth season. These unique conditions do not occur along the new twigs of the current growth season, and, therefore, the wood produced in the youngest twigs during their first year (when they are only a few weeks/months old) has a diffuse-porous wood pattern (Cochard and Tyree 1990; Lo Gullo et al. 1995; Rodriguez-Zaccaro et al. 2019). It should be noted that this diffuse-porous wood pattern in the youngest twigs/branches of ring-porous trees is induced by high-auxin stimulation produced by young growing leaves, and it is not due to the influence of "cambial age" as suggested by Rodriguez-Zaccaro et al. (2019) (read about the vague concept of "cambial age" in Chap. 12).

Therefore, there is no need for "older cambia" to produce the wide earlywood vessels typical to ring-porous wood pattern (Rodriguez-Zaccaro et al. 2019), but only the requirement for the unique endogenous conditions of extremely low-auxin stimulation (from dormant-looking buds before swelling) during early spring, allowing the slow vessel widening process typically forming the wide earlywood vessels in the deciduous ring-porous trees (Aloni 1991, 2001, 2013a).

Summary

- Vessel elements have evolved independently from tracheids in several diverse groups of plants. Polar auxin flow likely regulated the differentiation of perforation in tracheary elements, which can be induced experimentally in pine seedlings by high-auxin concentration.
- Fibers and vessels have originated from tracheids. During vascular evolution, the original hormonal mechanism that induced the differentiation of tracheids, a combination of both auxin and gibberellin, has become more specific in higher plants. Auxin by itself induces short vessel elements, while gibberellin, in the presence of auxin, has become the specific signal which induces elongated fibers.
- The development of ring-porous wood from diffuse-porous xylem in temperate deciduous hardwood trees is an adaptation of ring-porous trees to selective pressures in limiting environments.
- The *limited-growth hypothesis* suggests that during the evolution of temperate deciduous hardwood trees, limiting environments limited vegetative growth, reduced leaf production, and decreased auxin synthesis, which caused an increase in the sensitivity of the cambium to extremely low levels of IAA stimulation, produced in dormant-looking buds, before bud break. These changes enable slow wide earlywood vessel widening in the beginning of the growth season.
- Diffuse-porous trees (*Populus euphratica*) and ring-porous trees (*Quercus ithaburensis*) can change their porosity under opposite environmental conditions. They produce a ring-porous pattern under stress conditions, whereas a diffuse-porous pattern under favorable environments.
- An extremely low-auxin concentration (0.003% NAA in lanolin) applied to disbudded shoots of the ring-porous tree *Melia azedarach* induced slow differentiation resulting in wide earlywood vessels in the deciduous ring-porous trees, whereas a high-auxin concentration (0.1% NAA in lanolin) induced rapid differentiation resulting in narrow earlywood vessels like in a diffuse-porous wood.
- In deciduous ring-porous tree, the sensitive cambium undergoes extremely fast reactivation before bud break, which occurs almost simultaneously in the branches and along the trunk. This is why the bark of deciduous ring-porous trees may be peeled a few days before any bud swelling can be observed in spring. Conversely, deciduous diffuse-porous trees require several weeks for a "wave" of cambial reactivation to extend from the twigs of a large tree downward to the base of its trunk.
- The first wide earlywood vessels of ring-porous trees are initiated 6 to 2 weeks before the onset of leaf expansion, allowing the earlywood vessels to expand

along a few weeks before bud opening and, therefore, become very wide. Conversely, in diffuse-porous species, the first earlywood vessels are initiated 2 to 7 weeks after the onset of leaf expansion, and due to their rapid differentiation, they remain relatively narrow.

- Diffuse-porous species start the growth season a few weeks earlier than ring-porous trees and have a longer growth season which is characterized by continuous production of young leaves during a few months. Conversely, ring-porous trees have been adapted to a shorter growth season in limiting environments, to leaf out later than diffuse-porous trees, produce young leaves for only a short period of a few weeks and later they have mainly mature leaves.

- Continuous development of new young-IAA-producing leaves along the growth season of diffuse-porous trees stimulates continuous production of new vessels along the entire growth season with relatively thin-wall fibers, whereas in ring-porous trees, the dominating mature leaves, which produce gibberellin, induce the development of numerous well-developed hard lignified fibers during most of the growth season with only a few narrow vessels, building a strong trunk wood, resistant to extreme stormy conditions.

- The wide earlywood vessels are very efficient in water transport but therefore vulnerable and tend to become occluded by tyloses or gum toward the end of their relatively short season. Conversely, diffuse-porous trees produce narrow earlywood vessels that operate in water transport along a few years.

References and Recommended Reading[1]

Aloni R (1979) Role of auxin and gibberellin in differentiation of primary phloem fibers. Plant Physiol 63: 609–614.

Aloni R (1987) Differentiation of vascular tissues. Annu Rev Plant Physiol 38: 179–204.

*Aloni R (1991) Wood formation in deciduous hardwood trees. In: *Physiology of Trees*. AS Raghavendra (ed). Wiley & Sons, New York, pp. 175–197.

Aloni R (1993) The role of cytokinin in organised differentiation of vascular tissues. Aust J Plant Physiol 20: 601–608.

Aloni R (1995) The induction of vascular tissues by auxin and cytokinin. In: *Plant Hormones: Physiology, Biochemistry and Molecular Biology*, PJ Davies (ed). Kluwer, Dordrecht, pp. 531–546.

*Aloni R (2001) Foliar and axial aspects of vascular differentiation - hypotheses and evidence. J Plant Growth Regul 20: 22–34.

Aloni R (2010) The induction of vascular tissues by auxin. In: *Plant Hormones: Biosynthesis, Signal Transduction, Action!* PJ Davies (ed). Kluwer Academic Publishers, Dordrecht, pp. 485–506.

*Aloni R (2013a) The role of hormones in controlling vascular differentiation. In: *Cellular Aspects of Wood Formation*, J Fromm (ed). Springer-Verlag, Berlin, pp. 99–139.

Aloni R (2013b) Role of hormones in controlling vascular differentiation and the mechanism of lateral root initiation. Planta 238: 819–830.

---

[1] Papers of particular interest for suggested reading have been highlighted (with *)

Aloni R, Alexander JD, Tyree MT (1997) Natural and experimentally altered hydraulic architecture of branch junctions in *Acer saccharum* Marsh. and *Quercus velutina* Lam. trees. Trees 11: 255–264.

Aloni, R, Aloni E, Langhans M, Ullrich CI (2006) Role of cytokinin and auxin in shaping root architecture: regulating vascular differentiation, lateral root initiation, root apical dominance and root gravitropism. Ann Bot 97: 883–893.

*Aloni R, Feigenbaum P, Kalev N, Rozovsky S (2000) Hormonal control of vascular differentiation in plants: the physiological basis of cambium ontogeny and xylem evolution. In: *Cell and Molecular Biology of Wood Formation*, RA Savidge, JR Barnett, R Napier (eds). BIOS Scientific Publishers, Oxford, pp. 223–236.

Aloni R, Langhans M, Aloni E, Dreieicher E, Ullrich CI (2005) Root-synthesized cytokinin in *Arabidopsis* is distributed in the shoot by the transpiration stream. J Exp Bot 56: 1535–1544.

Aloni R, Peterson CA (1997) Auxin promotes dormancy callose removal from the phloem of *Magnolia kobus* and callose accumulation and earlywood vessel differentiation in *Quercus robur*. J Plant Res 110: 37–44.

Aloni R, Raviv A, Peterson CA (1991) The role of auxin in the removal of dormancy callose and resumption of phloem activity in *Vitis vinifera*. Can J Bot 69: 1825–1832.

Aloni R, Schwalm K, Langhans M, Ullrich CI (2003) Gradual shifts in sites of free-auxin production during leaf-primordium development and their role in vascular differentiation and leaf morphogenesis in *Arabidopsis*. Planta 216: 841–853.

Aloni R, Zimmermann MH (1983) The control of vessel size and density along the plant axis - a new hypothesis. Differentiation 24: 203–208.

Bailey IW (1944) The development of vessels in angiosperms and its significance in morphological research. Am J Bot 31: 421–428.

Bailey IW, Tupper WW (1918). Size variation in tracheary cells: I. A comparison between the secondary xylems of vascular cryptogams, gymnosperms and angiosperms. Proc Am Acad Arts Sci 54: 149–204.

Bannan MW (1958) An occurrence of perforated tracheids in *Thuja occidentalis* L. New Phytol 57: 132–134.

Barbez E, Kubeš M, Rolčík J, Béziat C, Pěnčík A, Wang B, Rosquete MR, Zhu J, Dobrev PI, Lee Y, Zažímalovà E, Petrášek J, Geisler M, Friml J, Kleine-Vehn J (2012) A novel putative auxin carrier family regulates intracellular auxin homeostasis in plants. Nature 485: 119–122.

Baum SF, Aloni R, Peterson CA (1991) The role of cytokinin in vessel regeneration in wounded *Coleus* internodes. Ann Bot 67: 543–548.

Bologna NG, Voinnet O (2014) The diversity, biogenesis, and activities of endogenous silencing small RNAs in *Arabidopsis*. Annu Rev Plant Biol 65: 473–503.

Bradford KJ, Trewavas AJ (1994) Sensitivity thresholds and variable time scales in plant hormone action. Plant Physiol 105: 1029–1036.

Cochard H, Tyree MT (1990). Xylem dysfunction in *Quercus*: Vessel sizes, tyloses, cavitation and seasonal changes in embolism. Tree Physiol 6: 393–407.

Coenen C, Lomax TL (1997) Auxin-cytokinin interactions in higher plants: old problems and new tools. Trends in Plant Sci 2: 351–356.

Dayan J, Voronin N, Gong F, Sun T-p, Hedden P, Fromm H, Aloni R (2012) Leaf-induced gibberellin signaling is essential for internode elongation, cambial activity, and fiber differentiation in tobacco stems. Plant Cell 24: 66–79.

Eklöf S, Åstot C, Blackwell J, Moritz T, Olsson O, Sandberg G (1997) Auxin-cytokinin interactions in wild-type and transgenic tobacco. Plant Cell Physiol 38: 225–235.

Evert RF (2006) *Esau's Plant Anatomy, Meristems, Cells, and Tissues of the Plant Body - their Structure, Function, and Development*. Wiley & Sons, Hoboken, NJ.

Evert RF, Eichhorn SE (2013) *Raven Biology of Plants*, 8th edn. Freeman, New York.

Feng S, Jacobsen SE, Reik W (2010) Epigenetic reprogramming in plant and animal development. Science 330: 622–627.

Fonti P, Solomonoff N, García-González I (2007) Earlywood vessels of *Castanea sativa* record temperature before their formation. New Phytol 173: 562–570.

Gerrienne P, Gensel PG, Strullu-Derrien C, Lardeux H, Steemans P, Prestianni C (2011) A simple type of wood in two early Devonian plants. Science 333: 837.

Harrison CJ, Morris JL (2017) The origin and early evolution of vascular plant shoots and leaves. Phil Trans R Soc B 373: 20160496.

Hou H-W, Zhou Y-T, Mwange K-N, Li W-F, He X-Q, Cui K-M (2006) ABP1 expression regulated by IAA and ABA is associated with the cambium periodicity in *Eucommia ulmoides* Oliv. J Exp Bot 57: 3857–3867.

*Indig FE, Aloni R (1989) An experimental method for studying the differentiation of vessel endings. Ann Bot 64: 589–592.

Jacobs WP (1952) The role of auxin in differentiation of xylem around a wound. Am J Bot 39: 301–309.

Kalev N, Aloni R (1998) Role of auxin and gibberellin in regenerative differentiation of tracheids in *Pinus pinea* L. seedlings. New Phytol 138: 461–468.

Kudo K, Nabeshima E, Begum S, Yamagishi Y, Nakaba S, Oribe Y, Yasue K, Funada R (2014) The effects of localized heating and disbudding on cambial reactivation and formation of earlywood vessels in seedlings of the deciduous ring-porous hardwood, *Quercus serrata*. Ann Bot 113: 1021–1027.

Kudo K, Utsumi Y, Kuroda K, Yamagishi Y, Nabeshima E, Nakaba S, Yasue K, Takata K, Funada R (2018) Formation of new networks of earlywood vessels in seedlings of the deciduous ring-porous hardwood *Quercus serrata* in springtime. Trees 32: 725–734.

Kudo K, Yasue K, Hosoo Y, Funada R (2015) Relationship between formation of earlywood vessels and leaf phenology in two ring-porous hardwoods, *Quercus serrata* and *Robinia pseudo-acacia,* in early spring. J Wood Sci 61: 455–464.

*Lavrič M, Eler K, Ferlan M, Vodnik D, Gričar J (2017) Chronological sequence of leaf phenology, xylem and phloem formation and sap flow of *Quercus pubescens* from abandoned karst grasslands. Front Plant Sci 8: 314.

*Lechowicz MJ (1984) Why do temperate deciduous trees leaf out at different times? Adaptation and ecology of forest communities. Am Nat 124: 821–842.

Li H, Taylor EL, Taylor TN (1996) Permian vessel elements. Science 271: 188–189.

*Liphschitz N (1995) Ecological wood anatomy: changes in xylem structure in Israeli trees. In: *Wood Anatomy Research 1995. Proc Inter Symp Tree Anatomy and Wood Formation*, W Shuming (ed), Tianjin, China. International Academic Publishers. Beijing, pp. 12–15.

Lo Gullo MA, Salleo S, Piaceri EC, Rosso R (1995). Relations between vulnerability to xylem embolism and xylem conduit dimensions in young trees of *Quercus cerris*. Plant Cell Environ 18: 661–669.

Matsumoto-Kitano M, Kusumoto T, Tarkowski P, Kinoshita-Tsujimura K, Václavíková K, Miyawaki K, and Kakimoto T (2008) Cytokinins are central regulators of cambial activity. Proc Natl Acad Sci USA 105: 20027–20031.

Moissiard G, Voinnet O (2004) Viral suppression of RNA silencing in plants. Mol Plant Pathol 5: 71–82.

Nieminen K, Immanen J, Laxell M, Kauppinen L, Tarkowski P, Dolezal K, Tähtiharju S, Elo A, Decourteix M, Ljung K, Bhalerao R, Keinonen K, Albert VA, Helariutta Y (2008) Cytokinin signaling regulates cambial development in poplar. Proc Natl Acad Sci USA 105: 20032–20037.

Palni LMS, Burch L, Horgan R (1988) The effect of auxin concentration on cytokinin stability and metabolism. Planta 174: 231–234.

Pérez-de-Lis G, Rozas V, Vázquez-Ruiz RA, García-González I (2018) Do ring-porous oaks prioritize earlywood vessel efficiency over safety? Environmental effects on vessel diameter and tyloses formation. Agric For Meteorol 248: 205–214.

Priestley JH, Scott LI (1936) A note upon summer wood production in the tree. Proc Leeds Phil Soc 3: 235–248.

*Puchałka R, Koprowski M, Gričar J, Przybylak R (2017) Does tree-ring formation follow leaf phenology in Pedunculate oak (*Quercus robur* L.)? Eur J For Res 136: 259–268.

Ríos G, Leida C, Conejero A, Badenes ML (2014) Epigenetic regulation of bud dormancy events in perennial plants. Front Plant Sci 5: 247.

Rodriguez-Zaccaro FD, Valdovinos-Ayala J, Percolla MI, Venturas MD, Pratt RB, Jacobsen AL (2019) Wood structure and function change with maturity: age of the vascular cambium is associated with xylem changes in current-year growth. Plant Cell Environ 42: 1816–1831.

Sachs T (1981) The control of patterned differentiation of vascular tissues. Adv Bot Res 9: 151–262.

Savidge RA (1996) Xylogenesis, genetic and environmental regulation. IAWA J 17: 269–310.

Sass-Klaassen U, Sabajo CR, den Ouden J (2011) Vessel formation in relation to leaf phenology in pedunculate oak and European ash. Dendrochronologia 29: 171–175.

Scarpella E, Helariutta Y (2010) Vascular pattern formation in plants. Curr Top Dev Biol 91: 221–265.

Schrader J, Baba K, May ST, Palme K, Bennet M, Bhalerao RP, Sandberg G (2003) Polar auxin transport in the wood-forming tissue of hybrid aspen is under simultaneous control of developmental and environmental signals. Proc Natl Acad Sci USA 100: 10096–10101.

*Suzuki M, Yoda K, Suzuki H (1996) Phenological comparison of the onset of vessel formation between ring-porous and diffuse-porous deciduous trees in a Japanese temperate forest. IAWA J 17: 431–444.

*Takahashi S, Okada N, Nobuchi T (2013) Relationship between the timing of vessel formation and leaf phenology in ten ring-porous and diffuse-porous deciduous tree species. Ecol Res 28: 615–624.

Trewavas AJ (1983) Is plant development regulated by changes in concentration of growth substances or by changes in the sensitivity to growth substances? TIBS 8: 354–357.

Tylewicz S, Petterle A, Marttila S, Miskolczi P, Azeez A, Singh RK, Immanen J, Mähler N, Hvidsten TR, Eklund DM, Bowman JL, Helariutta Y, Bhalerao RP (2018) Photoperiodic control of seasonal growth is mediated by ABA acting on cell-cell communication. Science 360: 212–215.

Tyree MT, Zimmermann MH (2002) *Xylem Structure and the Ascent of Sap*, 2nd edn. Springer-Verlag, Berlin.

Uggla C, Moritz T, Sandberg G, Sundberg B (1996) Auxin as a positional signal in pattern formation in plants. Proc Nat Acad Sci USA 93: 9282–9286.

Wareing PF (1951) Growth studies in woody species IV. The initiation of cambial activity in ring-porous species. Physiol Plant 4: 546–562.

*Wheeler EA, Baas P (1991) A survey of the fossil record for dicotyledonous wood and its significance for evolutionary and ecological wood anatomy. IAWA Bull ns 12: 275–332.

Wheeler E, Baas P (2019) Wood evolution: Baileyan trends and functional traits in the fossil record. IAWA J 40: 488–529.

Yoda K, Wagatsuma H, Suzuki M, Suzuki H (2003) Stem diameter changes before bud opening in *Zelkova serrata* saplings. J Plant Res 116: 13–18.

Zhang R, Zhang X, Wang J, Letham DS, McKinney SA, Higgins TJV (1995) The effect of auxin on cytokinin levels and metabolism in transgenic tobacco tissue expressing an *ipt* gene. Planta 196: 84–94.

How Vascular Differentiation in Hosts Is Regulated by Parasitic Plants and Gall-Inducing Insects

20

Parasitism of plants and insects on host plants is common. They are discussed in the same chapter, because these parasitic organisms use comparable hormonal mechanisms based on the same hormonal signals that enable both parasitic plants and gall-inducing insects to take over and directly control vascular differentiation of their hosts, for serving their own nutrient needs, and also provide a protecting shelter for the insects.

It is relatively easy for a plant to become a parasite, as, naturally, the parasite produces all the exact phytohormonal signals as manipulating tools to overcome a potential host. Therefore, a parasitic plant can establish itself on a wide variety of host species, while galling insects are more specific in choosing their host. As will be discussed below, galling insects can synthesize high concentrations of both major plant hormones, namely, auxin and cytokinins, which provide them with powerful tools to modify plant morphogenesis and specifically to regulate vascular differentiation of their hosts.

20.1 Xylem and Phloem Connections Between Parasitic and Host Plants

Parasitic plants evolved from non-parasitic plants. Parasitism is a highly successful easy life strategy, and, therefore, it is not a surprise that there are about 4000 species of parasitic plants in approximately 20 families of flowering plants (Kuijt 1969; Barkman et al. 2007; Westwood et al. 2010). DNA sequence data from the three plant genomes (the mitochondrial, chloroplast, and nuclear genomes) reveals that there are at least a dozen independent origins of the parasitic lifestyle in different angiosperms families, indicating numerous ancestral plant lineages, which demonstrates their parallel independent evolution (Barkman et al. 2007; Bromham et al. 2013). Theoretical and experimental evidence suggest that the rates of molecular evolution could be raised in parasitic organisms compared to non-parasitic taxa, and

© Springer Nature Switzerland AG 2021
R. Aloni, *Vascular Differentiation and Plant Hormones*,
https://doi.org/10.1007/978-3-030-53202-4_20

parasitic lineages have a faster rate of molecular evolution than their non-parasitic relatives (Bromham et al. 2013).

Parasitic plants differ in the extent to which they depend on their hosts for nutrients. **Hemiparasites** are facultative xylem-feeding green plants that photosynthesize their nutrition and depend on their host only for water and minerals. The more specialized are the xylem and phloem-feeding obligate parasites known as **holoparasites**, which are non-photosynthetic and depend on their hosts for all nutrition (Irving and Cameron 2009). Parasitism has shaped molecular plastid genome evolution during the transition from autotrophy to a non-photosynthetic parasitic lifestyle. The functional reduction process resulted in plastid gene losses during periods of relaxed selection (Wicke et al. 2016).

Parasitic plants use root exudates for stimulating their successful germination immediately near a host root. As discussed earlier (see Sect. 3.8), plant roots exudate **strigolactones** (SLs) to attract arbuscular mycorrhizal fungi, which are plant-fungus symbionts that facilitate the uptake of soil nutrients (Akiyama et al. 2005). Parasitic plants adapted themselves to use these SLs signals (Cook et al. 1972), as well as other exudates like sesquiterpene lactone (Raupp and Spring 2013) to stimulate and ensure their germination adjacent to the root of their host.

Parasitic plants attached themselves to the vascular system of their host plant and act as a sink, extracting water and nutrients from their host through a specialized feeding structure called **haustorium**, which is a specialized root-like organ that serves as a bridge between the parasite and the host (Joel 2013; Shimizu et al. 2018). Thus, the parasite plant forms a natural type of grafting (see Sect. 10.2) that develops vascular connections between the two organisms. Parasitic plants can attach themselves and graft to a wide variety of unrelated species (Musselman 1980; Westwood et al. 2010; Kokla and Melnyk 2018).

Parasitic plants, like western hemlock dwarf mistletoe (*Arceuthobium tsugense*) that infect western hemlock (*Tsuga heterophylla*) trees, in temperate coniferous forests of the western United States, decrease tree growth. Under stress conditions of warm weather and drought, the parasitic plants increase the vulnerability of the trees, causing productivity losses and mortality events (Bell et al. 2020).

Devastating parasitic weeds, such as *Orobanche*, *Striga*, and *Phelipanche* (of the Orobanchaceae family), cause serious damage to host plants and tremendous losses to economically important crops, of a few billion US dollars per year worldwide (Scholes and Press 2008; Parker 2009).

Hemiparasites depend on water and minerals from their host but produce assimilates. To understand how the vessels of a hemiparasitic plant connect to the vessels of the host, the connection sites of *Loranthus acacia* that germinates and develops on branches of the *Acacia raddiana* tree, were analyzed (Aloni 2015). The study revealed that the parasitic and host plants can form a continuous vessel system connected by **simple perforation plates** enabling rapid and uninterrupted water and mineral uptake into the parasite (Fig. 20.1). The vessels of the parasite are narrower than the vessels of the host and therefore safer against embolism under stress conditions, which can give an advantage to the parasite. The narrow vessels of the parasite indicate that the parasite produces higher auxin concentration than the host

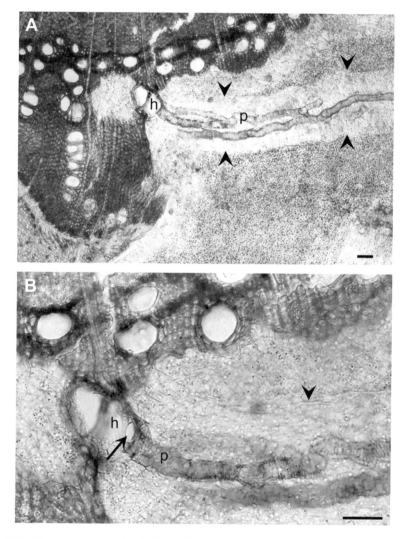

Fig. 20.1 Cross sections showing the haustorium penetration (from the *right side*) of the parasitic plant *Loranthus acacia* into the stem of its host tree *Acacia raddiana*. The *border lines* of the haustorium are marked with arrowheads. The xylem of the host is located in the *upper left side*. (**A**) The actively transporting vessels in both the host and the parasite were stained *red* by safranin transported through these vessels; the dye was applied to the host and moved upward into the parasite leaves (that became red) by transpiration. The 80 μm cross section was not stained after sectioning. (**B**) Magnified view of the same section focusing on a naturally occurring vessel perforation (marked by *arrow*), which forms a continuous open water-transporting lumen from the relatively wide vessel of the host to the narrow vessel of the parasite. h, host vessel; p, parasite vessel. Bars = 100 μm (From Aloni 2015)

plant (see Chap. 13). The continuity found in vessels of the *L. acacia* parasite and its host *A. raddiana* is similar to the continuous phloem reported by Dörr and Kollmann (1995) who found fully differentiated sieve plates of conducting elements that connected the phloem tissues of the holoparasite *Orobanche crenata* with its host *Vicia narbonensis,* forming direct symplastic connection through interspecific open sieve pores. Likewise, the sieve plate formed between *O. cumana* and sunflower sieve elements ensures a direct, symplastic contact between host and parasite and allows the parasite to obtain assimilates from the host (Krupp et al. 2019). However, additional studies show also other situations, e.g., in terminal haustoria of *Alectra vogelii*, the parasite phloem was separated from the host phloem by parenchyma (Dörr et al. 1979). In haustoria of the root-parasitic plant *Phelipanche aegyptiaca*, the phloem-conducting cells in haustorium retain nuclei and are not mature sieve elements. However, the translocation of symplasmic tracers from the host to the parasite demonstrated that they can function in assimilates translocation (Ekawa and Aoki 2017).

Bar-Nun et al. (2008) studied the role of IAA and auxin transport inhibitors applied to the host root on the infection of *Arabidopsis thaliana* by *Orobanche aegyptiaca*. They found that during germination and early development of the *O. aegyptiaca* parasite, it behaves as a root, namely, it acts as a sink to the auxin arriving from the stem of the host. Application of IAA, or auxin transport inhibitors, all resulted in a similar drastic reduction of *Orobanche* infection. These results demonstrate that the *O. aegyptiaca* parasite manipulates the host by acting as a sink for auxin. Disruption of auxin action or auxin flow at the contact site could be a novel basis for controlling infection by *Orobanche* parasites, supporting Bar-Nun et al. (2008) hypothesis that chemical disruptions of auxin transport and activity could influence the infection of the host by the parasite.

The hormonal mechanism of *Loranthus acacia* germination and growth on branches of *Acacia raddiana* trees is completely opposite. As the *L. acacia* parasite acts as a typical auxin-inducing shoot, from early germination and growth, the parasite is the source of the shoot signals, which descend into the host branch toward the host root. To determine if the perforations at the parasite-host junction (Fig. 20.1) are open and enable uninterrupted water flow from the host to the parasite, the vessel continuity was analyzed with diluted latex paint in water that was applied to the host stem (as described by Aloni and Griffith 1991). In addition, vessel continuity was also analyzed (under water to detect pressurized air bubbles) with pressurized air applied to the host (following Aloni and Griffith 1991). The results of both techniques were analyzed by making serial cross sections in the parasitic stem (from its young internodes downward), both showing the flow of latex paint particles and pressurized air flow into the parasite, confirming that the perforations are open and form a continuous open lumen through both the host and parasite vessels. Furthermore, auxin application (1% IAA in lanolin) on the stems of young *L. acaciae* parasite induced substantial increase in vessel differentiation, in the basipetal direction, in both the parasite and continuously in its *A. raddiana* host, demonstrating that the IAA can flow from the parasite to the host and induce continuous vessel differentiation also in the host (Aloni 2015). To understand how the simple open perforation differentiates between the vessel elements of the parasite and the host (Fig. 20.1b), there is need to acknowledge that both the parasitic plant and its host produce exactly the same hormonal signal, namely, indole-3-acetic acid. The polar moving stream of IAA from the parasite merges with a polar stream of IAA in the host. At their junction, the higher induced local auxin concentration promotes the differentiation of a perforation between the vessel element of the parasite and the vessel element of the host (see Sect. 19.1). Taken together, these observations and experimental results show that auxin is a transmissible signal from the parasite through the haustorium into the host and that it can induce continuous vessels between the two organisms, with open perforations at their vessel junctions.

It should be clarified that the differentiation of continuous vessels extending from the parasite to the host occurred simultaneously at a similar early stage of differentiation of both vessels. A new differentiating vessel descending from the parasite could not induce a perforation (an opening) in the cell wall of already dead vessel element in the host (Aloni 2015). Therefore, the report of Krupp et al. (2019) on the development of phloem connection between the holoparasite, (non-photosynthetic) *Orobanche cumana*, and its sunflower host is of great interest as it is different from the finding in the xylem. In their elegant study, they could distinguish between the host and the parasite sieve elements by differences found in the structure of their plastid, which enabled them to trace the exact contact site between the phloem of both species. Krupp et al. (2019) detected newly formed plasmodesmata between the host sieve-tube elements and parenchymatic cells of the parasite, thus showing that undifferentiated cells of the parasite could form plasmodesmata connections to already differentiated sieve elements of the sunflower host.

It is not a surprise that parasitic plant is so successful in establishing themselves on a wide range of host plants of unrelated species, as all these organisms use exactly the same hormonal signals. Once a signal from a parasitic plant penetrated into the host, it induces and regulates the host systems that are limited in their ability to antagonize the long-distance controlling signal originating in the parasite. The most important system for successful development of a parasitic plant is the vascular system, which enables the parasitic plant to feed on the host. Auxin and its polar movement is the primary signal that induces both phloem and xylem in flowering plants (see Chap. 5). As auxin can move through a graft union (see Sect. 10.2) and induces vascular continuity of both phloem and xylem along its pathway, parasitic plants successfully induce a continuous operating phloem and xylem systems connecting the parasite to the host plant.

A high auxin-to-cytokinin ratio contributed to haustorium development of the hemiparasitic *Santalum album*, and the concentrations of indole-3-acetic acid, zeatin, and zeatin riboside substantially increased after parasite-host attachment and penetration of the haustorium to the host. In addition, an increase was recorded in the GA-like substances and abscisic acid in the parasite tissues (Zhang et al. 2012), indicating that haustorium growth is regulated by increased hormonal stimulation that enhances and activates the haustorium to penetrate and grow into the host tissues and successfully induce the vascular connections required for absorbing water and nutrients from the host. It is likely that the increase in hormonal synthesis found in the *S. album* haustorium helps to produce mechanical force and enzymatic activity needed for penetration into the host (Zhang et al. 2012). Similarly, an increase in cytokinin concentration during infection was found in both the hemiparasitic plant *Phtheirospermum japonicum* and the *Arabidopsis thaliana* host (Spallek et al. 2017).

The parasite *Phtheirospermum* induces an increase in the size (hypertrophy) of the host vascular tissues, which leads to better physical attachment between the two organisms, resulting in increased sink strength at the haustorium connecting sites, to improve nutrient withdrawal by parasitic plants. This induced hypertrophy in *Arabidopsis*, required cytokinin signaling genes (*AHK3,4*) but not cytokinin biosynthesis genes (*IPT1,3,5,7*) in the host. Wild-type hosts with hypertrophy, that enabled better attachment for the parasite, were smaller than *ahk3,4* mutant hosts resistant to hypertrophy, suggesting that the induced hypertrophy of the host vascular tissues improves the efficiency of parasitism. Taken together, these results demonstrate that the interspecies movement of a parasite-derived hormone modified both host's vascular differentiation, root development, and fitness, which enable parasites to control their hosts (Spallek et al. 2017).

The attacked host plants use their defense responses to limit the invasion of the haustorium. Host plants use the defense hormones **jasmonic acid** (JA) and **salicylic acid** (SA) which are both stress hormones (Okada et al. 2015) that enable defense responses. For example, the attack of the parasite *Striga hermonthica* upregulates JA response gene expression in susceptible and resistant sorghum cultivars, and SA-induced defenses promote and determine sorghum resistance to *S. hermonthica* (Hiraoka and Sugimoto 2008). The limited studies that have addressed plant responses to parasitism by other plants probably suggest that both JA and SA can mediate effective host defenses (Smith et al. 2009).

20.2   Vascular Differentiation Controlled by Gall-Inducing Insects

Gall-inducing insects are parasitic organisms capable of controlling development and differentiation of plant tissues to serve their own needs. There are more than 15,000 gall-inducing insects from many families of different insect orders (Rohfritsch and Shorthouse 1982). The plants do not profit from the galls, which cause minor or sometimes serious damages to the host. The specific relationships of the insects with their host plants constitute one of the most elaborate and complicated insect-plant interactions, involving manipulation of the plant tissues as a food source and as protective habitation. The insects can induce galls on all plant organs of many families of angiosperms, gymnosperms, and ferns. The galls present best examples of plasticity of the plant responses to the galling insects (Rohfritsch 1992; Ferreira et al. 2019). The insects that induce the galls manipulate plant morphogenesis to form abnormal swelling outgrowths that might be considered as "**new plant organs**" which are induced by stimuli originating in the invading insects (Shorthouse et al. 2005). The insect-induced galls are not tumorous outgrowths, as the insects stimulate the development of an unusual type of organized tissues, which might become desiccated after the parasitic inhabitants leave the gall that served them as a feeding incubator. Galls on plants can be induced also by fungi, bacteria, and viruses. The well-studied development of the gall-inducing bacterium *Agrobacterium tumefaciens* will be discussed in Chap. 21.

The insect-induced galls are usually species-specific structures, and their anatomy and metabolism are strongly related to the gall inducer species and to the host plants (Ferreira et al. 2019 and references therein). The specialized group of gall-inducing aphids are host specific in the gall stage. Most aphid species alternate between trees (the primary host), where the gall is induced, and grasses or shrubs (secondary hosts). The galling aphids present some of the most complex and diverse life histories in the insect world (Wool 2004; Shorthouse et al. 2005).

There are numerous galls induced by different insect species with striking anatomical diversity and patterns. The gall's inhabitants induce novel networks of vascular tissues inside the gall, which supplies a nutritive tissue with nutrients on which the larvae feed, or aphids get their supply directly from the sieve tubes of the phloem tissue they induced, or obtain it from a specific fungus which proliferates in the larva chambers (Fig. 20.2) and on which the larvae feed, rather than on plant tissues, or in addition to them (Dorchin et al. 2002; Wool 2004).

Analysis of vascular differentiation in the galls induced by the midges *Izeniola obesula* on the host plant *Suaeda monoica* (Dorchin et al. 2002), provides evidence for the source of the vascular-inducing signal and the nature of the signal. Figure 20.2B shows that the earliest vessel elements (that will supply water to the larva) mature first in the larva chamber, showing the source of signaling and demonstrating that at this site the concentration of the inducing signal is the highest. Earlier observations (Aloni 2004) demonstrated that initial vessel-element maturation occurs at sites of high local auxin concentrations, e.g., near the auxin-producing hydathodes or at vascular junctions (where two auxin streams merge) in developing leaves (see Sect. 2.4; Fig. 2.21B, C); early vessel maturation, which occurs in discontinuous xylem patterns, starts immediately beneath the auxin-producing tips of developing flower organs (at the sites of high auxin concentrations) in the tip of a young gynoecium and in young stamen primordia (Fig. 2.22A, B). During gall development, the discontinuous vessel elements, which first mature at the larva chamber, become connected to form a continuous vessel from the larva chamber to an original bundle of the host stem (Fig. 20.2D).

In addition, the differentiation of a circular vessel at the larva chamber from stem pith parenchyma, characterized by low tissue polarity, also indicates that auxin is the stimulating signal originating in the larva chamber. Circular vessels are induced by auxin in sites where auxin flux is expected to follow in circular routes (Sachs and Cohen 1982), which normally occurs in tissue cultures (Aloni 1980), at branch junctions (see Chap. 14, Figs. 14.1 and 14.2) (Lev-Yadun and Aloni 1990; Aloni et al. 1997), suppressed buds (Aloni and Wolf 1984) and in tumorous galls (Chap. 21; Fig. 21.10) (Aloni et al. 1998).

Both the differentiation of circular vessels and the early maturation of vessel elements at the larva chamber are evidenced that the insect larva is the source of the vessel-inducing signal. The merging of the novel vessels from the larva chamber

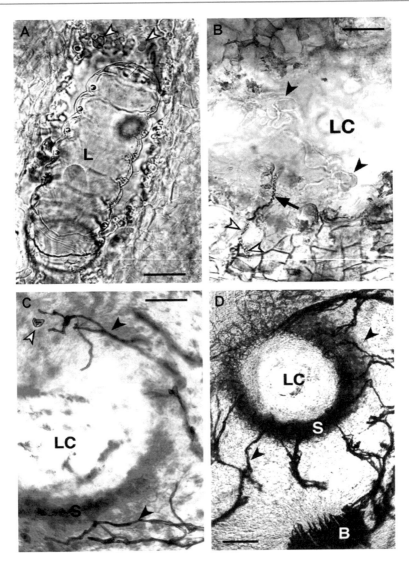

Fig. 20.2 Tissue differentiation in stem galls induced by the midges *Izeniola obesula* on the host plant *Suaeda monoica*, observed in cleared, hand-cut cross sections. Showing: (**A**) In 1-week-old gall, young larva (L) in its chamber. (**B**) A mature vessel element (*arrow*) and an early stage of vessel element development (*white arrowhead*), starting from the *larva chamber* (LC), that containing fungal mycelia (*black arrowhead*) on which the larva feeds. (**C**) A circular vessel (*white arrowhead*) differentiated near the origin of the larva-induced vessels (*black arrowheads*) descending from the larva chamber towards the stem's vascular bundles. (**D**) In mature gall, showing larva chamber protected by surrounding sclerenchyma (S) and supplied with water by network of novel vessels (*arrowheads*) induced by the larva, which merges into the original stem's vascular *bundle* (*B*). Bars = 25 μm (**A, B**), 50 μm (**C**), 100 μm (**D**). (From Dorchin et al. 2002)

into the stem's original bundles indicates that the polar auxin flow is the inducing signal originating from larva activity in its chamber. Consequently, the above vascular analysis strongly suggests that the gall-inducing larva produces auxin. This assumption was in accordance with early finding that phytohormones were found in the saliva and extracts of many types of gall inducers (reviewed in Hori 1992). However, all the early plant hormonal studies were done using bioassays, which have limited validity.

The larvae of the fruit fly *Eurosta solidaginis* which induces ball galls on the stems of *Solidago altissima* contain high levels of both IAA and cytokinins (Mapes and Davies, 2001a, b). The phytohormones in these studies were identified and quantified by gas chromatography-mass spectrometry, with internal standards. High concentrations of the following cytokinins: zeatin (Z), zeatin riboside (ZR), isopentenyladenosine (iPA), and isopentenyladenine (iP) were detected in the larvae of the fruit fly *E. solidaginis*. The larvae and galls contained elevated levels of cytokinins up to 53 times more than those in stem tissues (Mapes and Davies 2001b).

No applications of plant hormones or related substances have been able to duplicate the development or structure of insect galls. The failure to mimic gall development by application of compounds has almost certainly resulted from the inability to effectively simulate both the precise location, hormonal concentrations, and continuous production by the insect within the plant tissue, suggesting that the larvae of *E. solidaginis* act as continuous source points of both IAA and cytokinins during the development of the galls (Mapes and Davies 2001b).

Similarly, the midge *Rhopalomyia yomogicola* that induces galls on *Artemisia princeps* contained high levels of IAA and cytokinins. The gall midge larvae synthesized IAA from tryptophan, suggesting that in addition to the hormonal role in regulating gall development and vascular differentiation, the plant hormones may also have a functional significance in maintaining the feeding part of the gall as fresh nutritive tissue (Tanaka et al. 2013).

The gall-inducing sawfly larvae *Pontania* sp. on its host plant *Salix japonica* contains about 1000 ng g^{-1} IAA, which is about 100 times higher concentration of auxin than that in the leaf tissue of its host (Yamaguchi et al. 2012). Also in this gall system, the sawfly larvae synthesize IAA from tryptophan. Sawfly larvae contain also high concentrations of t-zeatin. In the glands of adult sawflies, the contents which are injected into leaves upon oviposition and are involved in the initial stages of gall formation contain an extraordinarily high concentration of t-zeatin riboside. The abnormally high concentration of this cytokinin in the glands strongly suggests that the sawfly larvae can synthesize cytokinins as well as IAA (Yamaguchi et al. 2012).

Based on the finding of Suzuki et al. (2014) that not only galling insects but also non-galling insects, like the silkworm *Bombyx mori*, can produce bioactive auxin (IAA) by *de novo* synthesis from tryptophan (Trp); Takei et al. (2019) elegantly showed that both **galling and non-galling insects synthesize indole-3-acetic acid** (IAA) from tryptophan (Trp) via two intermediates, indole-3-acetaldoxime (IAOx) and indole-3-acetaldehyde (IAAld). They have succeeded to isolated an enzyme that catalyzes the last step "IAAld → IAA" from a silk-gland extract of the silkworm. The enzyme is BmIAO1 which causes the nonenzymatic conversion of Trp to IAAld and the enzymatic conversion of IAOx to IAA, suggesting that BmIAO1 alone is responsible for IAA production in *B. mori* and that this enzyme is possibly a key player in IAA production in a broad range of insects (Takei et al. 2019).

The above studies (Mapes and Davies, 2001a, b; Tanaka et al. 2013; Yamaguchi et al. 2012; Suzuki et al. 2014; Takei et al. 2019) demonstrate that **galling insects can synthesize high concentrations of both major plant hormones, namely, auxin and cytokinins**. The ability of gall-inducing insects to produce these two basic controlling plant hormones provides them with powerful tools to manipulate the development of plant tissue and specifically to induce vascular differentiation for their benefit. These two working tools enable insects to shape plant morphogenesis and form countless types of striking galls to supply their food and provide their protective habitation.

In addition to the vascular tissues induced by the galling insects inside their gall (e.g., Fig. 20.2B–D), growing galls induce increased vascular differentiation below them, which is promoted by the auxin produced by the insects. A remarkable increase in xylem formation is evident under large galls induced by aphids (Figs. 20.3 and 20.4). Such galls may contain a few hundreds aphids (Wool 2004), which make the gall extremely large. During the life cycles of the gall-inducing aphids, *Slavum wertheimae* on *Pistacia atlantica* trees (Fig. 20.3) (Aloni et al. 1989) and *Baizongia pistaciae* on *Pistacia palaestina* trees (Fig. 20.4) (Aloni 1991), from the fertilized aphid egg emerges the nymph in the spring and induces the gall on a very young, unfolded leaf. Within the gall, which serves as a reproductive incubator for the aphids, a single genotype is reproduced parthenogenetically, and a few hundred aphids are formed before dispersal (Wool 2004). The parthenogenesis during most of their life cycle is interrupted by a single stage of sexual reproduction on the primary host shoot, when the aphids leave the gall. Both aphid species substantially induce more xylem (likely induced by their auxin and cytokinins stimulation) below their galls, characterized by the production of wide vessel in the latewood (Figs. 20.3D and 20.4B), while in the intact control branches, the latewood contain fewer narrow vessels (Figs. 20.3C and 20.4C). The wide latewood vessels of both species developed under the galls were very long and extended from 30 to 50 cm (these maximum vessel length data was determined under water with pressurized air bubbles, R Aloni unpublished results). These wide and long vessels are very efficient water transporting conduits for the growing aphid population. However, when the aphids leave the gall at the end of the season, these wide latewood vessels become embolized and cause serious damage to the gall-bearing branches, demonstrating the parasitic nature of these insects that during the season function as sinks for nutrients and when they leave the galls, the gall-bearing branches tend to dry out (R. Aloni, unpublished results).

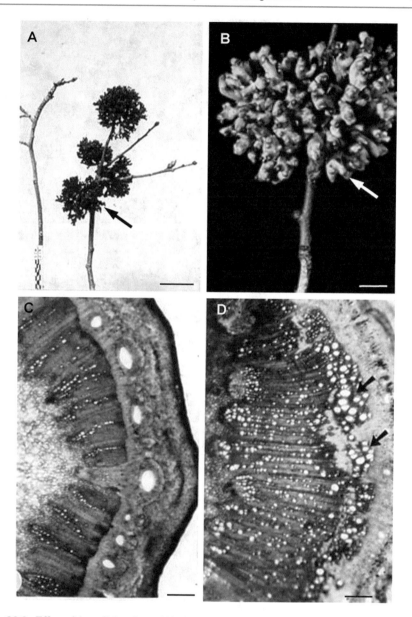

Fig. 20.3 Effect of the gall-forming aphid *Slavum wertheimae* on secondary xylem differentiation below the gall, in branches of *Pistacia atlantica* trees. (**A**) *P. atlantica* branch before bud break, carrying four coral-like ('cauliflower') galls (*right*), and a control branch with no galls (*left*). (**B**) Close-up view of the coral-like gall (*arrow*). (**C**) and (**D**) are transverse sections from 1-year-old branches of *P. atlantica*. Both sections are at the same magnification. (**C**) is taken from a control branch, free of galls. (**D**) is taken 20 mm below a coral-like gall induced by *S. wertheimae*. More xylem is evident immediately below the gall and is characterized by numerous wide vessels. In the latewood, there are some very wide vessels (*arrows*). Bars = 250 μm (**C, D**), 10 mm (**B**), 30 mm (**A**). (From Aloni et al. 1989)

Fig. 20.4 Effect of the gall-forming aphid *Baizongia pistaciae* on secondary xylem differentiation, shown in transverse sections made in 3-year old branches, 100 mm below the gall (**B**), and at an equivalent location in an ungalled branch (**C**) of the same *Pistacia palaestina* tree. The earlywood vessels of the current year are marked with *black arrows* in both branches (**B**, **C**). (**A**) *P. palaestina* branch with the typical elongated gall shape (*arrow*) produced by *B. pistaciae* aphids. (**B**) Substantially more xylem was induced below the gall (by hundreds of aphids living inside this gall) with wide vessels (*white arrow*) in the latewood. (**C**) Normal xylem with narrow latewood vessels (*white arrow*) differentiated in an adjacent ungalled branch. Bars = 250 μm (**B**, **C**), 25 mm (**A**). (From Aloni 1991)

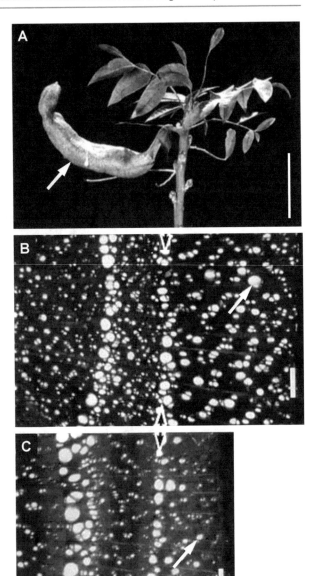

Summary

- Parasitic plants and insect inducing galls use their host to supply their needs. Both groups are successful in manipulating their hosts due to their ability to produce the two major phytohormones, auxin and cytokinins, in higher concentrations than the hormonal levels in their host tissues and therefore effectively control the development of their host and the differentiation of their vascular tissues.

- Parasitic plants are successful in establishing themselves on a wide range of host plants of unrelated species, as all these organisms use exactly the same set of hormonal signals. Once a hormonal signal from a parasitic plant penetrates into the host, it regulates and controls the host tissues by the long-distance auxin signal originating in the parasite.
- Parasitic plants can induce in their hosts a continuous and functional vessel system, which connect both organisms through simple open perforations.
- Likewise, the xylem and phloem-feeding parasites (holoparasites) can also form direct symplastic connection through interspecies open sieve pores for supplying their water and nutrients.
- Conversely, Insect inducing galls are usually species-specific. These specific relationships of galling insects with their host plants constitute one of the most elaborate and complicated insect-plant interactions, where the insects manipulate the plant tissues to provide their food and become their protective habitation.

References and Recommended Readings[1]

Akiyama, K., Matsuzaki K, Hayashi H (2005) Plant sesquiterpenes induce hyphal branching in arbuscular mycorrhizal fungi. Nature 435: 824–827.

Aloni R (1980) Role of auxin and sucrose in the differentiation of sieve and tracheary elements in plant tissue cultures. Planta 150: 255–263.

Aloni R (1991) Wood formation in deciduous hardwood trees. In: *Physiology of Trees*. AS Raghavendra (ed). Wiley & Sons, New York, pp. 175–197.

Aloni R (2004) The induction of vascular tissue by auxin. In: *Plant Hormones: Biosynthesis, Signal Transduction, Action!* PJ Davies (ed). Kluwer Academic Publishers, Dordrecht, Boston, London, pp. 471–492.

*Aloni R (2015) Ecophysiological implications of vascular differentiation and plant evolution. Trees 29: 1–16.

Aloni R, Alexander JD, Tyree MT (1997) Natural and experimentally altered hydraulic architecture of branch junctions in *Acer saccharum* Marsh. and *Quercus velutina* Lam. trees. Trees 11: 255–264.

Aloni R, Griffith M (1991) Functional xylem anatomy in root-shoot junctions of six cereal species. Planta 184: 123–129.

Aloni R, Katz DA, Wool D (1989) Effect of the gall-forming aphid *Slavum wertheimae* on the differentiation of xylem in branches of *Pistacia atlantica*. Ann Bot 63: 373–375.

Aloni R, Wolf A (1984) Suppressed buds embedded in the bark across the bole and the occurrence of their circular vessels in *Ficus religiosa*. Am J Bot 71: 1060–1066.

Aloni R, Wolf A, Feigenbaum P, Avni A, Klee HJ (1998) The *Never ripe* mutant provides evidence that tumor-induced ethylene controls the morphogenesis of *Agrobacterium tumefaciens*-induced crown galls on tomato stems. Plant Physiol 117: 841–847.

Bar-Nun N, Sachs T, Mayer AM (2008) A role for IAA in the infection of *Arabidopsis thaliana* by *Orobanche aegyptiaca*. Ann Bot 101: 261–265.

---

[1] Papers of particular interest for suggested reading have been highlighted (with *)

Barkman TJ, McNeal JR, Lim SH, Coat G, Croom HB, Young ND, Depamphilis CW (2007) Mitochondrial DNA suggests at least 11 origins of parasitism in angiosperms and reveals genomic chimerism in parasitic plants. MC Evol Biol 7: 248.

*Bell DM, Pabst RJ, Shaw DC (2020) Tree growth declines and mortality were associated with a parasitic plant during warm and dry climatic conditions in a temperate coniferous forest ecosystem. Glob Change Biol 26: 1714–1724.

*Bromham L, Cowman PF, Lanfear R (2013) Parasitic plants have increased rates of molecular evolution across all three genomes. BMC Evol Biol 13: 126.

Cook CE, Whichard LP, Wall ME (1972) Germination stimulants. II. The structure of strigol - a potent seed germination stimulant for witchweed (*Striga lutea* Lour.). J Am Chem Soc 94: 6198–6199.

Dorchin N, Freidberg, A, Aloni R (2002) Morphogenesis of stem gall tissues induced by larvae of two cecidomyiid species (Diptera: Cecidomyiidae) on *Suaeda monoica* (Chenopodiaceae). Can J Bot 80: 1141–1150.

Dörr I, Kollmann R (1995) Symplasmic sieve element continuity between *Orobanche* and its host. Bot Acta 108: 47–55.

Dörr I, Visser JH, Kollmann R (1979) On the parasitism of *Alectra vogelii* Benth. (Scrophulariaceae) – III. The occurrence of phloem between host and parasite. Z Pflanzenphysiol 94: 427–439.

Ekawa M, Aoki K (2017) Phloem-conducting cells in haustoria of the root-parasitic plant *Phelipanche aegyptiaca* retain nuclei and are not mature sieve elements. Plants (Basel) 6: E60.

*Ferreira BG, Álvarez R, Bragança GP, Alvarenga DR, Pérez-Hidalgo N, Isaias RMS (2019) Feeding and other gall facets: patterns and determinants in gall structure. Bot Rev 85: 78–106.

Hiraoka Y, Sugimoto Y (2008) Molecular responses of sorghum to purple witchweed (*Striga hermonthica*) parasitism. Weed Sci 56: 356–363.

Hori K (1992) Insect secretion and their effect on plant growth, with special reference to hemipterans. In: *Biology of insect-induced galls*. JD Shorthouse, O Rohfritsch (eds). Oxford University Press, New York, pp. 157–170.

Irving LJ, Cameron DD (2009) You are what you eat: interactions between root parasitic plants and their hosts. Adv Bot Res 50: 87–138.

Joel DM (2013) Functional structure of the mature haustorium. In: *Parasitic Orobanchaceae – parasitic mechanisms and control strategies*, DM Joel, J Gressel, LJ Musselman (eds). Springer, Berlin, pp. 25–60.

Kokla A, Melnyk CW (2018) Developing a thief: haustoria formation in parasitic plants. Dev Biol 442: 53–59.

Krupp A, Heller A, Spring O (2019) Development of phloem connection between the parasitic plant *Orobanche cumana* and its host sunflower. Protoplasma 256: 1385–1397.

Kuijt J (1969) *The Biology of Parasitic Flowering Plants*. University of California Press, Berkeley.

Lev-Yadun S, Aloni R (1990) Vascular differentiation in branch junction: circular patterns and functional significance. Trees 4: 49–54.

Mapes CC, Davies PJ (2001a) Indole-3-acetic acid in the ball gall of *Solidago altissima*. New Phytol 151: 195–202.

Mapes CC, Davies PJ (2001b) Cytokinins in the ball gall of *Solidago altissima* and in the gall forming larvae of *Eurosta solidaginis*. New Phytol 151: 203–212.

Musselman LJ (1980) The biology of *Striga, Orobanche*, and other root parasitic weeds. Annu Rev Phytopathol 18: 463–489.

Okada K, Abe H, Arimura G (2015) Jasmonates induce both defense responses and communication in monocotyledonous and dicotyledonous plants. Plant Cell Physiol 56: 16–27.

Parker C (2009) Observations on the current status of *Orobanche* and *Striga* problems worldwide. Pest Manag Sci 65: 453–459.

Raupp FM, Spring O (2013) New sesquiterpene lactones from sunflower root exudate as germination stimulants for *Orobanche cumana*. J Agric Food Chem 61: 10481–10487.

Rohfritsch O (1992) Patterns in gall development. In: *Biology of insect-induced galls*. JD Shorthouse, O Rohfritsch (eds), Oxford University Press, New York, pp. 60–86.

Rohfritsch O, Shorthouse JD (1982) Insect galls. In: *Molecular Biology of Plant Tumors*, G Kahl, JS Schell (eds). Academic Press, New York, pp. 131–152.

Sachs T, Cohen D (1982) Circular vessels and the control of vascular differentiation in plants. Differentiation 21: 22–26.

Scholes JD, Press MC (2008) *Striga* infestation of cereal crops - an unsolved problem in resource limited agriculture. Curr Opin Plant Biol 11: 180–186.

Shimizu K, Hozumi A, Aoki K (2018) Organization of vascular cells in the haustorium of the parasitic flowering plant *Cuscuta japonica*. Plant Cell Physiol 59: 715–723.

*Shorthouse JD, Wool D, Raman A (2005) Gall-inducing insects - nature's most sophisticated herbivores. Basic Appl Ecol 6: 407–411.

Smith JL, De Moraes CM, Mescher MC (2009) Jasmonate- and salicylate-mediated plant defense responses to insect herbivores, pathogens and parasitic plants. Pest Manag Sci 65: 497–503.

*Spallek T, Melnyk CW, Wakatake T, Zhang J, Sakamoto Y, Kiba T, Yoshida S, Matsunaga S, Sakakibara H, Shirasu K (2017) Interspecies hormonal control of host root morphology by parasitic plants. Proc Natl Acad Sci USA 114: 5283–5288.

Suzuki H, Yokokura J, Ito T, Arai R, Yokoyama C, Toshima H, Nagata S, Asami T, Suzuki Y (2014) Biosynthetic pathway of the phytohormone auxin in insects and screening of its inhibitors. Insect Biochem Mol Biol 53: 66–72.

*Takei M, Kogure S, Yokoyama C, Kouzuma Y, Suzuki Y (2019) Identification of an aldehyde oxidase involved in indole-3-acetic acid synthesis in *Bombyx mori* silk gland. Biosci Biotechnol Biochem 83: 129–136.

*Tanaka Y, Okada K, Asami T, Suzuki Y (2013) Phytohormones in Japanese mugwort gall induction by a gall-inducing gall midge. Biosci Biotechnol Biochem 77: 1942–1948.

Westwood JH, Yoder JI, Timko MP, dePamphilis CW (2010) The evolution of parasitism in plants. Trends Plant Sci 15: 227–235.

Wicke S, Müller KF, dePamphilis CW, Quandt D, Bellot S, Schneeweiss GM (2016) Mechanistic model of evolutionary rate variation en route to a nonphotosynthetic lifestyle in plants. Proc Natl Acad Sci USA 113: 9045–9050.

*Wool D (2004) Galling aphids: specialization, biological complexity, and variation. Annu Rev Entomol 49: 175–192.

Yamaguchi H, Tanaka H, Hasegawa M, Tokuda M, Asami T, Suzuki Y (2012) Phytohormones and willow gall induction by a gall-inducing sawfly. New Phytol 196: 586–595.

Zhang X, Teixeira da Silva JA, Duan J, Deng R, Xu X, Ma G (2012) Endogenous hormone levels and anatomical characters of haustoria in *Santalum album* L. seedlings before and after attachment to the host. J Plant Physiol 169: 859–866.

Cancer and Vascular Differentiation

21

21.1 Role of Vascular Differentiation in Plant Tumor Development

Plants might develop tumors (Ullrich and Aloni 2000; Aloni and Ullrich 2008; Ullrich et al. 2019). When young seedlings are infected by the soil bacterium *Agrobacterium tumefaciens*, they develop tumors (Fig. 21.1), might not survive, and die. However, plant tumors do not ultimately kill their host but reduce plant growth and can cause damage to crop yield of agricultural plants, by reducing the quality and quantity of fruits, which may cause severe economic damages. Therefore, there is a major challenge to develop methods for preventing cancer formation, suppressing tumor growth, recovery from cancer and avoiding tumor damages (Ullrich et al. 2009, 2019; Kuzmanović et al. 2018). When a cancer develops on a large tree, although the tumor may reach considerable sizes of more than 100 mm and sometimes even about 1000 mm, it may not kill the tree. The tumor tissues compete with healthy tissues on water and nutrient supply that may substantially impair plant growth. Therefore, understanding the control mechanisms of tumor development is a basic tool and major request for finding cure solutions for preventing cancer damages in plants. In addition, understanding the hormonal role of plant tumors has the potential of inventing therapy solutions to cure also cancers of human beings as will be discussed in this chapter.

The terms **cancer**, or tumor, defines a collection of diseases with the common feature of uncontrolled growth and structures. When a plant tumor, or a solid animal and human cancer, is analyzed anatomically, the tumorous tissues are characterized by unorganized tissues of unpredicted patterns of different cell sizes or shapes at various polarities and cell division activity (Fig. 21.2). Vascular differentiation in a plant tumor produces functional tissues, as will be shown below.

The most studied and common tumor in plant, which can develop on hundreds of different sensitive dicotyledonous species, is called **crown gall**, which is induced by the soil bacteria *Agrobacterium tumefaciens* and *A. vitis* (Cavara 1895; Smith and Townsend 1907; Burr and Otten 1999; Aloni and Ullrich 2008; Ullrich et al. 2019).

© Springer Nature Switzerland AG 2021
R. Aloni, *Vascular Differentiation and Plant Hormones*,
https://doi.org/10.1007/978-3-030-53202-4_21

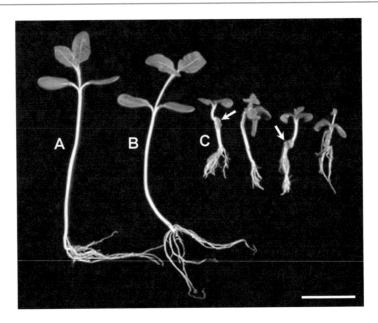

Fig. 21.1 Effects of wounding and exposure to *Agrobacterium tumefaciens* on tumor development in sunflower (*Helianthus annuus*) seedling, 2 weeks after infection. (**A**) Intact control not
exposed to the bacterium. (**B**) Unwounded seedling exposed to *A. tumefaciens* did not develop
tumor and continue normal growth. (**C**) Wounded seedlings exposed to *A. tumefaciens* developed
tumors (*arrows*) which substantially retarded plant growth. The seedling with the crown galls
degenerated and died after another 2–3 weeks. Bar = 50 mm

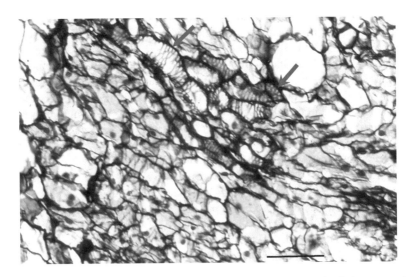

Fig. 21.2 Photomicrograph of a transverse section in a typically unorganized plant tumor tissue
in an *Agrobacterium tumefaciens*-induced crown gall on sunflower, showing vessel elements
(*arrows*) with lignified secondary cell wall (stained red by safranin) and parenchyma cells in various orientations and sizes, due to random cell divisions. Bar = 100 μm

The vectors which transfer the bacteria to plants are usually insects and sometimes birds, which wound the plant and infect the injured tissues. The bacteria do not induce crown gall tumors without wounding (Fig. 21.1B), or environmentally induced cracks that develop at the base of the stem, due to extreme changes of temperatures and moisture at the soil surface. Cellular changes are required in dividing plant cells in order to generate cancer as will be discussed below. The important discovery that from a large *Agrobacterium tumefaciens* tumor inducing plasmid, **Ti-plasmid**, a DNA sequence of about 20 kb, **T-DNA**, is stably incorporated into the higher plant genome (Van Larebeke et al. 1974; Chilton et al. 1977) was a breakthrough in molecular biology and biotechnology. Molecular biological research exploded upon the discovery that disarmed *agrobacterium* (from which the oncogenic genes were removed) can be used as gene vectors between pro- and eukaryotes. The T-DNA-located oncogenes have been identified. The most prominent ones which induced the tumor are those encoding enzymes of auxin (*iaaM, iaaH*), cytokinin (*ipt*), and opine biosynthesis (*nos/ocs*) (Weiler and Schröder 1987). Opine is used by the infecting bacteria as a specific source of their nutrient supply.

The expression of the oncogenes *iaaM* and *iaaH* for auxin synthesis, via a pathway from tryptophan over indoleacetamide (IAM), is well documented in bacteria. Usually, there is a requirement for a permanent presence of the phytohormone-producing bacteria in the gall for its development. These oncogenic genes were found in *Pseudomonas syringae* subsp. *savastanoi*, which causes gall tumors in olive (Fig. 21.3A) and oleander (Yamada et al. 1985; Ullrich et al. 2019), and in *Erwinia herbicola* pv. *gypsophilae*, which induces tumorous galls in *Gypsophila* (Manulis et al. 1998; Chalupowicz et al. 2006). Likewise, the cytokinin synthesis pathway via isopentenyltransferase (*ipt*) has been found in microorganisms. The synergism with auxin might lead to gall formation, as induced by *Rhodococcus fascians* (Thimann and Sachs 1966; Vereecke et al. 2000). However, in *Agrobacterium tumefaciens* galls (Thomashaw et al. 1986), after infection and the T-DNA is stably incorporated into the higher plant genome, there is no need for the presence of the *A. tumefaciens* bacteria in the gall for its additional development.

Plant tumors induced by *A. tumefaciens* were considered unorganized or partly organized masses (Sachs 1991 and references therein). However, uncovering the entire three-dimensional architecture of vascular tissues (Figs. 21.3B, 21.4, and 21.5) in crown gall induced by *A. tumefaciens*, in thick sections cleared with lactic acid and stained with lacmoid (Aloni et al. 1995) [a method which simultaneously revealing the three-dimensional structure of both vessels (by staining their lignified walls dark blue) and sieve elements (by staining the callose of the sieve plates sky blue) (Fig. 21.5) stained according to Aloni and Sachs (1973)], unveiled a quite sophisticated vascular network of continuous vascular strands extending from the host plant up to the tumor surface (Figs. 21.3B and 21.4). The development of these strands indicates synthesis of auxin by the *A. tumefaciens*-transformed plant cells located immediately beneath the surface of the fast growing tumor (Figs. 21.3B and 21.4) (Aloni et al. 1995). The crown galls developed on a castor bean (*Ricinus communis*) stem contained two types of vascular strands: tree-like-branched bundles which developed towards the tumor surface in the fast growing regions and globular

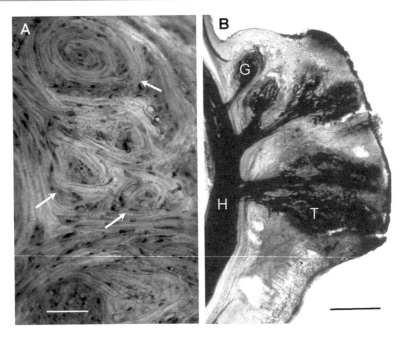

Fig. 21.3 Photomicrographs of vascular tissues in tumors shown in longitudinal radial sections both stained with lacmoid (according to Aloni 1979). (**A**) Circular patterns (*arrows*) of vessels and fibers in olive (*Olea europaea*) hardwood gall induced by the bacterium *Pseudomonas syringae* subsp. *savastanoi*. (**B**) Typical patterns of vascular differentiation in a 3-week-old crown gall tumor induced by *A. tumefaciens* on castor bean (*Ricinus communis*) showing both globular (G) and tree-like (T) vascular bundles, emerging from a main bundle of the host (H). Note that the lower side of the host bundle is thicker than its upper side, due to additional vascular tissues induced by the tumor. Bars = 200 μm (**A**); 20 mm (**B**). (From Ullrich et al. 2019 (**A**); and from Aloni et al. 1995 (**B**))

bundles with circular vessels in the slowly developing parts (Figs. 21.3B and 21.4), which have an inner xylem surrounded by phloem (Aloni et al. 1995). The xylem and phloem bundles (Fig. 21.5A) are interconnected by a dense network of phloem anastomoses (Fig. 21.5B) (Aloni et al. 1995), which are also common in healthy stems (Aloni and Sachs 1973; Aloni and Peterson 1990; Aloni and Barnett 1996) (see Sect. 2.2.2).

On the other hand, growing crown galls dramatically affect the structure of the developing host tissues at the tumor-host junction (Fig. 21.6B, C). The tumor causes in the host stem pathologic xylem adjacent to the tumor, which is characterized by narrow vessels adjacent to the tumor (Fig. 21.6C) that restrict water flow to the shoot organs above the gall (Aloni et al. 1995; Schurr et al. 1996), whereas the gall promotes more transporting xylem immediately below the tumor (*arrow* in Fig. 21.7B) with regular vessel diameters; therefore, the tumor itself is well sup-plied with water (Fig. 21.7). These anatomical findings led us to suggest the ***gall constriction hypothesis*** (Aloni et al. 1995) for explaining the mechanism of hydrau-lic constriction in the tumor-stem junction, necessary for efficient competition for

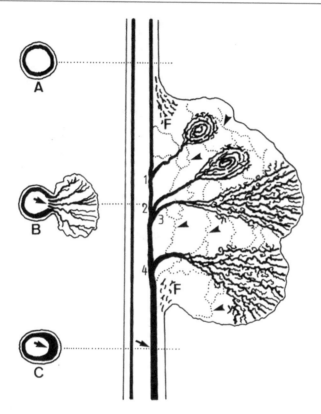

Fig. 21.4 Longitudinal schematic diagram showing the pattern of vascular bundles (*thick lines*), and phloem anastomoses (*dotted lines*) in *Agrobacterium tumefaciens*-induced crown gall on a *Ricinus communis* stem. In addition, three schematic diagrams of transverse sections, illustrating the symmetric structure of a healthy vascular system in the host stem above the tumor (**A**), the pathologic host xylem (*arrow*) with giant rays in a median position (**B**), and the asymmetric xylem differentiation below the tumor (**C**). Numbers 1 and 2 mark the connecting sites of globular vascular bundles to the host vascular system, while 3 and 4 mark the base of tree-like branched vascular bundles, both consisting of xylem and phloem. *Arrowheads* mark some of the phloem anastomoses between the bundles (shown in Fig. 21.5B); F marks regenerative phloem fibers with unique anatomical ramifications (shown in Chap. 2, Fig. 2.25B, C) restricted to the upper and lower basal regions of the tumor. (From Aloni et al. 1995)

water and mineral supply between the developing gall and the host shoot. The hypothesis proposes that growing galls retard the development of their host shoot by a signal that reduces vessel expansion and thus limits vessel width in the host stem. It was, therefore, suggested that this controlling signal is the hormone **ethylene** which is known to substantially reduce vessel width in plants (Yamamoto et al. 1987), also induces large vascular rays (see Chap. 15), and inhibits fiber differentiation, as was evident in the pathologic xylem tissue adjacent to the tumor (Fig. 21.6B, C). This is a typical example that demonstrates how only by understanding the implication of the pathologic xylem anatomy inflicted by the tumor in the host stem;

Fig. 21.5 Photomicrographs of longitudinal views of phloem and xylem in 8-week-old *Agrobacterium tumefaciens*-induced crown galls developed on *Ricinus* cleared in lactic acid and stained with lacmoid (according to Aloni and Sachs 1973). (**A**) Showing a portion of a large vascular bundle build of vessels (*arrowheads*) and sieve tubes, identified by turquoise blue staining of callose on their sieve-plates (*arrow*). (**B**) Phloem anastomoses between the bundles consisting of sieve tubes only (*arrows*). Bar = 100 μm. (From Aloni et al. 1995)

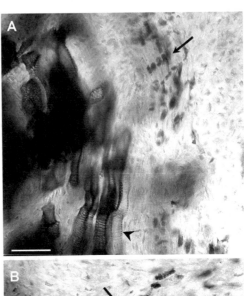

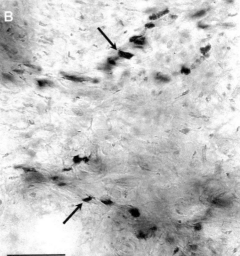

we could propose the *gall constriction hypothesis* and suggest the possible role of ethylene in tumor development (Aloni et al. 1995).

The integration and expression of the T-DNA of the bacterial Ti plasmid within the plant nuclear DNA (Zambryski et al. 1989) substantially elevate **auxin** concentrations in crown gall tissues, which may be 500 times higher in the tumor than in control tissues (Weiler and Spanier 1981). As there are no specific oncogenes for ethylene synthesis in the bacterial T-DNA, we proposed that the high auxin levels, which are known to enhance ethylene synthesis (Yang and Hoffman 1984; Abeles et al. 2012), promote the synthesis of ethylene in crown galls (Aloni et al. 1995). It is also possible that elevated **cytokinin** levels, which may be 1500 times higher in the tumor than in control tissues (Weiler and Spanier 1981; Akiyoshi et al. 1983),

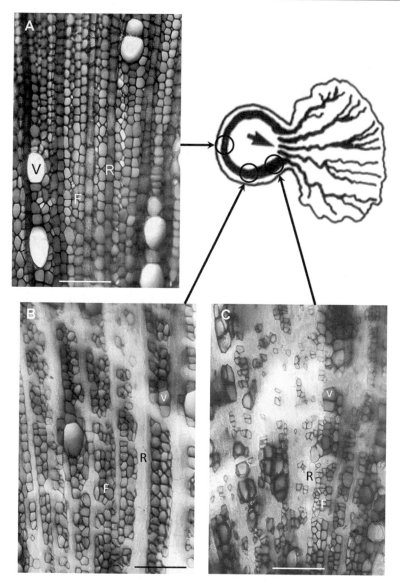

Fig. 21.6 Influence of 8-week-old *Agrobacterium tumefaciens*-induced crown gall on vascular differentiation in a *Ricinus* host stem. Showing transverse sections of healthy xylem formed away from the tumor (**A**), in comparison with pathological xylem close to the tumor (**B**), and very close to the crown gall (**C**) (the location of each section is marked by a circle on the schematic diagram). The vessels (V) and rays (R) had a normal structure away from the tumor (**A**). Narrow vessels and fibers (F) with large unlignified rays differentiated close to the grown gall (**B**). Maximum effect of the crown gall on the host was evident adjacent to the tumor, where vessels were very narrow and the rays were giant and unlignified (**C**). Bars = 100 μm. (From Aloni et al. 1995)

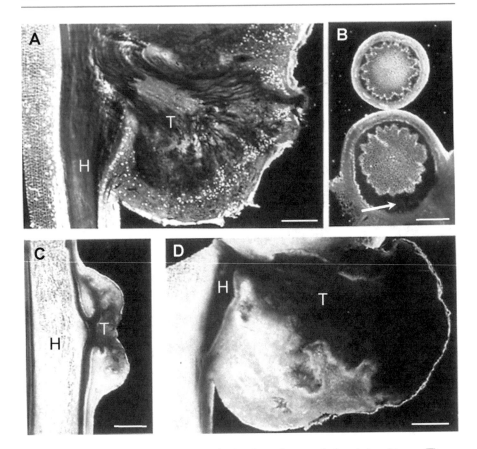

Fig. 21.7 Structure and function of xylem in *Agrobacterium tumefaciens*-induced tumor (T) on a stem of the *Ricinus communis* host (H). (**A**) Longitudinal section in an 8-week-old tumor stained with lacmoid. The xylem of the tumor is connected to the host and extends nearly up to the tumor periphery. (**B**) Stem cross sections above and below a 3-week-old tumor stained with lacmoid. The xylem formed immediately below the tumor (*arrow*) is twice as large as that above tumor (compare schematic diagrams A vs C in Fig. 21.4). (**C**) Path of water that were labelled with the red acid fuchsin, moving from the host stem into a 1-week-old tumor. Note that the host xylem below the tumor is thicker than above it. (**D**) Path of water uptake, labelled with red acid fuchsin, applied from the host stem into a 3-week-old tumor, extending up to the tumor periphery. The isolated red patches were found to be connected to the main water conduits. All the sections shown in the photomicrographs were viewed under dark field microscopy. Bars = 2 mm. (From Schurr et al. 1996)

might also promote ethylene synthesis, because cytokinin is also known to stimulate ethylene production (Mattoo and White 1991; Cary et al. 1995). In addition, it is possible that there is a synergistic effect within crown-gall tissues whereby a combination of the elevated levels of both auxin and cytokinin boosts ethylene production in the tumor.

The *gall constriction hypothesis* (Aloni et al. 1995) was experimentally confirmed (Aloni et al. 1998; Wächter et al. 1999) by showing that tumor-induced ethylene is a limiting and controlling factor of crown gall morphogenesis. By using an almost ethylene-insensitive tomato (*Lycopersicon esculentum*) mutant, the *Never ripe* (*Nr*), and its isogenic wild-type (WT) parent, Aloni et al. (1998) demonstrated that infection by *A. tumefaciens* results in high rates of ethylene evolution from the developing crown galls. **Tumor-induced ethylene** is a limiting and controlling factor of tumor morphogenesis; very high ethylene concentrations were produced continuously by a growing crown gall during a few weeks in tomato (Aloni et al. 1998), up to 140 times more ethylene than in wounded, but not infected control stems of *Ricinus communis* reaching a maximum at 5 weeks after infection (Wächter et al. 1999). Our experimental results (Aloni et al. 1998; Wächter et al. 1999) confirm the results of Romano et al. (1993), which showed that many high concentration "auxin" effects promoted by *iaaM* were actually ethylene effects. However, we should note that auxin also has a direct effect on vascular differentiation within the tumor, as auxin is a limiting and controlling factor in the differentiation of both vessels and sieve tubes (Jacobs 1952; Roberts et al. 1988; Aloni 1995).

Ethylene promoted the typical unorganized callus shape of the gall (Fig. 21.8A), which maximized the tumor surface on wild-type stems, whereas the galls on the *Nr* stems had a smooth appearance with minimum surface area (Fig. 21.8B) (Aloni et al. 1998). In addition, *Agrobacterium tumefaciens*-induced galls produce ethylene that substantially reduces vessel diameter in the host stem of wild-type tomato (Fig. 21.9A) compared to wide vessels in the ethylene-insensitive mutant (Fig. 21.9B) (Aloni et al. 1998). The combination of increased tumor surface (Fig. 21.8A) and decreased vessel diameter in the host stem beside the tumor (Figs. 21.6B, C and 21.9A) ensures water supply priority to the growing tumor (Fig. 21.8C) over the shoot on the wild-type host. Conversely, on the ethylene-insensitive (*Nr*) plants, many of the grown gall tumors (Fig. 21.8D) did not develop and remain small and some degenerated, likely because they could not compete with the well-supplied *Nr* host shoots due to the stem's wide stem vessels (Fig. 21.9B) that enable water uptake to the upper stem foliage.

Interestingly, fibers are nearly absent from tumor tissue of *Ricinus communis* (Aloni et al. 1995) and tomato (Fig. 21.10A), probably due to either low gibberellin levels in the tumor, which is a limiting factor for fiber differentiation (Aloni 1979), or more likely due to high ethylene concentrations that inhibit fiber differentiation (Yamamoto et al. 1987; Aloni 2013). Conversely, in woody galls that develop on olive trees, which are induced by *Pseudomonas syringae* subsp. *savastanoi* (Yamada et al. 1985), the mature galls contain fibers (Fig. 21.3A). The differentiation of many well-developed fibers in crown gall tumors developed on the *Nr* mutant (Fig. 21.10B), which is insensitive to ethylene, supports the latter suggestion that the almost absence of fibers in the soft tumors is due to high ethylene concentrations in these galls. This may cause the crumbly structure of herbaceous plant tumors. However, regenerative phloem and xylem fibers with unique ramifications developed at the base of the tumor (Chap. 2; Fig. 2.25B, C) where marginal streams of gibberellin from the stem penetrate into the gall (Aloni et al. 1995). Due to the interrupted intrusive growth of the tumorous fiber tips by unorganized cell division and random growth of the neighboring parenchyma cells (typical to cancer tissues, Fig. 21.2), promoting initiations

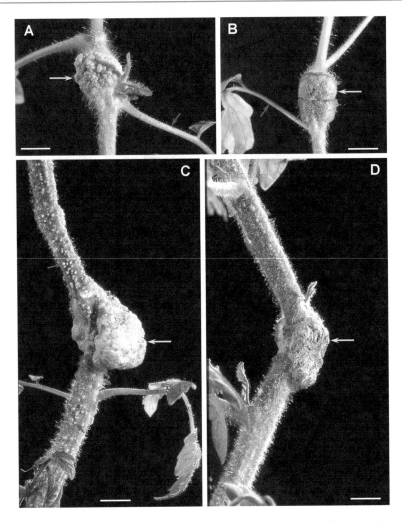

Fig. 21.8 Comparison of *A. tumefaciens*-induced crown galls (marked by *yellow arrows*) on wild-type tomato (*Lycopersicon esculentum*) (**A** and **C**) and the ethylene insensitive *Nr* mutant (**B** and **D**) stems. (**A**) Front view of a 3-week-old tumor developed on a wild-type plant showing the typical unorganized callus shape (resulting in enlarged surface) of a young crown gall and the epinastic response of the leaves (*red arrow*) both above and below the tumor (typically induced by ethylene). (**B**) Front view of a 3-week-old tumor developed on the *Nr* mutant, characterized by a smooth minimal surface and leaves in the normal orientation (*red arrow*). Note that the lower half of the gall is protected by epidermis. (**C**) Side view of a 2-month-old tumor on a wild-type stem with numerous adventitious roots (white spots marked by *red arrows*) developed both above and below the crown gall. (**D**) Side view of a 2-month-old tumor on the *Nr* mutant, showing a fibrous hard gall and a stem almost free of adventitious roots. All photographs are at the same magnification, bars = 10 mm. (From Aloni et al. 1998)

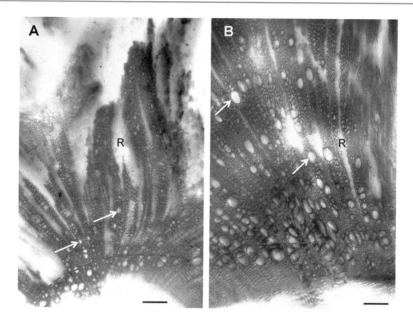

Fig. 21.9 The effects of 6-week-old *A. tumefaciens*-induced crown galls on xylem differentiation in tomato host stems, in transverse sections cleared with lactic acid and stained with lacmoid. The crown galls were located above the micrographs, and the white region at the lower part of the photographs is the cleared parenchymatic pith. The border between the xylem formed after infection with *A. tumefaciens* (upper part of each micrograph) and the intact xylem developed before the treatments (lower part) is delineated by *a broken red line*. (**A**) Limited differentiation of pathologic xylem with very narrow vessels (*arrows*) and wide rays (R) characterize the wild-type host. (**B**) Massive xylem with wide vessels (*arrows*) and almost normal rays (R) characterized the host stem formed adjacent to the *tumor* on *the Nr* mutant. Bars = 200 µm. (From Aloni et al. 1998)

of side outshoot growth sprouting from the tumorous fiber body (Fig. 2.25B, C). Owing to unorganized cell polarities in tumor tissues, the longitudinal polar auxin flow is regularly interrupted and therefore IAA tends to move in circular patterns resulting in circular and spiral structures of vascular tissues, which are very common and typical features of cancer tissues (Figs. 21.3 and 21.10).

On the surface of developing crown galls, the epidermis becomes severely disrupted by the actively growing galls and the parenchyma cells inside the gall are left with little or no structural barrier to water loss (Fig. 21.7D) (Schurr et al. 1996). Therefore, the transpiration rate of the tumors increases up to tenfold higher than that of non-infected *Ricinus* stem. Under dark conditions, the transpiration rate of a 3-week-old tumors on *Ricinus* was 28 mmol m^{-2} s^{-1}, that of leaves of control plants is about 12 mmol m^{-2} s^{-1}, while that of leaves on plants with tumors was only 6 mmol m^{-2} s^{-1} (Schurr et al. 1996; Wächter et al. 1999). These physiological changes are due to alterations in water transport structures of both the increased unprotected gall surface and the reduction in the width of tumor-adjacent vessels in wild-type plants (Fig. 21.6B, C). Under the same stress conditions, the leaves of wild-type tomato with tumor became yellow and dried out, whereas the tumor infected

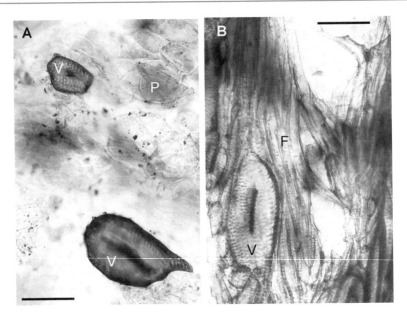

Fig. 21.10 Typical patterns of vascular tissues in 6-week-old *A. tumefaciens*-induced crown galls on tomato stems observed in thick, longitudinal radial sections cleared with lactic acid and stained with lacmoid. (**A**) Circular vessels surrounded by parenchyma cells in a wild-type stem. (**B**) Circular vessel surrounded by fibers in the *Nr* mutant. Circular vessels (V); fibers (F); parenchyma (P) cells. Bars = 100 μm. (From Aloni et al. 1998)

ethylene-insensitive *Nr* plants remain green, active, and healthy (Fig. 21.11) (Aloni et al. 1998).

These results demonstrate that in addition to the well-defined roles of auxin and cytokinin, there is a critical role for the gaseous hormone ethylene in determining crown-gall morphogenesis. The continuous production of normal wide vessels in the stem of the ethylene-insensitive *Nr* host plants (Fig. 21.9B) ensures water supply priority to the host shoot over the tumor. A limited supply of water and nutrients to the initiating crown gall tumors might therefore reduce or even inhibit cancer development. Evidently, the use of ethylene-insensitive agricultural plants prevents tumor development, and even if there is an infection, tumor growth is limited (Aloni et al. 1998) and does not reduce plant growth, fruit quality, and quantity on infected plants. The selection for ethylene insensitive or low ethylene sensitivity fruit trees, which are usually tumor free, is the simplest and most effective method for solving the severe economic damage induced by the most devastating and common cancer on wild-type plants. In modern agriculture, lines of insensitive-ethylene fruit plants are being used and should be practiced worldwide for preventing cancer damages in crop plants for obtaining high agricultural yields.

Fig. 21.11 The effects of 4-week-old *A. tumefaciens*-induced crown galls located at the base (*yellow arrow*) of the stems on shoot development and leaf senescence in tomato plants, grown under identical conditions. (**A**) Retarded wild-type shoot (*left*) and a typically taller *Nr* shoot with large leaves (*right*). Note that the older leaves in the wild-type plant started to turn yellow and senesce. (**B**) Moderate water stress caused leaf senescence in the wild-type shoot (*left*), whereas most of the leaves remained green and healthy on the *Nr* shoot (*right*). Bars = 50 mm. (From Aloni et al. 1998)

21.2   Comparison Between Plants and Animal Tumors

Comparison between plant tumors and animal tumors reveals major differences, as well as striking similarities and analogies (Ullrich et al. 2019). Plant tumors are benign, i.e., they usually do not ultimately kill their host, although they can cause considerable economic damage. On the other hand, the majority of human tumors are malignant with a lethal outcome.

In human beings and animals, neovascularization of tumors increases the probability of metastatic spread (Doonan and Hunt 1996; Gaspar 1998), whereas there is no metastatic spread from a plant tumor to new sites, because plants do not have a circulatory system and their cells are attached to each other by rigid cellulosic cell wall structure and therefore somatic plant cells are immobile. Consequently, the physical removal of the entire tumor tissues will eradicate the cancer and cure the plant.

Solid tumor growth in animals and human beings depends on angiogenesis, namely, the formation of new capillaries from pre-existing vasculature by migration and proliferation of endothelial cells. Prerequisite of solid tumor development in animals is the development of their vascular system. Solid tumors that fail to induce the formation of new blood vessels remain very limited (Folkman et al. 1989; Folkman 1990, 1995; Folkman and Shing 1992; Burri et al. 2004; Liang et al. 2011; Fus and Górnicka 2016). Actually, animal and human tumors overexpress

angiogenic growth factors, of which TNF-α (tumour necrosis factor), FGF (fibro-
blast growth factor), VEGF (vascular endothelial growth factor), and PDGF
(platelet-derived growth factor) are considered to be major mediators of angiogenesis in many different tumor types (Risau 1990; Peterson et al. 2012; Fus and
Górnicka 2016). Human squamous carcinoma cells induce a dense network of
blood vessels that supplies the tumor stroma with nutrients, water, and oxygen. It is
therefore possible that angiogenesis inhibition would promote its ultimate goal, the
recovery of people from different types of solid cancers. Anti-angiogenic therapy
has rapidly evolved into an integral component of current standard anti-cancer treatment and fulfilled its expected promise. Early detection of tumors combined with
anti-angiogenic treatments, would direct anti-angiogenic therapy toward the conversion of cancer to a dormant, chronic, manageable disease. (Folkman 1971, 1990,
1995; Folkman et al. 1989; Folkman and Shing 1992; Abdollahi and Folkman 2010;
Liang et al. 2011).

In analogy, plant tumors produce auxin and cytokinin that induce a dense and
continuous vascular system from the host xylem and phloem tissues up to the tumorous gall surface (Figs. 21.3B, 21.4 and 21.7) (Aloni et al. 1995; Schurr et al. 1996).
Tumor development on a leaf or a small branch does not allow much growth due to
their vascular restriction that limits supply to the gall which remains small, while
tumor growth on the trunk of a large tree induces a well-developed vascular system
allowing the gall to reach an extremely large size.

Wounding is a precondition of T-DNA transfer from the Ti plasmid of *A. tumefaciens* to the plant genome (Fig. 21.1). Similarly in animals, tumor initiation can be
promoted by wounding. Wounding of a tumor in animals promotes faster growth
and spread of the cancer tissue due to rapid regeneration of the tumor tissue. In
plants, wounding stimulates cell division, during which the T-DNA incorporates
into the plant's DNA (Citovsky et al. 2007) and thus enables integration of the pro-
karyotic genes into the nucleus of the dividing plant cell. Furthermore, in both animals and plants, gradients of growth factors are established, in plants by the polar
auxin transport, inducing and regulating tumor vascular differentiation. Surprisingly,
in human squamous cell carcinoma, a tenfold accumulation of auxin was found at
the early differentiating stage, with an unknown function (Shimojo et al. 1997).

Although there is no need for the presence of *Agrobacterium tumefaciens or
A. vitis* bacteria in the induced crown gall after the T-DNA was stably incorporated
into the plant genome (Thomashaw et al. 1986). In all the other plant tumors, there
is a requirement for the permanent presences (between the cancer cells) and activity
of the tumor-inducing-bacteria to continuously produce auxin and cytokinins, which
promote and maintain tumor growth (Manulis et al. 1998; Ullrich and Aloni 2000;
Chalupowicz et al. 2006; Ullrich et al. 2019).

In human beings, bacteria and human cancer cells can cooperate to create tumorous niches. Microbe-cancer cell interactions contribute to cancer progression. Some
bacteria, like *Bacillus* sp. and *Escherichia coli*, produce peptides that can promote
metastasis. The microbes enhance cancer cell fitness by promoting proliferation as
well as protecting cancer cells from the control of the immune system (Whisner and
Aktipis 2019). A recent elegant analysis of the **tumor microbiome** in seven cancer

types, including breast, lung, ovary, pancreas, melanoma, bone, and brain tumors, demonstrated that each tumor type has a distinct microbiome composition and that breast cancer has a particularly rich and diverse microbiome. The intratumor bacteria are found mostly inside the cancer cells (intracellular) and are present in both the cancer and immune cells (Nejman et al. 2020).

21.3   The Plant Hormone Jasmonate, Antibiotics and Bioactive Secondary Metabolites Are Promising Candidates for Human Cancer Therapy

Jasmonate was recorded in early stages of crown gall tumor formation (Veselov et al. 2003; Aloni and Ullrich 2008). The naturally occurring plant hormones jasmonates, including their volatile methyl jasmonate (MeJA), which activate defense-related genes against insects and pathogens (Howe 2004; Wang et al. 2019), induce cell cycle arrest, apoptosis, and non-apoptotic cell death selectively in animal cancer cells and, therefore, are promising candidate to cure of human cancer (Rotem et al. 2005). Jasmonates induced cytochrome c release and swelling in mitochondria isolated from cancer cells but not from normal ones. Thus, the selectivity of jasmonates against cancer cells is rooted at the mitochondrial level and probably exploits differences between mitochondria from normal *versus* cancer cells (Goldin et al. 2007).

Several research groups have reported in recent years that members of the plant defense hormone family of jasmonates, and some of their synthetic derivatives exhibit anti-cancer activity in vitro and in vivo. Jasmonates increased the life span of EL-4 lymphoma-bearing mice and exhibited selective cytotoxicity toward cancer cells while sparing normal blood lymphocytes, even when the latter were part of a mixed population of leukemic and normal cells drawn from the blood of chronic lymphocytic leukemia (CLL) patients (Flescher 2007).

Thus, jasmonates join a growing number of old and new cancer chemotherapeutic compounds of bioactive plant originating secondary metabolites (Ullrich et al. 2019). Three mechanisms of action have been proposed to explain the anti-cancer activity of jasmonates. These include the following:

(1) The bio-energetic mechanism jasmonates induce severe ATP depletion in cancer cells via mitochondrial perturbation.
(2) The re-differentiation mechanism jasmonates induce re-differentiation in human myeloid leukemia cells via mitogen-activated protein kinase (MAPK) activity.
(3) The reactive oxygen species (ROS)-mediated mechanism-jasmonates induce apoptosis in lung carcinoma cells via the generation of hydrogen peroxide, and pro-apoptotic proteins of the Bcl-2 family.

Additionally, jasmonates can induce death in drug-resistant cells. The drug resistance was conferred by either p53 mutation or P-glycoprotein (P-gp) over-expression. Hence, the jasmonate family is a promising a novel anti-cancer agents present new hope for the development of cancer therapeutics, which should attract further scientific and pharmaceutical interest (Flescher 2007; Zhang et al. 2016; Ullrich et al. 2019).

Interestingly, also at the mitochondrial level and strict dependence on mitochondrial biogenesis, a recent study shows that mitochondrially targeted **antibiotics** are efficient in eradicating cancer stem cells (CSCs), across many tumor types under in vitro experimental conditions. The study demonstrated that 4–5 different classes of FDA-approved antibiotics (azithromycin, doxycycline, tigecycline, pyrvinium pamoate, and the anti-parasitic drug chloramphenicol) can be used to selectively target cancer stem cells, across several tumor types, in 12 different cancer cell lines, across 8 different tumor types (breast, ductal carcinoma in situ, ovarian, prostate, lung, pancreatic, melanoma, and glioblastoma (brain)) (Lamb et al. 2015). The mitochondrially targeted antibiotics should also be considered for prevention studies specifically focused on the prevention of tumor recurrence and distant metastasis. Additionally, early clinical trials with the doxycycline and azithromycin antibiotics (intended to target cancer-associated infections, but not cancer cells) have already shown positive therapeutic effects in cancer patients, although their ability to eradicate cancer stem cells was not yet appreciated.

The discovery that antibiotics can be used to selectively target cancer stem cells (Lamb et al. 2015) supports a promising concept for cancer therapy of combining antibiotics with other human cancer therapeutic treatments for increasing the recovery success from tumor. The recent finding that different types of cancer have developed diverse and specific tumor microbiome composition inside the cancer cells (Nejman et al. 2020), bacteria that protect the cancer cells from the immune system (Whisner and Aktipis 2019), may explain why antibiotics can eliminate the bacteria-cancer interactions and thus enable the immune system to control the unprotected cancer stem cells. Generally, the use of generic antibiotics for anti-cancer therapy should significantly reduce the costs of patient care, making treatment more accessible, especially in the developing world.

Finally, the exploitation of **bioactive secondary metabolites** originating in plants to treat human tumors bears a tremendous therapeutic potential. Isolated phytochemicals and their synthetic derivatives extract from plants may offer new therapy options to decrease human tumor incidence and mortality (Ullrich et al. 2019).

Summary

- Although *Agrobacterium tumefaciens-induced* plant tumors are characterized by unorganized tissues of unpredictable cell patterns, the tumors develop functional collateral bundles built of continuous and functional sieve tubes and vessels, connected by phloem anastomoses.
- In comparison to the majority of human tumors that are malignant with a lethal outcome, plant tumors are benign, i.e., they do not ultimately kill their host, but they can cause considerable economic damage in agricultural crops.

- In human beings and animals, neovascularization in tumors increases the probability of metastatic spread, whereas there is no metastatic spread from a plant tumor to new sites, because somatic plants cells are attached to each other by rigid cellulosic cell walls and therefore they are immobile.
- *Tumor-inducing bacteria* cause tumorous galls following wounding, during which a DNA sequence of about 20 kb (T-DNA), from a large *Agrobacterium tumefaciens* tumor-inducing plasmid (Ti-plasmid), is stably incorporated into the higher plant genome.
- The T-DNA-located oncogenes *iaaM*, *iaaH* for auxin synthesis, and *ipt* for cytokinin production boost hormonal concentrations in the tumorous gall tissues, which may be 500 times higher for auxin and 1500 times higher for cytokinin concentrations than in control tissues, promoting cancer development and vascular differentiation.
- Tumorous crown gall tissues produce also high ethylene that may be 140 higher than in control tissues, likely resulting from the very high auxin and cytokinin concentrations.
- Ethylene produced by the tumors promotes the typical unorganized callus shape of the tumor which maximizes tumor surface. This causes a substantially increase in transpiration rate through the gall's disrupted tumor epidermis (due to actively growing parenchyma cells inside the gall). In addition, the ethylene substantially reduces vessel diameter in the host stem. The combination of increased tumor surface and decreased vessel diameter in the host stem beside the tumor ensures water supply priority to the growing tumor over the host shoot.
- The use of ethylene-insensitive agricultural plants prevents tumor development, and even if there is an infection, tumor growth is limited and does not reduce plant growth, fruit quality, and quantity on infected plants.
- The selection for ethylene insensitive or low ethylene sensitivity fruit trees, which are usually tumor free, is the simplest and most effective method for solving the severe economic damage induced by the most devastating and common cancer on agricultural plants.
- The plant hormone jasmonate (JA) was recorded in early stages of crown gall tumor formation activates defense-related genes against insects and pathogens. JAs can be used for cancer therapy in human due to their selectivity against cancer cells, which is rooted at the mitochondrial level, as JAs probably exploits differences between mitochondria from normal *versus* cancer cells.
- Also at the mitochondrial level and strict dependence on mitochondrial biogenesis; mitochondrially antibiotics are efficient in eradicating cancer stem cells, across many tumor types under in vitro experimental conditions.
- Different types of cancer have developed diverse and tumor type-specific microbiome composition inside the cancer cells and the immune cells. These intracellular bacteria protect the cancer cells from the immune system. Antibiotics that eliminate the bacteria-cancer interactions enable the immune system to control the unprotected cancer stem cells.
- Comparison between the development of plant and solid tumors of animal has shown an analogous requirement for neovascularization in both, presaging

possible strategies for prevention and therapy. Anti-angiogenic therapy in human has rapidly evolved into an integral component of current standard anti-cancer treatment, fulfilling its earlier proposed promise. Likewise, therapy with secondary plant metabolites is a promising direction.

References and Recommended Readings[1]

Abdollahi A, Folkman J (2010) Evading tumor evasion: current concepts and perspectives of anti-angiogenic cancer therapy. Drug Resist Updat 13: 16–28.

Abeles FB, Morgan PW, Saltveit ME, Jr (2012) *Ethylene in Plant Biology*, 2nd ed, Academic Press, San Diego, CA.

Akiyoshi DE, Morris RO, Hinz R, Mischke BS, Kosuge T, Garfinkel DJ, Gordon MP, Nester EW (1983) Cytokinin/auxin balance in crown gall tumors is regulated by specific loci in the T-DNA. Proc Natl Acad Sci USA 80: 407–411.

Aloni R (1979) Role of auxin and gibberellin in differentiation of primary phloem fibers. Plant Physiol 63: 609–614.

Aloni R (1995) The induction of vascular tissues by auxin and cytokinin. In: *Plant Hormones: Physiology, Biochemistry and Molecular Biology*, PJ Davies (ed). Kluwer Academic, Dordrecht, pp. 531–546.

Aloni R (2013) The role of hormones in controlling vascular differentiation. In: *Cellular Aspects of Wood Formation*, J Fromm (ed). Springer-Verlag, Berlin, pp. 99–139.

Aloni R, Barnett JR.1996. The development of phloem anastomoses between vascular bundles and their role in xylem regeneration after wounding in *Cucurbita* and *Dahlia*. Planta 198: 595–603.

Aloni R, Peterson CA (1990) The functional significance of phloem anastomoses in stems of *Dahlia pinnata* Cav. Planta 182: 583–590.

Aloni R, Sachs T (1973) The three-dimensional structure of primary phloem systems. Planta 113: 345–353.

*Aloni R, Pradel KS, Ullrich CI (1995) The three-dimensional structure of vascular tissues in *Agrobacterium tumefaciens*-induced crown galls and in the host stems of *Ricinus communis* L. Planta 196: 597–605.

Aloni R, Ullrich CI (2008) Biology of crown gall tumors. In: *Agrobacterium*. T Tzfira, V Citovsky (eds). Springer, New York, pp. 565–591.

*Aloni R, Wolf A, Feigenbaum P, Avni A, Klee HJ (1998) The *Never ripe* mutant provides evidence that tumor-induced ethylene controls the morphogenesis of *Agrobacterium tumefaciens*-induced crown galls on tomato stems. Plant Physiol 117: 841–847.

Burr TJ, Otten L (1999) Crown gall of grape: biology and disease management. Annu Rev Phytopathol 37: 53–80.

Burri PH, Hlushchuk R, Djonov V (2004) Intussusceptive angiogenesis: its emergence, its characteristics, and its significance. Dev Dyn 231: 474–88.

Cary AJ, Liu W, Howell SH (1995) Cytokinin action is coupled to ethylene in its effects on the inhibition of root and hypocotyl elongation in *Arabidopsis thaliana* seedlings. Plant Physiol 107: 1075–1082.

Cavara F (1895) Aperçu sommaire de quelques maladies de la vigne parues en Italie en 1894. Rev Inter Viticult Oenologie 6: 447–449.

Chalupowicz L, Barash I, Schwartz M, Aloni R, Manulis S (2006) Comparative anatomy of gall development on *Gypsophila paniculata* induced by bacteria with different mechanisms of pathogenicity. Planta 224: 429–437.

---

[1] Papers of particular interest for suggested reading have been highlighted (with *)

Chilton M, Currier T, Farrand S, Merlo D, Sciaky D, Montoya A, Gordon M, Nester E (1977) Stable incorporation of plasmid DNA into higher plant cells: the molecular basis of crown gall tumorigenesis. Cell 11: 263–271.

Citovsky V, Kozlovsky SV, Lacroix B, Zaltsman A, Dafny-Yelin M, Vyas S, Tovkach A, Tzfira T (2007) Biological systems of the host cell involved in Agrobacterium infection. Cell Microbiol 9: 9–20.

Doonan J, Hunt T (1996) Why don't plants get cancer? Nature 380: 481–482.

*Flescher E (2007) Jasmonates in cancer therapy. Cancer Lett 245: 1–10.

*Folkman J (1971) Tumor angiogenesis: therapeutic implications. N Engl J Med 285: 1182–1186.

Folkman J (1990) What is the evidence that tumors are angiogenesis dependent? J Natl Cancer Inst 82: 4–6.

Folkman J.1995. Tumor angiogenesis. In: *The Molecular Basis of Cancer*. J Mendelsohn, MP Howley, MA Israel, LA Liotta (eds). WB Saunders, Philadelphia, Pennsylvania, pp. 206–232.

Folkman J, Shing, Y (1992) Angiogenesis. J Biol Chem 267: 10931–10934.

Folkman J, Watson K, Ingber D, Hanahan D (1989) Induction of angiogenesis during the transition from hyperplasia to neoplasia. Nature 339: 58–61.

Fus ŁP, Górnicka B (2016) Role of angiogenesis in urothelial bladder carcinoma. Cent Euro J Urol 69: 258–263.

Gaspar T (1998) Plants can get cancer. Plant Physiol Biochem 36: 203–204.

*Goldin N, Heyfets A, Reischer D, Flescher E (2007) Mitochondria-mediated ATP depletion by anti-cancer agents of the jasmonate family. J Bioenerg Biomembr 39: 51–57.

Howe GA (2004) Jasmonates. In: *Plant Hormones: Biosynthesis, Signal Transduction, Action!* PJ Davies (ed). Kluwer Academic Publishers, Dordrecht, Boston, pp. 610–634.

Jacobs WP (1952) The role of auxin in differentiation of xylem around a wound. Am J Bot 39: 301–309.

Kuzmanović N, Puławska J, Hao L, Burr TJ (2018) The ecology of *Agrobacterium vitis* and management of crown gall disease in vineyards. Curr Top Microbiol Immunol 418: 15–53.

*Lamb R, Ozsvari B, Lisanti CL, Tanowitz HB, Howell A, Martinez-Outschoorn UE, Sotgia F, Lisanti MP (2015) Antibiotics that target mitochondria effectively eradicate cancer stem cells, across multiple tumor types: treating cancer like an infectious disease. Oncotarget 6: 4569–4584.

Liang G, Butterfield C, Liang J, Birsner A, Folkman J, Shing Y (2011) Beta-35 is a transferrin-derived inhibitor of angiogenesis and tumor growth. Biochem Biophys Res Commun 409: 562–566

Manulis S, Haviv-Chesner A, Brandl MT, Lindow SE, Barash I (1998) Differential involvement of indole-3-acetic acid biosynthetic pathways in pathogenicity and epiphytic fitness of *Erwinia herbicola* pv. *gypsophilae*. Molecular Plant-Microbe Interactions 11: 634–642.

Mattoo AK, White WB (1991) Regulation of ethylene biosynthesis. In: *The Plant Hormone Ethylene*, AK Mattoo, JC Suttle (eds). CRC, Boca Raton, FL, pp. 21–42.

*Nejman D, Livyatan I, Fuks G, Gavert N, Zwang Y, Leore T, Geller LT, Rotter-Maskowitz A, Weiser R, Mallel G, Gigi E, Meltser A, Douglas GM, Kamer I, Gopalakrishnan V, Dadosh T, Levin-Zaidman S, Avnet S, Atlan T, Cooper ZA, Arora R, Cogdill AP, Khan MdAW, Ologun G, Bussi Y, Weinberger A, Lotan-Pompan M, Golani O, Perry G, Rokah M, Bahar-Shany K, Rozeman EA, Blank CU, Ronai A, Shaoul R, Amit A, Dorfman T, Kremer R, Cohen ZR, Harnof S, Siegal T, Yehuda-Shnaidman E, Gal-Yam EN, Shapira H, Baldini N, Langille MGI, Ben-Nun A, Kaufman B, Nissan A, Golan T, Dadiani M, Levanon K, Bar J, Yust-Katz S, Barshack I, Peeper DS, Raz DJ, Segal E, Wargo JA, Sandbank J, Shental N, Straussman R (2020) The human tumor microbiome is composed of tumor type-specific intracellular bacteria. Science 368: 973–980.

Peterson JE, Zurakowski D, Italiano JE Jr, Michel LV, Connors S, Oenick M, D'Amato RJ, Klement GL, Folkman J (2012) VEGF, PF4 and PDGF are elevated in platelets of colorectal cancer patients. Angiogenesis 15: 265–273.

Risau W (1990) Angiogenic growth factors. Prog Growth Factors Res 2: 71–79.

Roberts LW, Gahan BP, Aloni R (1988) *Vascular Differentiation and Plant Growth Regulators.* Springer-Verlag, Berlin.

Romano CP, Cooper ML, Klee HJ (1993) Uncoupling auxin and ethylene effects in transgenic tobacco and *Arabidopsis* Plants. Plant Cell 5: 181–189.

Rotem R, Heyfets A, Fingrut O, Blickstein D, Shaklai M, Flescher E (2005) Jasmonates: novel anticancer agents acting directly and selectively on human cancer cell mitochondria. Cancer Res 65: 1984–1993.

*Sachs T (1991) Callus and tumor development. In: *Pattern Formation in Plant Tissues,* T Sachs. Cambridge university press, Cambridge, pp. 38–55.

Schurr U, Schuberth B, Aloni R, Pradel KS, Schmundt D, Jähne B, Ullrich CI (1996) Structural and functional evidence for xylem-mediated water transport and high transpiration in *Agrobacterium tumefaciens*-induced tumors of *Ricinus communis.* Bot Acta 109: 405–411.

Shimojo E, Yamaguchi I, Murofushi N (1997) Increase of indole-3-acetic acid in human esophageal cancer tissue. Proc Japan Acad Ser B – Physic Biol Sci 73: 182–185.

Smith EF, Townsend CO (1907) A plant tumor of bacterial origin. Science 25: 671–673.

Thimann KV, Sachs T.1966. The role of cytokinins in the 'fasciation' disease caused by *Corynebacterium fascians.* Am J Bot 53: 731–739.

Thomashow MF, Hugly S, Buchholz WG, Thomashow LS (1986) Molecular basis for the auxin-independent phenotype of crown gall tumor tissues. Science 231: 616–618.

*Ullrich CI, Aloni R (2000) Vascularization is a general requirement for growth of plant and animal tumours. J Exp Bot 51: 1951–1960.

*Ullrich CI, Aloni R, Saeed MEM, Ullrich W, Efferth T (2019) Comparison between tumors in plants and human beings: Mechanisms of tumor development and therapy with secondary plant metabolites. Phytomedicine 64: 153081.

Ullrich CI, von Eitzen-Ritter M, Jockel A, Efferth T (2009) Prevention of plant crown gall tumor development by the anti-malarial artesunate of *Artemisia annua.* J Cultiv Plants 61: 31–36.

Van Larebeke N, Engler G, Holsters M, Van den Elsacker S, Zenen I, Schell J (1974) Large plasmid in *Agrobacterium tumefaciens* essential for crown gall-inducing activity. Nature 252: 255–264.

Vereecke D, Burssens S, Simón-Mateo C, Inzé D, Van Montagu M, Goethals K, Jaziri M (2000) The *Rhodococcus fascians*– plant interaction: morphological traits and biotechnological applications. Planta 210: 241–251.

Veselov D, Langhans M, Hartung W, Aloni R, Feussner I, Götz C, Veselova S, Schlomski S, Dickler C, Bächmann K, Ullrich CI (2003) Development of *Agrobacterium tumefaciens* C58-induced plant tumors and impact on host shoots are controlled by a cascade in production of jasmonic acid, auxin, cytokinin, ethylene, and abscisic acid. Planta 216: 512–522.

Wächter R, Fischer K, Gäbler R, Kühnemann F, Urban W, Bögemann GM, Voesenek LACJ, Blom CWPM, Ullrich CI (1999) Ethylene production and ACC-accumulation in *Agrobacterium tumefaciens*-induced plant tumours and their impact on tumour and host stem structure and function. Plant Cell Environ 22: 1263–1273.

*Wang J, Wu D, Wang Y, Xie D (2019) Jasmonate action in plant defense against insects. J Exp Bot 70: 3391–3400.

Weiler EW, Schröder J (1987) Hormone genes and crown gall disease. Trends Biochem Sci 12: 271–275.

Weiler EW, Spanier K (1981) Phytohormones in the formation of crown gall tumors. Planta 153: 326–37.

*Whisner CM, Aktipis CA (2019) The role of the microbiome in cancer initiation and progression: how microbes and cancer cells utilize excess energy and promote one another's growth. Curr Nutr Rep 8: 42–51.

Yamada T, Palm CJ, Brooks B, Kosuge T (1985) Nucleotide sequences of the *Pseudomonas savastanoi* indoleacetic acid genes show homology with *Agrobacterium tumefaciens* T-DNA. Proc Natl Acad Sci USA 82: 6522–6526.

Yamamoto F, Angeles G, Kozlowski TT (1987) Effect of ethrel on stem anatomy of *Ulmus americana* seedlings. IAWA Bull ns 8: 3–9.

Yang SF, Hoffman NE (1984) Ethylene biosynthesis and its regulation in higher plant. Annu Rev Plant Physiol 35: 155–189.

Zambryski P, Tempé J, Schell J (1989) Transfer and function of T-DNA genes from *Agrobacterium* Ti and Ri plasmids in plants. Cell 56: 193–201.

*Zhang M, Su L, Xiao Z, Liu X, Liu X (2016) Methyl jasmonate induces apoptosis and pro-apoptotic autophagy via the ROS pathway in human non-small cell lung cancer. Am J Cancer Res 6: 187–199.

Index

A

ABA signaling, 80, 81
ABC flowering model, 2
ABC transporters, 77
Abscisic acid (ABA), 74, 79–81, 101, 207,
    208, 210, 216, 254–255, 278, 279, 297
Acaciae raddiana, 294–296
Acacia mangium, 268
Acer rubrum, 266
Acer saccharum, 237, 238, 281
Acidity (pH), 58
Adult wood, 215–218, 220
Aerenchyma, 28, 78
Ageing parenchyma cells, 191, 195
Agrobacterium tumefaciens, 38, 309–317, 319,
    322, 324, 325
Agrobacterium tumefaciens-induced tumors,
    309–311, 316, 324, 325
Agrobacterium vitis, 309, 322
Alectra vogelii, 296
Altered phloem development (APL)
    transcription factor, 105–106, 123, 124
1-Aminocyclopropane-1-carboxylate synthase
    (ACS), 179, 182
1-Aminocyclopropane-1-carboxylic acid
    (ACC), 78, 179, 182, 269
Amyloplasts, 265
Anastomoses, 24–26, 191
Angiogenesis, 321, 322
Angiosperms, 11, 14, 34, 43, 45, 72, 108, 111,
    200, 209, 217, 218, 220, 232, 265–270,
    273, 282, 293, 298
Animal tumors, 321
Annual growth rings, 215, 226
Annular secondary wall thickenings, 152, 192
Anthers, 32, 33, 166–171, 173
Antibiotics, 324, 325
Aphids (feeding on the stem's sieve
    tubes), 200

Apical buds, 40, 61, 97, 115, 131–135, 137,
    141, 188, 191, 200, 217–219, 240,
    242, 285
Apical dominance, 132–135, 137, 168, 173
Apical meristems, 7, 164
Apoplast, 58, 74
Arabidopsis thaliana, 2, 9, 11, 27, 59, 69, 74,
    136, 143, 144, 147, 148, 150, 153, 154,
    156–159, 164, 165, 167, 168, 171, 172,
    174, 186, 190, 195, 200, 201, 296, 298
Arceuthobium tsugense, 294
ARR5::GUS expression, 9, 76, 77, 134, 136
Artemisia princeps, 301
Auxin and sugar, 102
Auxin canalization, 63
Auxin concentrations, 29, 33, 37, 40, 62, 64,
    76, 102–104, 106, 107, 110, 116, 117,
    123, 145, 146, 166, 185, 188, 192, 199,
    210, 216, 220, 224–227, 230, 232, 233,
    240, 242, 248, 249, 252, 255, 257, 267,
    283–285, 294, 297
Auxin gradients, 73, 110, 111, 141, 200, 226,
    227, 231–233, 240, 252, 255, 256, 283
Auxin-gradient hypothesis (six-point
    hypothesis), 110, 226
Auxin immunolocalization, 60, 72, 150
Auxin induces lateral roots, 177
Auxin influx carriers AUX1/LAX, 66
Auxin maximum, 58, 59, 63, 106, 107, 124,
    141, 143, 144, 148–151, 155, 157–159,
    169, 179, 181, 182
Auxin movement in a wave-like pattern,
    209, 233
Auxin pathway model, 67
Auxin producing sites, 155, 157–159
Auxin-resistance/like-AUX (AUX/LAX)
    protein family, 58
Auxin-resistance1 (AUX1),
    64–67

© Springer Nature Switzerland AG 2021
R. Aloni, *Vascular Differentiation and Plant Hormones*,
https://doi.org/10.1007/978-3-030-53202-4

Auxin sensitivity, 65, 73, 76, 277, 282, 283, 286
Auxin signaling, 12, 30, 31, 73, 83, 97, 101, 111, 123, 142, 203, 206, 207, 210, 224, 231
Auxin synthesis in leaf primordial, 141, 145–155
Auxin transport, 59, 62–65, 67, 71, 73, 104, 105, 123, 132, 141, 144, 149, 154, 167, 171, 173, 231, 266, 267, 296
Auxin transport in wave patterns, 72–73, 199–200, 227
Auxin transport pathways, 154
AUX1 auxin carrier, 58, 64–66

B
Bacillus thuringiensis, 261
Baizongia pistaciae, 302, 304
Bicollateral vascular bundles, 12, 16, 24
Bioactive gibberellins, 74, 75, 119, 122, 204, 208, 217, 220
Bioactive hormones, 60, 62
Bioactive secondary metabolites, 324
Blockage of the plasmodesmata by ABA, 208, 210, 279
Bomby xmori (silkworm), 301
Bottleneck, 238, 240, 242
Branch junctions, 40, 237, 239, 240, 281, 299
Brassinosteroids (BRs), 82–83, 122, 269
Brefeldin A, 64, 65, 74
Brown algae (Phaeophyta), 101, 102, 124
Bud swelling, 283, 284, 287

C
Callose (β-1,3-glucan), 16–19, 26, 38, 47, 71, 103, 106
Callus, 10, 35, 102, 103, 105, 106, 193, 194, 196, 245, 317, 318, 325
Cambial activity, 75, 76, 79, 83, 97, 113, 121, 187, 188, 192, 195, 199–210, 216, 225, 238, 245, 256, 266, 268, 269
Cambial age, 215, 221, 286
Cambial dormancy, 79, 208, 210
Cambial fusiform initials, 78, 110, 192, 216, 217
Cambial ray initials, 210
Cambial reactiviation, 284
Cambial region, 11, 72, 202–204, 227, 267
Cambial variants, 201–203, 210
Cambial zone, 11, 17, 71, 72, 203, 206, 227, 266, 269

Cambium, 3, 4, 9–11, 13, 16–18, 25, 28, 30, 35, 36, 38, 40, 45, 47, 56–58, 67–76, 78, 79, 81, 83, 97, 102, 110, 114, 116, 121, 122, 131, 188–191, 195, 199–210, 215–218, 220, 221, 224, 227, 231, 233, 238, 245–249, 252, 253, 259, 265–267, 269, 276, 278, 279, 282–284, 286, 287
Cambium circumference, 200–203, 210, 231, 233
Canalization hypothesis, 62–63
Cancer, 1, 5, 82, 309–326
Cancer stem cells (CSCs), 324, 325
Cancer therapy, 5, 82, 324, 325
Carpels, 164, 165, 169
Cederus sp., 259
Cell determination, 160
Cell grafting, 194, 196
Cell size and density, 223–233, 251, 252
Cellulose, 43, 45, 46, 107, 265, 269
Cellulose microfibrils, 44, 107, 265
Cell walls, 7, 11, 14, 18, 20, 22–24, 29, 41, 45, 47, 48, 58, 83, 84, 107, 122, 215, 217, 223, 252, 265, 269, 273, 275, 283, 284, 297, 310, 321, 325
Cereals, 40, 241, 242
Circular vascular tissues, 237–242
Circular vessels, 40, 237–240, 242, 299, 300, 312, 320
Circumnutating growth, 187
Coleus blumei, 8, 14, 22, 38, 39, 102, 106, 109, 113–115, 187–189, 195, 224
Collateral vascular bundles, 12, 15
Companion cell (CC), 14, 47, 225
Compression wood, 42–44, 217, 265, 267, 269
Conductive efficiency, 24
Conduit size and density, 223, 226, 227
Conifers, 11, 30, 36, 42, 56, 76, 79, 108, 110, 111, 187, 195, 210, 216, 217, 220, 231, 245, 246, 252, 259, 261, 262, 269
Conjugated auxin, 58, 60–62, 134, 150, 159, 166–168, 171
Connective auxin transport, 71
Cooperative tree community, 3, 192, 195
Cordiera concolor, 208
Core wood, 215, 217, 220
Crown galls, 38, 39, 82, 309–323, 325
Crystalline cellulose fibrils, 43
Cucurbita maxima, 191, 195
Cucurbita moschata, 164
Cucurbita pepo, 113
Cucurbita sp., 24, 26
Cunninghamia lanceolata, 216
Cupressus sp., 259
Cycling and vesicle trafficking, 123

Cycling of AUX1, 64–66
Cycling of PIN1, 64
Cytokinin (CK), 5, 9, 56, 66, 71, 76–79, 85,
    97, 106, 112, 113, 117, 118, 120, 122,
    124, 132–137, 179–182, 185, 188, 190,
    193, 204, 206, 207, 216, 269, 279, 282,
    283, 298, 301, 311, 314, 316, 320,
    322, 325
Cytokinin-dependent root apical
    dominance, 98, 135

D

Dahlia pinnata, 17, 18, 24–26, 105
Deciduous hardwood trees, 73, 76, 277, 279,
    282, 287
Deformed vessels, 29, 45, 47, 48,
    107, 204–205
Developmental plasticity, 79, 142, 192, 207
Diarch root, 27
Dicotyledons (dicots), 13, 17, 28, 67, 81, 145,
    156, 177, 179, 181, 182, 190, 223, 238,
    252, 309
Differentiating vessels, 34, 63, 68, 71, 72, 177,
    182, 201, 224, 247, 283
Diffuse-porous wood, 279, 282, 286, 287
Discontinuous vascular patterns, 31
Dormancy, 7, 12, 16, 35, 71, 79, 207, 208,
    245, 279
Dormant buds, 40
Downward "wave" pattern, 145, 159
Dracaena draco, 36, 110
DR5::GUS expression, 62
Drought stress, 40, 207

E

Earlywood, 22, 110, 111, 224, 237, 279,
    285, 286
Earlywood vessels, 24, 280–288, 304
Efficiency vs. safety, 24, 238, 251
Efflux carriers, 63, 66, 71, 81, 83, 266, 269
Embolism, 19, 20, 23, 24, 40, 111, 238, 241,
    242, 294
Endodermis, 27, 74, 81, 179, 180, 182, 266
Environmental adaptation, 4, 40, 48
Environmental stresses, 29, 48, 107, 252
Ephedra campylopoda, 276, 278
Erwinia herbicola pv. Gypsophilae, 311
Escherichia coli, 322
Ethrel (2-chloroethylphosphonic acid), 247,
    249, 261
Ethylene (C₂H₄), 10, 56, 57, 78–79, 82, 84,
    85, 122, 177, 179–182, 206, 207, 210,

245–249, 261, 262, 268, 269, 313, 314,
    316–318, 320, 325
Eucalyptus globulus, 218, 219
Eucalyptus gomphocephala, 268
Eucalyptus regnans, 231
Eucalyptus sp., 97, 220, 256, 268
Eucommia ulmoides, 79
Eurosta solidaginis, 301

F

Fertilized ovaries, 172, 174
Fiber differentiation, 47, 74–76, 84, 109,
    113–118, 120, 122, 124, 164, 192,
    204–205, 209, 210, 268, 269, 313, 317
Fibers, 11, 14, 19, 22–24, 38, 39, 44–47, 57,
    74–76, 84, 107–109, 113–122, 124,
    164, 173, 200, 201, 204–205, 209, 210,
    216, 218, 226, 232, 242, 251–253, 268,
    269, 276–280, 286–288, 312, 313, 315,
    317, 320
Floral organs, 164, 165, 167–169, 171
Florigen, 84, 163, 164, 173
Flowering, 27, 74, 76, 84, 132, 163–165, 168,
    173, 217, 218, 220, 293, 297
Flowering locus T (FT) gene, 84, 163
Flowers, 2, 31, 33, 57, 61, 62, 74, 77, 82, 84,
    134, 163–175, 218, 299
Foliar and axial organs, 106
Fragmented patterns, 187
Fraxinus mandshurica, 268
Free auxin, 60–62, 67, 72, 147–150,
    159, 167–170
Freely-ending veinlets, 2, 152, 157–159, 172
Fruit development, 171, 174
FT protein, 163, 164, 173
Fusiform initials, 10, 246

G

GA₁, 74, 75, 121, 220
GA₃, 74, 108, 109, 115–120, 204–206, 268,
    277, 278
GA₄, 74, 75, 220
GA₇, 74
GA₁₂, 75, 121, 220
GA₂₀, 74, 75, 121, 220
Gall constriction hypothesis, 312, 314, 316
Gall-inducing insects, 293, 298, 302
Galling aphids, 299
Galling insects, 293, 298, 301, 302, 305
GA 2-oxidase, 119, 121, 124
GA 20-oxidase, 75, 118, 119, 121, 124
GA precursors, 74, 121

GA transport, 74
GA transporters, 74
Gelatinous fibers (G-fibers), 42–46, 265, 266,
    268, 269
Gelatinous layer (G-layer), 43–46, 265, 266,
    268, 269
Gene expression, 55, 62, 148, 149, 168, 170,
    174, 193, 206, 216, 273, 298
Gibberellic acid (GA₃), 74, 115, 116, 118, 225,
    266, 277
Gibberellins (GAs), 74–76, 82, 122, 203,
    217, 218
GNOM (GN), 64
Gradients, 14, 17, 37, 62, 73, 110, 112, 141,
    159, 200, 206, 208, 209, 216, 220, 224,
    226, 227, 231–233, 248, 249, 252, 255,
    256, 322
Graft hybridization, 194, 195
Grafting, 3, 81, 163, 164, 192–196, 218, 219,
    221, 294
Graft transmissible, 74
Graft union, 164, 192–194, 196, 218, 219, 297
Guaiacyl (G), 107, 108
Gymnosperms, 10, 36, 43, 67, 108, 265, 267,
    269, 298
Gypsophila sp., 311

H
Hardwood, 11, 24, 57, 245, 269, 312
Haustorium, 294–298
Hedera helix (English ivy), 217
Helianthus annuus, 117–120
Hemicelluloses, 107
Hemiparasites, 294
Herbs, 209, 210
Hibiscus cannabinus, 247, 252, 253, 257
Holoparasites, 294, 305
Hordeum vulgare, 40
Hormonal crosstalk and interactions, 84–85
Host plant (primary host *vs.* secondary
    host), 299
Humulus lupulus (hop), 22, 46
Hydathodes, 2, 31–35, 106, 107, 146–152,
    155–157, 159, 299
Hydraulic architectures, 241, 242
Hydraulic performance, 223, 232
Hydraulic safety zones, 3
Hydraulic segmentation, 111, 237–240, 242
Hypotheses, 2, 3, 44, 62, 103, 110, 111, 115,
    133, 145, 148, 177, 224–227, 231–233,
    240, 246, 252, 255, 257, 274, 282,
    283, 296
Hypothesis of tracheid diameter regulation,
    110, 224

I
IAA catabolism, 231, 233
IAA conjugation, 231, 233, 282
IAA, indole-3-acetic acid, auxin, 2, 57, 111,
    132, 200, 297, 301
Immunolocalization, 62, 77, 147, 166
Indole-3-acetic acid (IAA), 2, 57–73, 79,
    102–106, 108–112, 115, 116, 119, 120,
    132, 141, 143–145, 147–159, 167–171,
    179–182, 186–188, 199–201, 203–206,
    208, 209, 217, 227, 228, 231–233, 240,
    252, 257, 266, 267, 276, 279, 283, 286,
    287, 296, 297, 301, 302
Influx carrier, 74, 81
Interfascicular cambium, 10, 120, 199, 206
Interxylary phloem, 17, 19
Intratumor bacteria (specific bacteria inside
    human cancer cells), 323
In vitro, 102, 122, 188, 194, 196,
    261, 323–325
In vivo, 102, 106, 112, 323
Ipomoea batatas, 27, 28, 178
Ipomoea purpurea, 163
IPT genes, 135
Isopentenyltransferase *(IPT)* genes, 311
Izeniola obesula, 299, 300

J
JA signaling, 81, 82
Jasmonates, 74, 81–82, 84, 207, 210, 261, 262,
    323, 324
Jasmonic acid (JA), 80–82, 298, 325
JA transporter, 81
Junctions, 33, 34, 40, 41, 145, 146, 185,
    193–195, 237–242, 255, 296, 297,
    299, 312
Juniperus sp., 259
Juvenile-adult transition, 215, 217, 218, 220
Juvenile wood, 215, 216, 218, 220, 221

K
Kinetin, 106, 112, 118–120, 188

L
Larix sp., 259
Larva chamber, 299, 300
Lateral root initiation (LRi), 98, 134, 135, 177,
    178, 181, 182
Lateral root initiation hypothesis, 177,
    179, 182
Latewood, 110, 111, 116, 224, 260, 279–281,
    285, 286, 302–304

Leaf apical dominance, 145, 148, 149, 152, 156, 159
Leaf blade, 74, 75
Leaf development, 57, 58, 97, 104, 110, 141, 142, 145, 146, 159, 199, 224
Leaf initiation, 58, 141, 159
Leaf margins, 144, 146, 147, 155
Leaf primordium, 2, 9, 31, 33, 36, 40, 59–61, 63, 106, 145, 146, 148–152, 154, 157–159, 186, 192
Leaf-venation hypothesis, 147, 148, 156, 157
Liana species, 200, 202–203
Lignification, 35, 107, 108, 115, 116, 220
Lignin, 35, 43, 44, 46, 75, 107, 108, 117, 118, 121, 220, 265
Lignin polymers, 107, 108
Limited-growth hypothesis, 282, 286, 287
Long-distance transport, 1, 14, 77, 80, 102, 194
Longitudinal and parallel venation patterns, 30
Loop pattern, 142, 171
Loranthus acacia, 294–296
Luffa cylindrica, 225

M
Marchantia polymorpha, 101
Mature wood, 215–217, 221
Melia azedarach, 246, 284, 285, 287
Meristemoids, 246
Meristems, 3, 5, 7–12, 31, 69–71, 77, 97, 163, 164, 167, 199–201, 206, 208, 210, 254–255
Mesophyll cells, 61, 83, 148
Metabolic stress, 45–48, 107
Metastasis, 322, 324
Metaxylem vessels, 15, 28–30, 40, 48, 238, 241
Methyl jasmonate (MeJA), 78, 81, 82, 261, 262, 323
Microbe-cancer cell interactions, 322
Microbiome, 323
MicroRNAs, 80, 220
Midvein (primary vein), 142–146, 148, 149, 151, 152, 155, 157–159, 186, 190
Monocotyledons (monocots), 110
Monolignols, 35, 107, 108
Monopteros (MP), 12, 63
Monoterpenoids, 259
Morphogen, 200, 209, 224, 227
Morphogenesis, 2, 31, 83, 143, 146, 152, 158–159, 167–170, 293, 298, 302, 317, 320
Morphogen gradients, 226
Mycorrhiza, 83

N
NAC-domain genes, 105
NAC-domain transcription factors, 122, 124
Narrow latewood vessels, 284, 304
Negative xylem pressure, 17, 20, 23
Neovascularization, 321, 325
Never ripe mutant (ethylene-insensitive), 261, 317, 320
Nicotiana tabacum, 11, 75, 83, 116, 117, 121, 167, 204–205
Nitrate (NO_3^-), 4, 74, 78, 80, 97, 98, 133
Non-galling insects, 301
Non-polar auxin route, 71
Non-polar gibberellin transport, 113–122, 205
Non-polar rapid auxin movement, 71, 200
NPA (1-N-naphthylphtalamic acid), auxin efflux inhibitor, 65, 157, 158

O
Old cambium, 4, 210, 283
Oleander (*Nerium oleander*), 311
Olive (*Olea europaea*), 312
Opposite wood, 42, 43, 265–267, 269
Organ communication, 131
Organ competition, 131
Organ junctions, 3, 111, 237, 242
Orobanchaceae, 294
Orobanche aegyptiaca, 296
Orobanche crenata, 296
Orobanche cumana, 296, 297
Orobanche sp., 294, 296
Oryza sativa, 40
Outer wood, 208, 215, 220
Oxalis stricta, 107
OxIAA (2-oxindole-3-acetic acid), 73, 231

P
Parallel evolution, 111, 273
Parasitic plants, 293–295, 297, 298, 304, 305
Parasitism, 1, 4, 293, 294, 298
Parthenocissus tricuspidata, 102
Perforation plates, 37, 111, 275, 279, 294
Perforations, 18, 23, 111, 119, 122, 238, 273–275, 287, 295–297, 305
Pericycle, 27, 29, 67, 70, 71, 77, 179–182
Permian plant, 275
Permian vessels (fossil evidence), 275
Petunia hybrid, 83
Petals, 82, 164–173, 190
pH, *see* Acidity
Phaseolus vulgaris, 228–230
Phelipanche aegyptiaca, 296
Phelipanche sp., 294

Phloem anastomoses, 17, 20, 24–26, 56, 71, 104, 105, 186, 187, 189, 191, 192, 194, 195, 312–314, 324
Phloem differentiation, 14, 31, 32, 45, 102–103, 115, 146, 194
Phloem only strands, 188
Phloem structure, 47
Phloem translocation, 164
Phosphate (PO₄-), 4, 18, 84, 97, 98
Phosphate deficiency, 97
Phtheirospermum japonicum, 298
Phyllotaxis, 141, 159, 218
Physcomitrella patens, 101, 123
Phytohormones (plant hormones), 55
Picea abies, 82
Picea mariana, 111, 231
PIN1 auxin carrier, 66
PIN3, 71, 104, 266, 267, 269
PIN4, 71, 104
PIN5, 154, 155
PIN6, 154, 155
PIN7, 71, 104
PIN8, 154, 155
Pinaceae, 82, 204, 259
Pin-formed (PIN), 58
Pinus contorta, 72, 199, 227, 228
Pinus halepensis, 259–261
Pinus longaeva, 3
Pinus pinea, 37, 110, 227, 274–277
Pinus radiata, 261
Pinus sylvestris, 71, 72, 267
Pistacia atlantica, 302, 303
Pistacia palaestina, 302, 304
Pisum sativum, 64, 133, 195, 200, 209
Pit, 18, 20, 22, 23, 223
Pit aperture, 20, 22, 23
Pith, 13, 16, 29, 30, 35, 42, 46, 202–203, 245, 246, 249, 299, 319
Pitted secondary wall thickenings, 192
Plant tumors, 309–311, 317, 321, 322, 324, 325
Plasma membrane, 35, 58, 63–66, 74, 77, 80–83, 97, 133, 154, 155, 266
Plasmodesmata, 14, 16, 17, 34, 76, 79, 132, 208, 210, 279, 297
Plasticity, 141, 142, 185, 190, 191, 195, 215, 298
Polar auxin transport, 11, 58, 59, 62, 63, 65, 66, 71, 104, 123, 155, 166, 167, 186, 247, 249, 286, 322
Polarities, 30, 31, 55, 58–63, 66, 112, 116, 142, 144, 146, 147, 167, 237, 299, 309, 319
Pollen development, 166–169, 173

Pollen grains, 167–169, 173
Polyarch monocot adventitious root, 27
Polycentric vascular rays, 248
Polygonum convolvulus, 107
Pontania sp., 301
Populus deltoids, 237
Populus euphratica, 282, 287
Populus robusta, 203
Populus sp., 12, 38, 75, 83, 188, 207, 266, 267, 269
Populus tremula × P. tremuloides, 206
Post-mortem lignification, 35, 108
Primary vascular tissues, 9, 27
Primary wall, 18, 22, 23, 35
Procambium, 8–12, 31, 32, 36, 38, 56, 58, 69, 70, 142, 143, 146, 148, 186, 245
Programmed cell death (PCD), 39, 101, 102, 107, 108, 122, 124
Protein transporters, 58
Protoxylem vessels, 15, 16, 20, 29, 30, 81, 123, 177–182, 254–255, 257
Pruning, 132, 240
Pseudolarix sp., 259
Pseudomonas syringae subsp.*savastanoi*, 311, 312, 317
Pseudotsuga menziesii, 261
Pseudotsuga sp., 259

Q
Quercus ithaburensis, 239, 282, 287
Quercus serrata, 283, 284, 286
Quercus velutina, 237, 238, 281
Quiescent center (QC), 9, 12

R
Radial files of vessels, 200
Radial pattern (radial vascular strands in roots), 27, 28
Radial patterns of vessels and sieve tubes, 200, 201, 210
Radial transport pathways, 245, 249
Rate of conduit differentiation, 227
Ray initials, 9, 10, 207, 245, 246
Ray initiation, 246, 247, 249
Rays, 9–11, 34, 35, 46, 78, 206, 245–249, 259–261, 268, 313, 315, 319
Ray spacing, 247
Reaction woods, 3, 40, 42–44, 207, 210, 215, 265–267, 269, 270
Reconnection of cut vascular systems (in grafting), 193
Redifferentiate, 110, 144

Regeneration, 31, 39, 56, 57, 76, 102, 106, 112, 113, 124, 132, 142, 143, 147, 152, 153, 185–195, 245, 322
Rejuvenations, 3, 76, 215, 217, 218, 220, 221
Reproductive phase, 76, 84, 163, 164, 217, 220
Resin ducts, 35, 79, 248, 259–262
Resin glands, 259, 260
Resin-secreting epithelial cells, 82, 259
Reticulated secondary wall thickenings, 119, 192
Reticulated vein patterns, 30
Rhapis excels, 40, 238
Rhodococcus fascians, 311
Rhopalomyia yomogicola, 301
Ricinus communis, 311–313, 316, 317
Ring-porous trees, 24, 73, 282–284, 286–288
Ring-porous tree evolution, 73, 282–288
Ring-porous wood, 76, 277, 279, 280, 282, 286, 287
Robinia pseudoacacia, 225, 280, 283
Root apical meristem (RAM), 7, 9
Root cap, 5, 7, 9, 76–78, 97, 98, 131, 135–137, 180–182, 206, 210
Root grafts, 192, 193
Root-parasitic weeds, 83–84
Roots, 1, 3, 4, 7–13, 17, 24, 27–31, 38, 40, 47, 48, 56, 58, 62–64, 67, 70, 71, 73, 74, 76–81, 83–84, 97, 98, 110–113, 116–118, 120, 123, 124, 131, 133–137, 141–143, 145, 155, 170, 171, 174, 177–182, 185, 186, 188, 190, 192–195, 199–201, 203, 204, 206, 207, 215–218, 220, 223, 224, 226, 227, 231–233, 237, 241, 242, 245, 248, 249, 252–257, 265, 266, 274, 282, 294, 296, 298, 318
Root signals, 83, 84, 217, 221
Rootstock (or stock), 193

S
Safety zones, 40, 238, 241, 242
Salicylic acid (SA), 298
Salix japonica, 301
Santalum album, 297, 298
Sapwood, 3, 24, 42, 78, 245, 249
Scalariform perforation plates, 275
Scions, 164, 193–195
Secale cereal, 241
Secondary cell wall biogenesis (SCWB), 84, 164, 173
Secondary vascular differentiation, 83, 131, 199–207, 269

Secondary vascular tissues, 11, 30, 199, 202–203, 206, 217, 245
Secondary veins, 142, 144, 145, 150, 152, 155, 157, 159, 171, 173
Secondary walls, 18, 20, 22, 29, 38, 43, 47, 48, 102, 107–109, 112, 113, 115, 122, 216, 220, 223, 224, 227, 231–233, 252–256, 283, 284
Secondary wall thickenings, 24, 152, 181, 192, 276
Secondary xylem (wood), 11, 34, 108, 113, 117–120, 199, 217, 220, 223, 226, 245, 266, 268, 281, 304
Seed development, 174
Segmentation hypothesis, 238, 240, 255
Sensitivity to auxin, 73, 277, 282, 283, 286
Sepals, 147, 164–166, 169, 171, 174
Sequoiadendron giganteum, 231
Sequoia sempervirens, 3, 4, 226, 231
Shoot apical dominance, 131–134, 137
Shoot apical meristem (SAM), 7–9, 12, 58, 59, 66, 217, 221
Shoot branching, 83, 97, 132, 133
Sieve elements (SEs), 14, 16, 17, 20, 26, 38, 47, 102, 103, 105, 124, 179, 186–188, 216, 223, 225, 232
Sieve plates, 16–20, 25, 26, 38, 47, 71, 103, 104, 106, 225, 296
Sieve tubes, 14, 17–20, 24–26, 31, 35, 46, 47, 57, 62, 63, 67–72, 76, 82, 104, 106, 107, 112, 124, 147, 179, 182, 187–189, 194, 195, 200, 201, 206, 224, 226, 237, 242, 248, 254–255, 299, 314, 317, 324
Simple perforation plate, 275
Six-point hypothesis, 110, 226, 240
Slavum wertheimae, 302, 303
Social status of a forest tree, 208–210
Softwood, 11, 36, 108
Solanum lycopersicum, 164
Solidago altissima, 301
Spiral secondary wall thickenings, 20, 21
Stamens, 33, 164–171
Statocytes, 76, 265, 266
Statoliths, 265, 266, 269
Stem cell niche (SCN), 7, 8, 11, 12
Stem cells, 3, 5, 7–12, 57, 76, 110, 185, 201, 203, 206
Stem growth oscillations, 187
Stigma, 32, 33, 166, 167, 169–171, 174, 175
Stresses, 24, 42, 48, 57, 78–81, 83, 97, 193, 195, 207, 210, 241, 242, 254–255, 261, 262, 268, 282, 287, 294, 298, 319, 321
Striga sp., 294

Striga hermonthica, 298
Strigolactone (SL), 80, 83–85, 97, 98, 132, 133, 137, 207, 210
Suaeda monoica, 248
Sucrose, 102–104, 106, 134
Sugar, 14, 35, 73, 76, 82, 102, 103, 132, 133, 137, 208
Sunflower (Helianthus annuus), 117, 120, 199, 310
Symbiotic fungi, 83–84
Symplastic connection, 296, 305
Symplastic pathways, 34
Synergism between hormones, 203
Syringa, 102, 103
Syringyl (S), 107, 108

T
TAA/YUC pathway, 58
Temperate deciduous hardwood trees, 73, 76, 277, 279, 282, 287
Tension wood, 42–44, 78, 83, 217, 237, 265–269
Tertiary veins, 142, 144, 151, 152, 157–160
Thuja plicata, 36, 37, 259, 260
TIBA (2,3,5-triiodobenzoic acid), auxin influx inhibitor, 66, 173, 266
Ti-plasmid (tumor-inducing plasmid), 311, 325
Tissue cultures, 76, 102, 103, 105, 106, 112, 122, 124, 152, 299
Tissue sensitivity, 277, 282, 283
Torus, 20, 22, 23
Total auxin, 62, 69, 72, 148
Totipotency, 185
Tracheary elements (TEs), 17, 18, 39, 41, 78, 83, 103, 105–107, 113, 122–124, 146, 154, 155, 188, 192, 207, 210, 223, 227, 233, 247, 249, 274, 275, 287
Tracheid differentiation, 110–111
Tracheid dimension, 110
Tracheids, 11, 17–20, 22–24, 36, 37, 40, 43, 44, 77, 79, 107–111, 187, 195, 204, 208–210, 216, 223–227, 231, 232, 238, 240–242, 245, 248, 259, 260, 265, 267, 273–279, 287
Tracheid with perforations, 18, 37, 111, 238, 273–275, 287
Transfer cells, 35
Transferred DNA of the tumor-inducing Ti plasmid (T-DNA), 311, 314, 322, 325
Transfusion tracheids, 36, 37, 56, 187, 195
Transition from juvenile to adult phases, 76

Traumatic resin ducts, 79, 82, 84, 259
Trichomes, 2, 71
Trifolium sp., 72
Triticum aestivum, 40
Tropaeolum majus, 163
Tryptophan (Trp), 301
Tryptophan aminotransferase of *arabidopsis* (TAA), 57, 167
Tsuga heterophylla, 294
Tsuga sp., 259
Tumor-host junction, 312
Tumor-induced ethylene, 316–317
Tumor microbiome, 322, 324
Tumorous fibers, 39
Tumors, 5, 39, 82, 261, 309–325
Tumor tissue, 38, 39, 310, 317, 322
Tumor type-specific microbiome, 325
Tyloses, 22–24, 35, 245, 280, 282, 288

V
Vascular adaptation, 251, 255
Vascular adaptation hypothesis, 251, 252, 255–257
Vascular bundle maturation, 32
Vascular bundles, 10, 12–14, 17, 24, 26, 30, 31, 37, 39, 61–63, 66–68, 104, 108, 112, 186, 187, 189, 191, 194–195, 199, 206, 224, 225, 254–255, 300, 312, 313
Vascular cambium, 10, 17, 57, 72, 83, 97, 106, 121, 185, 199, 200, 203, 221, 226–228
Vascular loops, 31, 142, 143, 146, 170–172
Vascular network patterns, 142
Vascular rays initiation and development, 246
Vascular regeneration, 3, 37, 56, 66, 151, 152, 160, 185, 187–190, 194, 195
Vascular-related NAC-domain (VND), transcription factor, 81, 106, 122, 124
Vegetative growth, 12, 76, 215, 217, 282, 287
Veinlets, 30, 33, 107, 142, 147, 151, 152, 157–160, 169–171, 173–175
Vein networks, 142, 151, 154, 157, 159
Veins, 2, 9, 12, 30, 33, 35, 63, 107, 142, 144–147, 149–155, 157–159, 170, 172, 173, 186, 187, 190, 207
Venation pattern formation, 145, 159
Venation patterns, 30
Vessel density, 223, 226, 227, 229, 232, 233, 256
Vessel diameter, 24, 34, 40, 48, 223, 225, 227, 229, 230, 232, 237, 238, 252, 255–257, 268, 280, 317, 325

Vessel differentiation, 29, 31, 32, 48, 72, 76, 80, 106, 111–113, 117, 123, 124, 146, 147, 156, 169, 173, 180, 181, 192, 195, 230, 276, 282–285, 296

Vessel elements, 21, 23, 33–35, 41, 71, 72, 104, 108, 111–113, 119, 122–124, 146, 147, 152, 153, 155, 156, 175, 177, 179–182, 185, 186, 189, 190, 192, 195, 206, 227, 237, 238, 240, 273, 275, 277, 279, 283, 286, 287, 297, 299, 300, 310

Vessel endings, 22, 238, 241, 242, 274

Vessel widening, 195, 224, 232, 256, 287

Vicia narbonensis, 296

VND6 and VND7 genes, 123

VND, transcription factor, 81, 106, 122, 124

Vulnerability, 24, 294

W

Wave-like pattern, 72–73, 199, 209, 227, 228, 231, 233

Weak acids, 74

Wide earlywood vessels, 280, 282–284, 286–288

Wood adaptation, 223

Woodiness, 209–210

Wood production, 45, 75, 122, 210, 220, 266

Woody angiosperms, 34, 76, 217, 218, 220, 245

Wound callose, 18, 106

Wounding effects on vascularization, 185–192, 195

Wounds, 11, 16, 18, 25, 26, 31, 35, 39, 56, 57, 63, 66, 81, 102, 106, 112–115, 124, 153, 154, 186–191, 194, 195, 238, 245, 261, 311

Wuschel-related homeobox (WOX) gene family, 11

X

Xylem differentiation, 14, 27, 31, 97, 102–104, 106, 121, 124, 144, 146, 148, 186, 194, 203, 204, 303, 304, 313, 319

Xylem fiber (termed libriform fibers), 38, 116, 117, 121

Xylem parenchyma, 17, 24, 34, 35, 67–72, 97, 108

Xylem structure, 242

Y

Young cambium, 58, 67–73, 209, 215, 217, 220, 282

YUCCA (YUC), 57, 167

Z

Zea mays, 15, 29, 48, 74, 190, 241, 252, 254–255, 257

Zeatin, 76, 106, 112, 119, 188, 297, 301

Zinnia elegans, 78, 83, 108, 122–124

Printed in the United States
by Baker & Taylor Publisher Services